Angiogenesis Protocols

Second Edition

John M. Walker
School of Life Sciences
University of Hertfordshire
Hatfield, Hertfordshire, AL10 9AB, UK

For other titles published in this series, go to
www.springer.com/series/7651

METHODS IN MOLECULAR BIOLOGY™

Angiogenesis Protocols

Second Edition

Edited by

Stewart Martin and Cliff Murray

University of Nottingham, Nottingham, UK
Source Biosciences plc., Nottingham, UK

☼ Humana Press

Editors
Stewart Martin
Nottingham University Hospital
Department of Clinical Oncology
City Hospital Campus
Nottingham
UK

Cliff Murray
Source Biosciences plc.
1 Orchard Place
Nottingham
Nottingham Business Park
UK

ISBN: 978-1-58829-907-9   e-ISBN: 978-1-59745-241-0
ISSN: 1064-3745   e-ISSN: 1940-6029
DOI: 10.1007/978-1-59745-241-0

Library of Congress Control Number: 2008936477

Cover illustration: Chapter 15, Figure 2.

Printed on acid-free paper

springer.com

Preface

It is 6 years since the first edition of Angiogenesis Protocols was published. It is surely a testament to the continuing interest in angiogenesis and the quality of the chapters that made up that important text that we now have a second edition in print.

The original concept behind this volume was to provide a single source for angiogenesis researchers that would provide a range of methods for cell isolation and assessing angiogenesis in vivo or in vitro. Inclusivity was key to this endeavour; a number of techniques should be described in detail, ranging in difficulty and resource requirements, so that most, if not all, interested laboratories could participate in this exciting research field, irrespective of levels of resource and expertise. We have endeavoured to remain true to that underlying principle in this new edition. The foundations remain firmly in place in the form of chapters on in vivo and in vitro assays for angiogenesis, techniques that now form part of the canon of angiogenesis literature. In addition, basic methods for isolation of the various cellular components of blood vessels continue to be included.

But, angiogenesis has inevitably moved on, and our understanding of the biology and physiology of blood and lymphatic vessels is that much more profound. In addition, we are now seeing the fruits of experimentation and clinical trials with first-generation antiangiogenic agents. Reflecting this new understanding of the biology of vessels, we have incorporated new chapters, from leaders in the field, including two chapters on lymphatic vessels, one on circulating endothelial progenitor cells, and one on angiogenic signalling pathways. In recognition of the growing importance of bioimaging and other noninvasive techniques that will have a profound impact on drug development and disease diagnosis and monitoring, we have also included chapters on imaging of angiogenesis and measurement of tissue blood flow.

By expanding the number of in vitro and other techniques to reflect advances in the field, we have inevitably had to sacrifice certain chapters that were in the first edition—a tough decision. We would like to think, however, that this volume will continue to provide not only a practical handbook for key techniques, but also an informative and enjoyable read for all those interested, no matter how directly, in angiogenesis.

Stewart Martin
Cliff Murray

Contents

SECTION V: IN VIVO TECHNIQUES

Contributors

ZUBAIR AHMED, BSc, PhD • *CRUK Molecular Angiogenesis Group, Divisions of Immunity and Infection and Cancer Studies, Institute of Biomedical Research, University of Birmingham, UK*

DAVID O. BATES, PhD • *Microvascular Research Laboratories, Bristol Heart Institute, Department of Physiology, University of Bristol, UK*

ANDREW V. BENEST, BSc • *Division of Vascular Oncology and Metastasis, DKFZ, Heidelberg, Germany.*

ECKART BERTELMANN, PhD • *Berlin Clinic for Ophthalmology, Charité University Medical Centre, Berlin, Germany*

ROY BICKNELL, BSc, PhD • *CRUK Molecular Angiogenesis Group, Divisions of Immunity and Infection and Cancer Studies, Institute of Biomedical Research, University of Birmingham, Birmingham, UK*

MIKE F. BURBRIDGE, PhD • *Cancer Research and Drug Discovery, Institut de Recherches Servier, Croissy sur Seine, France*

DR RAPHAELE BUSER • *Department of Oncology, University Medical Center, Geneva, Switzerland raphaele.buserllinares@medecine.unige.ch*

LYNN M. BUTLER, PhD • *Department of Physiology, Division of Medical Sciences, The Medical School, The University of Birmingham, UK*

NATALIE CHARNLEY, MBCHB, MRCP, FRCR. • *University of Manchester, Wolfson Molecular Imaging Centre, Manchester, UK*

ADRIAN T. CHURCHMAN • *Cardiovascular Division, School of Medicine King's College London, London, UK*

STEVEN CLASPER, BSc, PhD • *MRC Human Immunology Unit, Weatherall Institute of Molecular Medicine, John Radcliffe Hospital, Headington, Oxford. UK.*

WILLIAM COURT, MSc • *Cancer Research UK Centre for Cancer Therapeutics, McElwain Laboratories, Institute of Cancer Research, Surrey, UK*

VINCENT J. CUNNINGHAM, PhD • *GSK Clinical Imaging Centre, Imperial College, Hammersmith Hospital, London UK*

STEPHANIE DONALDSON, MSc, MSCI • *North Western Medical Physics, Christie Hospital, Manchester, and Imaging Science and Biomedical Engineering, University of Manchester, UK*

SUZANNE A. ECCLES, PhD • *Cancer Research UK Centre for Cancer Therapeutics, McElwain Laboratories, Institute of Cancer Research, Surrey, UK*

MÔNICA ALVES NEVES DINIZ FERREIRA, PhD • *General Pathology, Institute of Biological Sciences, Federal University of Minas Gerais, Brazil*

STEPHEN B. FOX, MD • *Department of Pathology, Peter MacCallum Cancer Centre, Melbourne, Australia*

IRA M. HERMAN, PhD • *Department of Physiology Center for Innovations in Wound Healing Research, Tufts University School of Medicine, Boston, Massachusetts, USA*

PETER W. HEWETT, PhD • *Vascular and Reproductive Biology, Institute for Biomedical Research, The Medical School, University of Birmingham, UK*

DAVID G. JACKSON, BA, PhD • *MRC Human Immunology Unit, Weatherall Institute of Molecular Medicine, John Radcliffe Hospital, Headington, Oxford. UK.*

FRANK M. KLENKE, MD • *Department of Orthopaedic Surgery, University of Berne, Switzerland*

ALICE LEUNG • *Department of Physiology Center for Innovations in Wound Healing Research, Tufts University School of Medicine, Boston, Massachusetts, USA*

MICHAEL LEUNIG, MD • *Department of Orthopaedic Surgery, Schulthess Klinik, Zurich, Switzerland*

KATHERINE M. MALINDA, PhD • *Blood and Vascular Review Group, Division of Extramural Research Activities, National Heart, Lung, and Blood Institute, Bethesda, Maryland, USA*

HELEN M. MCGETTRICK, PhD • *Department of Physiology, Division of Medical Sciences, The Medical School, The University of Birmingham, UK*

LUCIA MORBIDELLI, PhD • *Section of Pharmacology, Toxicology and Chemotherapy, Depart ment of Molecular Biology, University of Siena, Siena, Italy.*

GERARD B. NASH, PhD • *Department of Physiology, Division of Medical Sciences, The Medical School, The University of Birmingham, UK*

RICCARDO E. NISATO, PhD • *Laboratory for Mechanobiology and Morphogenesis (LMBM), Integrative Biosciences Institute (IBI), Faculty of Life Sciences, EPFL, Lausanne, Switzerland*

SILVIA PASSOS ANDRADE, PhD • *Departments of Physiology and Biophysics, Institute of Biological Sciences, Federal University of Minas Gerais, Brazil*

LISA PATTERSON, BSc • *Cancer Research UK Centre for Cancer Therapeutics, McElwain Laboratories, Institute of Cancer Research, Surrey, UK*

MICHAEL S. PEPPER, MBCHB, PhD, MD • *Department of Immunology, Faculty of Health Sciences, University of Pretoria, and Netcare Institute of Cellular and Molecular Medicine, South Africa*

M. LOURDES PONCE, PhD • *Section on Retinal Diseases and Therapeutics, National Eye Institute, Bethesda Maryland, USA.*

PAT PRICE, FRCR, MD • *Academic Department of Radiation Oncology, Christie Hospital, Manchester, UK*

VIVIEN E. PRISE, BSc • *Gray Cancer Institute, Mount Vernon Hospital, Northwood, Middlesex UK*

SHARON SANDERSON, PhD • *Cancer Research UK Molecular Angiogenesis Laboratory, Weatherhall Institute of Molecular Medicine, John Radcliffe Hospital, Oxford, UK*

MICHAEL C. SCHMID, PhD • *Moores UCSD Cancer Center, University of California, San Diego, California, USA*

AXEL SCKELL, MD • *Department of Trauma and Reconstructive Surgery, Charité University Medical Center, Berlin, Germany*

STEVEN D. SHNYDER, PhD • *Institute of Cancer Therapeutics, University of Bradford, Bradford, UK*

RICHARD C.M. SIOW , BSc, PhD • *Cardiovascular Division, School of Medicine King's College London, London, UK.*

GILLIAN M. TOZER, BSc, MSc, PhD • *Academic Unit of Surgical Oncology, University of Sheffield, School of Medicine and Biomedical Sciences, Sheffield, UK*

JUDITH A. VARNER, PhD • *Moores UCSD Cancer Center, University of California, San Diego, California, USA*

DAVID C. WEST, PhD • *School of Biological Sciences, University of Liverpool, Liverpool, UK*

MARINA ZICHE, MD • *Section of Pharmacology, Toxicology and Chemotherapy, Dept. Molecular Biology, University of Siena, Siena, Italy.*

Section I

Review Articles

Chapter 1

Angiogenic Signalling Pathways

Zubair Ahmed and Roy Bicknell

Abstract

Hypoxia is widely recognised as a key driving force for tumor angiogenesis by its induction of vascular endothelial growth factor (VEGF) and other direct-acting angiogenic factors. We describe the effect of hypoxia on gene expression and downstream angiogenic signalling; however, the angiogenic process is complex, and many other signalling pathways beyond VEGF are implicated in the formation of new vessels. These include extra-cellular signalling pathways such as the notch/delta, ephrin/Eph receptor, roundabout/slit, and netrin/UNC (uncoordinated) receptor families as well as intracellular proteins such as hedgehog and sprouty. The remarkable diversity in angiogenic signalling pathways provides many opportunities for therapeutic intervention, and anti-angiogenesis is currently a major area of oncology research.

Key words: Endothelial cells, guidance molecules, hypoxia, tumor, vascular endothelial growth factor.

1. Introduction

Angiogenesis is a complex developmental process involving (*inter alia*) basement membrane degradation, endothelial cell proliferation, migration, and tube formation. Angiogenesis plays a central role in normal development and wound healing and in the aetiology of many diseases, such as psoriasis, diabetic retinopathy, and cancer. Since angiogenesis plays an essential role in tumor growth and invasion, anti-angiogenesis has been pursued for over 20 years as a route to novel cancer therapies. Many anti-angiogenic therapies have now been described that inhibit not only tumor growth but also cancer cell dissemination *(1–4)*. More recently, efforts have

S. Martin and C. Murray (eds.), *Methods in Molecular Biology, Angiogenesis Protocols, Second edition, Vol. 467*
© Humana Press, a part of Springer Science+ Business Media, LLC 2009
DOI: 10.1007/978-1-59745-241-0_1

been made to identify the regulators of vascular patterning and vessel guidance. In this context, the patterning of the nervous system has shed light on mechanisms of angiogenesis and has identified common regulators of both neuronal and vessel guidance. Both systems utilise complex branching networks of blood vessels or nerve cells to penetrate all regions of the body and provide a bidirectional flow of information. The two networks are often patterned similarly in peripheral tissues, with nerve fibres and blood vessels following parallel routes *(5)*. Neuronal and endothelial precursor cells also follow similar routes of migration during embryogenesis and respond to similar mesenchymal cues *(6)*, while genetic studies have indicated that both systems regulate each other's development. The latter is shown through the release of neurotrophic factors such as artemin and neurotrophin 3 by vessels *(7, 8)* and pro-angiogenic factors such as vascular endothelial growth factor (VEGF) by neurones *(9)*.

The behavioural similarities of the axonal growth cone and endothelial tip cell are complemented by the fact that both use a common repertoire of ligand/receptor signalling systems that couple the guidance cues detected in the external environment to the cytoskeletal changes necessary for guided cell migration. To date, four families of neuronal guidance molecules have been shown also to be involved in vascular development: the Ephrins, Semaphorins, Slits, and Netrins. These guidance cues can act to either attract or repel the tip cell, and different cues act over either short or long ranges, depending on whether they are diffusible or cell or matrix associated. *In vivo*, developing vessels navigate through tissue corridors by combinations of attractive cues made by cells in the corridor and repulsive signals expressed in the surrounding tissues. Thus, a greater understanding of the multiple guidance cues involved in vascular patterning could have clinical relevance in developing therapeutic agents to inhibit or stimulate angiogenesis in a variety of pathological conditions.

2. Hypoxia and Angiogenesis

Hypoxia is the result of tissues or tumors outgrowing their blood supply and thus being deprived of oxygen. The cellular response to hypoxia is mediated *via* transcription factors called hypoxia-inducible factors (HIFs). Hypoxia-inducible factor 1 (HIF-1) is a heterodimer of two DNA-binding proteins, HIF-1α and the aryl hydrocarbon nuclear translocator (ARNT or HIF-1β). Under normal oxygen tension, HIF-1α is rapidly degraded as a result of enzymatic prolyl-hydroxylation, but as the oxygen tension drops below 2%, HIF-1α is no longer degraded and translocates to the nucleus. Once in the nucleus, HIF-1α dimerises with HIF-1β, initiating a complex transcriptional response in genes having specific hypoxia response elements (HREs). Genes that arc up-regulated by HIF-1α include VEGF, proteins involved in the glycolytic pathway, and those involved in invasion (urokinase receptor, plasminogen activator 1) *(10)*.

As noted, HIF activation is regulated by stabilisation of the oxygen-sensitive α-subunit and its subsequent translocation to the nucleus, where it forms a functional complex with ARNT and transcriptional co-activators such as CBP/p300 *(11)*. Under normal oxygen concentrations HIFα binds the ubiquitin ligase, the Von Hippel-Lindau protein (pVHL), which targets HIFα for proteasomal degradation *(12)*. All three known HIFs interact with pVHL, an interaction that is conserved between species, requiring iron and oxygen-dependent hydroxylation of defined proline residues within the oxygen-dependent degradation domain of HIFα by prolyl hydroxylases. Other HIFs include HIF-2α, which is thought to be critical for embryonic development although its role in adult angiogenesis remains poorly defined compared to HIF-1α, and HIF-3α. In the tumor endothelium, HIF-1α is commonly up-regulated, while HIF-2α is expressed highly in stromal cells such as macrophages *(13)*, pointing to a different hypoxic response in different cell types.

VEGF is also strongly induced by HIF-1α *via* HREs in the 5′ and 3′ end of the gene, while many oncogenes also activate the transcription of VEGF and separately enhance function or expression of HIF-1α. It is clear that angiogenic genes downstream of HIF-1α are also likely to be important in tumor angiogenesis. Aside from VEGF, other direct-acting angiogenic factors induced by hypoxia include endothelins 1 and 2, adrenomedullin and angiogenin *(14)*. Oxygen regulated protein 150 (ORP 150), a VEGF chaperone, is also regulated by hypoxia *(15)*, as are connective tissue growth factor *(16, 17)*, leptin *(18)*, stromal cell-derived factor 1 (CXCL12) *(19)*, migratory inhibitory factor *(20)*, and placenta growth factor (PLGF) *(21)*. In addition, numerous polypeptide angiogenic factors, intermediary metabolites of glycolysis that are induced under anaerobic conditions, such as lactate and pyruvate, are angiogenic *(22)*. Given the wide range of angiogenic pathways regulated by hypoxia, the search for drugs targeting HIF is currently receiving a lot of attention.

2.1. Vascular Endothelial Growth Factor Signalling

VEGF was originally described as vascular permeability factor (VPF) that was released by tumor cells and promoted vascular leakage *(23)*. VEGFs are among the most important players in the regulation of blood and lymphatic vessel formation during embryonic development and wound healing. They are also involved in the maintenance of vessel homeostasis in adults *(24)*. VEGF is also an important mediator of neovascularization associated with diverse human diseases such as cancer as well as degeneration of the cells in the macula and neovascularization in the choroids in age-related macular degeneration *(25)*.

The most studied member of the VEGF family is VEGF-A, which signals and regulates vessel morphogenesis through VEGF receptors 1 and 2 (VEGFR1 and VEGFR2). Three VEG-FRs have been identified: VEGFR1, VEGFR2, and VEGFR3. VEGFR2 (also known as KDR, kinase domain insert region) is thought to be the main receptor that mediates proliferation of endothelial cells following stimulation by VEGF-A. VEGFR1, also known as flt-1, is a high-affinity VEGFR that occurs as a splice variant giving rise to a secreted soluble extra-cellular domain, sflt1, which is a potent antagonist of VEGF. VEGFR3 (flt4) is present in the vasculature during development, but in adults is restricted to the lymphatics. VEGFRs are tyrosine kinases that dimerise and can signal through mitogen-activated protein (MAP) kinases (MAPKs) and akt. The kinase domain is split into two functional domains, which gave rise to the original name KDR. Each receptor binds several VEGF ligands but with different specificities. For example, VEGF-A binds to VEGFR1 and -2 and sflt1, VEGF-C binds to VEGFR2 and VEGFR3, while PLGF1 and -2 bind to flt-1 soluble and receptor forms. VEGF-E is a viral VEGF molecule that binds VEGFR2 only (Fig. 1.1). Multiple splice variants of VEGF-A exist and affect its ability to bind heparin. Splicing and proteolytic processing of VEGF-C and -D and other members of the family have been reported.

As noted, VEGFR1 is required for normal vessel development during embryogenesis. Thus, homozygous deletion of VEGFR1 in mice is embryonic lethal at E8.5, which is caused by severe malformation of the vascular system *(26)*. A VEGFR1 splice variant lacking the intracellular tyrosine kinase and the transmembrane domains (sVEGFR1 or sflt1) has been shown to be deficient in signalling but is expressed in many tissues during embryonic development and apparently acts as a decoy for VEGF ligands *(27, 28)*. The kinase activity of VEGF1 is thought to be required for vessel development and to play an essential role during pathological angiogenesis and in wound healing through potentiation of VEGFR2 signalling *(29, 30)*.

VEGFR2 mediates VEGF's effects on endothelial cell migration, proliferation, differentiation, and survival in addition to vessel permeability and dilation. This is achieved through signalling via its kinase domains. Phosphorylation of tyrosine residues appears to be a key mediator in the signalling pathway of VEGFR2. Phosphorylation occurs at many sites, including Tyr1054 and Tyr1059, which are required for maximal kinase activity *(31)*. PLC-gamma has been shown to directly bind to phosphorylated Tyr1175, mediate activation of the MAPK/extra-cellular-signal-regulated

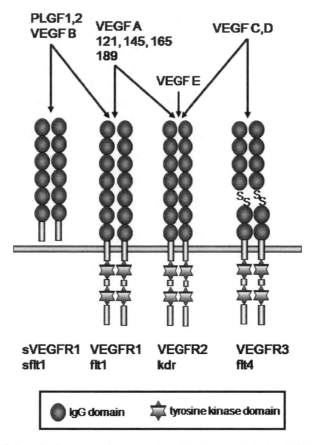

PLGF1,2
VEGF B

VEGF A
121, 145, 165
189

VEGF C,D

VEGF E

sVEGFR1   VEGFR1   VEGFR2   VEGFR3
sflt1   flt1   kdr   flt4

● IgG domain   ★ tyrosine kinase domain

Fig. 1.1. Schematic diagram to show vascular endothelial growth factor (VEGF) receptor (VEGFR) families and their ligands. Mammalian VEGFs bind to three VEGFR tyrosine kinases, which leads to the formation of VEGFR homodimers and heterodimers. There is some promiscuity in ligand–receptor interaction such that VEGF-C and -D are able to interact with VEGFR2. VEGF-A can also interact with both VEGFR1 and VEGFR2. Similarly, PLGF1 and -2 and VEGF-B can interact with soluble VEGFR1 (sVEGFR) and VEGFR1.

kinase 2 (ERK1/2) and endothelial cell migration (32). Mice expressing mutant forms of Tyr1173Phe VEGFR2 die early during embryonic development due to vascular defects, similar to vegfr2–/– null mice (33). VEGFR2 phosphorylation site Tyr951, a binding site for T-cell-specific adaptor (TSAd), has been shown to be important in endothelial cell migration (34, 35), and Tsad–/– mice show reduced vascularisation and growth of tumors (34).

VEGFR3 is synthesised as a 195-kDa precursor protein, proteolytically processed and thought to regulate lymphangiogenesis. VEGFR3 expression is found early in embryonic endothelial cells and later restricted to developing veins and lymphatics and not arteries (36, 37). In VEGFR3 null mice, vascular remodelling and maturation is abnormal in larger vessels with defective lumens that allow fluids to accumulate in the pericardial cavity, causing embryonic death *(38)*. VEGFR3 is also expressed on blood vessels in the near vicinity of tumors as well as being present in several benign and malignant tumor cells *(39, 40)*. VEGFR3 is activated by VEGF-C and -D, and proteolytically processed variants may also interact with VEGFR2. VEGFR3 can also promote cell migration and survival in the endothelium of the lymphatic system by protein kinase C (PKC)-dependent activation of MAPK *(41, 42)*.

Several mutations and polymorphisms in the VEGF pathway have been reported in Juvenile haemangioma, which shows a spontaneous growth of lesions in the early years of life followed by spontaneous regression. Analysis revealed missense mutations in the kinase domain or KDR of VEGFR2 and -3 in juvenile haemangioma *(43)*. Polymorphisms in the promoter region have also been analysed, and a −460/+405 polymorphism increased basal activity and led to a fivefold enhancement in responsiveness to phorbol esters in juvenile haemangioma *(44)*.

2.2. PLGF1 and VEGFR1

In angiogenesis, the main focus of attention has been VEGFR2. However, recent reports have implicated the varied role of VEGFR1. Adult transgenic mice deficient in PLGF1 showed a defective response to ischaemia, wound healing, inflammation, and cancer, but development of the embryonic vasculature was normal *(45)*. The effects could not be corrected by other ligands binding to flt1 such as VEGF-B, suggesting that there is a specific interaction between PLGF1 and VEGFR1. However, the deficit could be corrected by the introduction of bone marrow, which implies a role for progenitor cells in PLGF-mediated angiogenesis. Antibodies specific for flt1 inhibited tumor angiogenesis and inflammatory angiogenesis by inhibiting mobilization of bone marrow-derived myeloid precursors. PLGF was also able to induce arteriogenesis in adult rabbits *in vivo* and more effectively than a ligand specific for VEGFR2 (VEGF-E) *(46, 47)*. Tumor cell lines that were transfected with PLGF and grown *in vivo* conferred an anti-apoptotic effect on macrophages and endothelial cells, encouraging macrophage infiltration and survival *(48)*.

The finding that cells co-expressing PLGF and VEGF produce heterodimers that are inactive on VEGFR2 has complicated matters, suggesting that localization of production may also be important in regulating the overall effects of PLGF *(49)*. In addition, macrophages often have an important role in sustaining angiogenesis, partly by producing VEGF themselves *(50)*. Taken together, these results suggest an endocrine effect of secreted PLGF recruiting endothelial, myeloid, or monocyte precursors from the marrow to sites of angiogenesis.

**2.3. Semaphorins,
Neuropillins,
and Plexins**

Semaphorins are secreted or membrane-associated glycoproteins that have been grouped into eight classes on the basis of their structural element and amino acid sequence *(51–53)*. Semaphorins found in invertebrates have been grouped into classes 1 and 2, while classes 3–7 are those found in vertebrates, and the final group is encoded by viruses *(52)* (Fig. 1.2). All Semaphorins however, have a conserved 400-amino acid "Sema" domain and range from 400 to 1000 amino acids in length. An immunoglobulin-like domain is also found in Semaphorin classes 2–4 and 7, while class 5 has seven thrombospondin domains.

Membrane-associated Semaphorins bind to plexins, whereas secreted Semaphorins bind to Neuropillins, which do not signal on their own but act as co-receptors for Plexin signalling. Genetic studies in mice and *Drosophila* have shown that Semaphorin signalling acts as a repulsive cue in axon guidance and cell migration *(54)*. However, other studies have shown that Semaphorins can also attract certain types of neurons, with cytosolic cGMP (cyclic guanosine monophospahte) being critical in determining whether a particular response is repulsive or attractive *(55–58)*.

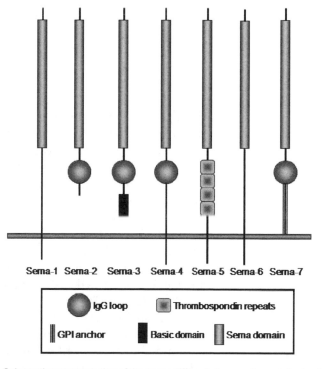

Fig. 1.2. Schematic representation of the seven different classes of semaphorins (Sema). Classes 1 and 2 are found in invertebrates; classes 3–7 are found in vertebrates. *IgG* immunoglobulin G. GPI = glycosylphosphatidylinositol

Among the Semaphorins (Sema), Sema3 variants have been most widely associated with angiogenesis and tumorigenesis. For example, Sema3A acts as a repellent for axons but has also been shown to inhibit migration of endothelial cells *(59, 60)*. In addition, both Sema3B and Sema3F have been identified as tumor suppressor genes that are frequently inactivated in small lung cancers *(61–64)*. Ectopic expression of Sema3F in a lung cancer cell line expressing the neuropillin-1 receptor resulted in tumor growth inhibition *(65–67)*, while the observation that Sema3B transcript expression is increased by p53 raises the possibility that Sema3B may influence cell growth associated with tumor progression *(68)*. Sema6A is temporally regulated during development of the nervous system *(69)*, and the soluble ectodomain is reported to cause growth cone collapse in sympathetic neurons and dorsal root ganglia neurons *(69)*. However, a recent study has shown that the ectodomain of Sema6A blocked VEGF-mediated endothelial cell migration *in vitro* by inhibition of Src, FAK, and ERK phosphorylation *(70)*. In *in vivo* Matrigel assays, the administration of Sema6A ectodomain inhibited both bFGF/VEGF and tumor cell line-induced neovascularisation *(70)*.

The neuropillin (NP) family was initially identified as receptors for repulsive axon guidance through class 3 Semaphorins *(71)*. NP1-deficient mice exhibit aberrant cardiovascular development and defective vessel branching during mid-gestation, suggesting a role for NPs in vascular patterning *(72)*. These vascular effects may not be due to disrupted Sema3A signalling as NP1 has also been shown to serve as a co-receptor for $VEGF_{165}$ but not other VEGF isoforms *(73)*. The secreted soluble form of NP1 was thought to function as a VEGF trap *(74)*; for example, a study showed that a natural truncated, soluble form of NP1 containing the A and B domains but not the C domain (which is required for receptor dimerisation) inhibited VEGF activity *(74)*. Tumors of rat prostate carcinoma cells expressing the soluble form of NP1 manifested extensive haemorrhage, damaged vessels, and the presence of apoptotic tumor cells *(75)*. Over-expression of NP1 in AT2.1 prostate cancer-derived cells led to faster tumor development without affecting the rate of proliferation of these cells *in vitro (60)*. This enhancement in tumor growth rate was associated with an increased density of blood vessels in the tumors, presumably due to effects on VEGF signalling since membrane-bound and soluble dimerised NP1 enhances VEGF signalling mediated by VEGFR2 *(60, 76)*.

NP2 also behaves as a splice form-specific VEGFR, and its VEGF binding characteristics closely mirror those of NP1. NP2 can bind $VEGF_{145}$, a VEGF splice form that is not recognised by NP1 *(77)*. During early embryogenesis, NP2 is expressed in veins, while NP1 expression is in arteries *(78)*. NP2 is later expressed in lymph vessels and plays a part in lymphangiogenesis *(79)* and as a receptor for the lymphangiogenesis-induced VEGF-C *(80)*. NP2-deleted mice are viable, displaying neuronal abnormalities, but the cardiovascular system is largely normal except for the presence of some abnormalities in peripheral lymph vessels *(79)*. However, mice with both NP1 and NP2 display much more acute cardiovascular defects than single-gene inactivation. These mice do not develop any blood vessels and die early *in utero,* confirming that NPs do have a role in vasculogenesis and developmental angiogenesis *(81)*. In agreement with these results, VEGF was shown to strongly inhibit retinal angiogenesis in mice lacking the NP2 *(82)*. These results clearly demonstrated that NP2 has an important regulatory role in VEGF function and of angiogenesis and that NP2 ligands may affect tumor angiogenesis and tumor lymphangiogenesis *(82)*.

The plexin family contains nine vertebrate members that are subdivided into four classes. Plexins are transmembrane proteins containing a cytoplasmic SP domain that includes tyrosine phosphorylation sites but no enzymatic activity. Their extra-cellular domains are distinguished by the presence of a Sema domain, a Met-related sequence domain and by glycine-proline-rich motifs *(83)*. Some Semaphorins bind directly to plexins; for example, semaphorin-4D binds to plexin-B1 *(84)*, semaphorin-3E to plexin-D1 *(85)*, and semaphorin-6D to plexin-A1 *(86)*. Type A plexins normally associate with NPs to form functional receptors for class 3 Sema *(87–90)*. Plexin-B1, on the other hand, forms complexes with the hepatocyte growth factor receptor, MET *(84)*. Sema6D potentiates the effects of VEGF as a result of complex formation between plexin-A1 and VEGFR2 *(86)*.

2.4. Notch and Delta   Originally discovered in *Drosophila*, the Notch signalling pathway is an example of a signalling system that plays multiple roles throughout development as well as affecting cell cycle progression and apoptosis. The complexity of Notch signalling is demonstrated by the presence of multiple Notch receptors and ligands, each having distinct expression profiles. Four Notch receptors (Notch 1–4) and five ligands (Jagged-1 and -2; Delta-1, -3, and -4) have so far been identified in this pathway *(91)*. Notch-delta signalling plays a crucial role in lateral inhibition; signalling from one cell expressing the ligand interacts with cells expressing the receptor. Notch activation results in cleavage and release of the intracellular domain, which translocates to the nucleus and activates transcription, thus blocking differentiation and allowing cells to proliferate and respond to later developmental ligands. The intracellular domain binds to the transcription factor RBP-Jk and then to a specific DNA sequence of the RBP-Jk binding site. This in turn up-regulates HES 1, 5, and 7 and other transcription factors such as HERP 1, 2, and 3, which have endothelial specificity *(92)*. All of the receptors and their ligands have been shown to be expressed in at least one vascular compartment (e.g., arteries, veins, capillaries, vascular smooth muscle cells, or pericytes) *(92)*. However, Notch-4 is specifically expressed in arterial vessels *(93)*, whereas other receptors are widely expressed in many cell types and tissues. Delta4 is also reported to be endothelial specific *(94, 95)*, and Notch signalling ligands and receptors are thought to be restricted to arterial vessels *(96)* and the venous vasculature *(91)*.

The importance of Notch in vascular development is highlighted by the genetic defect that arises in patients with mutations in the pathway, such as the Alagille syndrome, with loss-of-function mutations in a ligand for Notch and Jagged 1, and cerebral autosomal-dominant arteriopathy with subcortical infarcts and leukoencephalopathy (also known as CADASIL) *(97)*. Mice lacking Jagged-1 or Delta-1 were embryonic lethals and displayed severe vascular abnormalities and haemorrhaging of the embryo *(98, 99)*. Notch-1 and -4 double mutants and others with an intracellular processing-deficient allele of Notch-1 and embryos specifically lacking endothelial Notch-1 show defects in angiogenesis and vascular development of the yolk sac and placental labyrinthine and within the embryo itself and are embryonic lethal *(100–102)*. In the double knockouts, however, vasculogenesis remains normal, but angiogenic remodelling of the primary vascular plexus on the yolk sac does not occur *(100)*. The transcriptional targets of Notch, Hey1 and Hey2, are necessary for angiogenesis since loss of Hey1 alone does not lead to phenotypic defects, while double Hey1 and Hey2 knockout causes *de novo* vessel formation to occur, but vessels remain small or absent and fail to develop further, resulting in embryonic death *(103)*. Because Notch signalling modulates other pathways, such as the PI3 kinase-AKT and NF-κB, these and other studies demonstrated that Notch target genes have a key role in the regulation of angiogenic processes during embryonic development. Contrastingly, constitutive expression of active Notch4 led to vascular defects during development *(93)*, while in the adult mouse, constitutive activation in endothelial cells led to various organ-specific arteriovenous malformations and death that could be reversed by repression of Notch4 *(104)*.

2.5. Ephrins/Eph

Ephrins and their receptors, Ephs, are involved in various developmental processes during embryogenesis *(105–107)* and were initially characterised as involved in neuronal development *(108, 109)*. Eph proteins are receptor tyrosine kinases, and Ephrins are their membrane-bound ligands, often reciprocally expressed at tissue compartment boundaries, and take part in bidirectional signalling *(110, 111)*. Ephrins and Eph receptors are divided into either A or B types, depending on the way in which they are anchored to the plasma membrane (Fig. 1.3). There is considerable promiscuity in receptor–ligand binding; however, type A ephrins

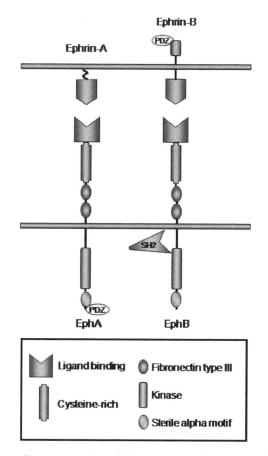

Fig. 1.3. Schematic representation of the structure of Eph and ephrins. Cell–cell contact occurs in a class-specific manner such that ephrinA ligands bind to EphA, while ephrinB ligands bind to EphB. Signalling occurs through binding of ephrins, which leads to receptor clustering and phosphorylation at specific sites. The two types of ephrins share similar structure, consisting of an extra-cellular globular domain, fibronectin type III repeats, cell membrane spanning domain, and tyrosine kinase domains.

preferentially bind Eph A receptors and type B ephrins bind type B Eph receptors.

The involvement of ephrins and Ephs is beginning to be unravelled in the development of the vasculature, and ephrin/Eph knockout mice have been reported to display a variety of vascular phenotypes. For example, knockdown of ephrinB2 or its cytoplasmic tail led to defects in vascular morphogenesis *(112)*, while disruption of either ephrinB2 or EphB4 led to failure in remodelling of the primary capillary plexus into properly patterned large and small vessels *(113, 114)*. These studies demonstrated the importance of ephrinB2/EphB4 signalling and their requirement for proper morphogenesis and patterning of the vascular system during development. The embryonic vasculature also expresses ephrinB1 in most arteries, while ephrin B1, EphB3, and EphB4 are expressed on veins *(115)*. Furthermore, ephrins and Ephs may mediate an interaction between the endothelium and pericyte interface since pericytes and smooth muscle cells adjacent to endothelial cells also express ephrinB2 and EphB2 *(115)*.

Several ephrins and Eph receptors are present on the vascular endothelium, including ephrinA1, which has a role in inflammatory angiogenesis induced by tumor necrosis factor-α *(116)*; ephrinB1, which promotes endothelial capillary-like assembly and attachment *in vivo(117)*; and ephrinB2, EphB3, and EphB4 *(115)*. In mouse xenografts of human MDA435 and K1767 Kaposi sarcoma cells and in the vasculature of human cancers, double immunostaining for CD34 showed ephrinA1 and EphA2 localisation throughout the endothelium *(118)*. A dominant negative EphA2 blocked capillary endothelial tube formation *in vitro(119)*, while other studies showed that soluble EphA2-Fc and EphA3-Fc inhibited tumor angiogenesis and growth *in vivo(119)*. Other studies have shown that blockade of EphA, ephrinB2, or EphB4 specifically inhibits VEGF-induced angiogenesis *(120, 121)*. These studies clearly demonstrated the mechanistic roles of ephrins/Ephs in tumor angiogenesis and provided novel anti-angiogenic and anti-tumor therapeutic targets.

2.6. Netrins and UNCs

The Netrins are a family of secreted proteins that are highly conserved throughout evolution and are the prototypical attractants. The family includes netrin-1, -2 (-3 in mice), and -4 and netrin-related molecules netrin-G1 and -G2. Netrins can act through the deleted in colorectal cancer (DCC) and the uncoordinated-5 (UNC5) receptor families. Netrin-1 is the most extensively studied gene in the netrin family and, in addition to its role in axon guidance, has been shown to be important in angiogenesis. Netrins-1, -2, and -4 have been shown to induce migration, proliferation, and tube formation in multiple cell lines, while knockdown of netrin-1a messenger RNA (mRNA) in zebrafish revealed a requirement for netrin signalling in the formation of parachordal vessels *(122)*. Netrins were also shown to activate blood vessel formation and accelerate revascularisation and reperfusion of ischaemic tissues in mice *(122)*.

In a study examining the expression of netrin receptors DCC, UNC5A, and UNC5B, only UNC5B expression was found in the vascular system during mouse embryogenesis *(123)*. UNC5B expression was also localized in arterial endothelium and in tip cells, indicating a role for UNC5B in tip cell guidance, while inactivation of UNC5B gene in mice led to embryonic lethality by E12.5, corresponding with abnormal branching of capillaries in the central nervous system *(123)*. In addition, morpholino knockdown of the orthologue of UNC5B or its ligand netrin-1a in zebrafish led to aberrant pathfinding of intersegmental vessels *(124)*. Furthermore, treatment of endothelial cells with netrin-1 resulted in tip cell filopodial retraction, an effect that was abolished in UNC5B-deficient mice *(123)*.

However, another study suggested that netrin-1 was pro-migratory, pro-adhesive, and pro-mitogenic on primary endothelial cells from human aorta and microvasculature *in vitro(125)*. The receptor implicated with this effect has not been identified, but the adenosine A2B receptor, which was reported to bind netrin-1 *(125)*, is expressed on endothelial cells, and may provide positive netrin-1 signals to endothelial cells *(124)*. The ability of netrins to promote growth of the vasculature has made them promising targets for therapeutic intervention and requires further investigation.

2.7. Roundabouts/Slits

Roundabouts (Robos) are single-pass transmembrane proteins that function as receptors for the Slit family of proteins and were thought to be restricted to cells of neuronal lineage. The slits and roundabouts mediate repulsive signalling in axon guidance, and Robo and Slit mutants exhibit defects in axon pathfinding at the ventral midline. To date, four Robos have been identified, with the structurally most divergent Robo4 (magic roundabout) showing endothelial-specific gene expression both *in vitro* and *in vivo(126)*. Although the role of Slit-Robo remains controversial in vascular guidance, interaction between Slit2 and Robo4 has been reported *(125)*; however, an interaction between Robo4 and any member of the Slit family has not been shown *(127)*. The effect of slit2 on endothelial cells is also controversial since one report claimed a promigratory effect *(128)* and another the opposite *(129)*.

Soluble Robo4 receptor has been reported to inhibit both *in vitro* endothelial cell migration and *in vivo* angiogenesis, although none of the slits have been shown to bind to this protein *(127)*. In zebrafish, morpholino knockdown of Robo4 resulted in temporal and spatial disruption of inter-somitic vessel development, with vessels sprouting from the aorta in the wrong direction and aborting prematurely. Such observations suggest that Robo4 may act as a guidance molecule in vascular development *(130)*. Another member of the Robo family, Robo1, has also been shown to be expressed on endothelial cells with a proposed role in Slit2-Robo1 signalling in promoting tumor angiogenesis *(128)*. Slit2 expression was localised in human malignant xenografts and in a range of cancer cell lines, while Robo1 expression was detected in mouse tumor endothelium. Tumor xenografts, which over-expressed Slit2, demonstrated increased tumor angiogenesis and accelerated tumor growth, while xenografts expressing soluble Robo1 showed reduced tumor growth and microvessel density. Given that Robo1 is able to form heterodimers with other Robos, it is possible that Robo1 and Robo4 interact on vascular sprouts or that Robo4 may modulate the response of Robo1 to slit2.

2.8. Hedgehog Signalling

Hedgehogs are a class of 19-kDa proteins that interact with heparin on the cell surface through N-terminal basic domains. They are tethered to the cell surface through cholesterol and fatty acid modification. Hedgehog signalling is crucial during development, and there exist three human homologues of the *Drosophila* hedgehog gene family: sonic hedgehog (shh), desert hedgehog (Dhh), and Indian hedgehog (Ihh). Signalling of hedgehogs occurs through interactions with the Patched1 receptor, which then activates transcription factors Gli1, Gli2, and Gli3. Shh is the most widely expressed of all the hedgehogs during development, and lack of Shh is embryonic lethal. Ihh is less widely expressed, and mice deficient in Ihh are able to survive until late gestation but eventually die as a result of skeletal and gut defects, while Dhh-deficient mice are viable but display peripheral nerve and male fertility defects.

Several studies have implicated a role for hedgehogs in angiogenesis. For example, hypervascularization of the neuroectoderm is observed with transgenic over-expression of Shh in the dorsal neural tube of zebrafish. As with Notch and Delta, both up- and down-regulation of Hedgehog protein give rise to defects in the vasculature. Shh administered to aged mice induced new vessel growth in ischaemic hind limbs *(131)*. An indirect involvement in angiogenesis has been demonstrated *in vitro* by the action of hedgehogs acting upstream of angiogenic factors *(131)*, with results later confirmed in zebrafish *(132)*. Other studies in zebrafish have shown that embryos lacking Shh activity fail to undergo arterial differentiation, characterised by the absence of arterial markers such as ephrinB2a, while injection of Shh mRNA induced ectopic ephrinB2a-specific vascular expression *(132)*. It appears that hedgehogs are primarily involved in arteriogenesis

2.9. Role of Sprouty in Angiogenesis

The *Drosophila Sprouty (Spry)* gene encodes a 63-kDa protein containing a cysteine-rich domain that is highly conserved in three human and four mouse homologues *(133–136)*. The *Drosophila* tracheal system and the mammalian lung are both formed by patterned branching morphogenesis, and branchless, a homologue of mammalian FGFs, is required for normal tracheal branch patterning *(137, 138)*. Branchless then activates Breathless, an FGF receptor homologue, and induces tracheal cell migration and branching *(137–140)*. Surprisingly, however, loss-of-function mutations in *Drosophila* caused enhanced tracheal branching *(136)*. Furthermore, inhibition of murine *Spry-2* by antisense oligonucleotides enhanced terminal branching of mouse cultured lung vessels *(134)*. These data suggest that Sprouty proteins negatively modulate branching morphogenesis in the *Drosophila* and mouse respiratory systems.

All four mammalian Sprys exhibit a restricted expression pattern in the embryo during early development, showing a dose correlation with sites of FGF signalling, suggesting a possible role of Spry in negative regulation of FGF signalling during mammalian and *Drosophila* development. Direct evidence for a role of Spry in angiogenesis was derived from a study in which mSpry4 was over-expressed in the developing endothelium of a mouse embryo using an adenoviral vector *(141)*. Embryos expressing mSpry4 showed less sprouting of smaller vessels from large ones and a primitive vasculature with poor branching and minimal sprouting of vessels, and the hearts of mSpry-expressing embryos were beating but incompletely developed *(141)*. Spry4 effects appear to be mediated through receptor tyrosine kinase pathways since there was a reduction in both basal and bFGF- or VEGF-induced MAPK phosphorylation in Spry4-expressing cells *(141)*. The MAPK pathway is important in the regulation of cell proliferation, migration, and differentiation during angiogenesis, and modulation of this pathway may account for some of the observed effects as a result of Spry over-expression.

3. Conclusions

Angiogenesis is one of the fastest growing fields in biomedical research, and advances in our basic understanding of the process and its clinical application have been relatively rapid. Angiogenesis is a complex, highly orchestrated process that plays a critical role in normal development and in the pathophysiology of common disease. Many different pathways converge on angiogenesis and include signalling by VEGF, Notch/delta, Ephrins/Ephs, semaphorins/plexins/neuropilins, slits/robos hedgehog, and sprouty (Fig. 1.4). Although substantial efforts have been made to understand the molecular regulators of vascular patterning and vessel guidance, only an outline of the interplay of mechanisms that coordinate these processes has been identified. Many similarities have been revealed between neuronal and vascular guidance. Indeed, vessels share the same evolutionary conserved guidance cues that enable axons to path-find their targets. It remains to be seen how these guidance cues orchestrate the development of new vessels.

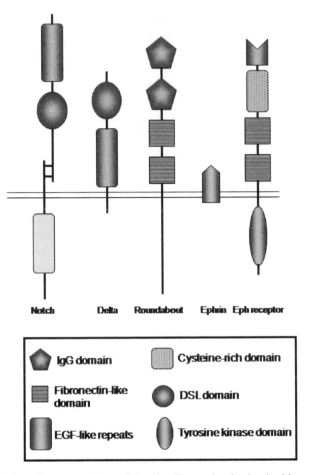

Fig. 1.4. Schematic representation of the signalling molecules involved in angiogenesis. The tumor endothelium has been shown to contain Notch, Delta, Roundabout, Ephrin, and their Eph receptors. It is yet unclear how these different molecules affect tumor angiogenesis; however, a greater understanding of the mechanisms of signalling may lead to better, more rationally designed antitumor/antiangiogenic therapy. *IgG* immunoglobulin G (Adapted from ref. *4*). DSL = Delta/Serrate/Lag2 domain and EGF = epidermal growth factor

References

1. Eskens, F. A. (2004) Angiogenesis inhibitors in clinical development; where are we now and where are we going? *Br J Cancer* 90, 1–7.

2. Tonini, T., Rossi, F., Claudio, P. P. (2003) Molecular basis of angiogenesis and cancer. *Oncogene* 22, 6549–6556.

3. Bicknell, R., Harris, A. L. (2004) Novel angiogenic signaling pathways and vascular targets. *Annu Rev Pharmacol Toxicol* 44, 219–238.

4. Sullivan, D. C., Bicknell, R. (2003) New molecular pathways in angiogenesis. *Br J Cancer* 89, 228–231.

5. Carmeliet, P. (2003) Blood vessels and nerves: common signals, pathways and diseases. *Nat Rev Genet* 4, 710–720.

6. Martin, P., Lewis, J. (1989) Origins of the neurovascular bundle: interactions between developing nerves and blood vessels in embryonic chick skin. *Int J Dev Biol* 33, 379–387.

7. Honma, Y., Araki, T., Gianino, S., et al. (2002) Artemin is a vascular-derived neurotropic factor for developing sympathetic neurons. *Neuron* 35, 267–282.

8. Kuruvilla, R., Zweifel, L. S., Glebova, N. O., et al. (2004) A neurotrophin signaling cascade coordinates sympathetic neuron development through differential control of TrkA trafficking and retrograde signaling. *Cell* 118, 243–255.

9. Mukouyama, Y. S., Shin, D., Britsch, S., Taniguchi, M., Anderson, D. J. (2002) Sensory nerves determine the pattern of arterial differentiation and blood vessel branching in the skin. *Cell* 109, 693–705.

10. Harris, A. L. (2002) Hypoxia--a key regulatory factor in tumor growth. *Nat Rev Cancer* 2, 38–47.

11. Wenger, R. H. (2002) Cellular adaptation to hypoxia: O2-sensing protein hydroxylases, hypoxia-inducible transcription factors, and O2-regulated gene expression. *FASEB J* 16, 1151–1162.

12. Ivan, M., Kondo, K., Yang, H., et al. (2001) HIFalpha targeted for VHL-mediated destruction by proline hydroxylation: implications for O2 sensing. *Science* 292, 464–468.

13. Leek, R. D., Talks, K. L., Pezzella, F., et al. (2002) Relation of hypoxia-inducible factor-2 alpha (HIF-2 alpha) expression in tumor-infiltrative macrophages to tumor angiogenesis and the oxidative thymidine phosphorylase pathway in Human breast cancer. *Cancer Res* 62, 1326–1329.

14. Pilch, H., Schlenger, K., Steiner, E., Brockerhoff, P., Knapstein, P., Vaupel, P. (2001) Hypoxia-stimulated expression of angiogenic growth factors in cervical cancer cells and cervical cancer-derived fibroblasts. *Int J Gynecol Cancer* 11, 137–142.

15. Ozawa, K., Kondo, T., Hori, O., et al. (2001) Expression of the oxygen-regulated protein ORP150 accelerates wound healing by modulating intracellular VEGF transport. *J Clin Invest* 108, 41–50.

16. Shimo, T., Kubota, S., Kondo, S., et al. (2001) Connective tissue growth factor as a major angiogenic agent that is induced by hypoxia in a human breast cancer cell line. *Cancer Lett* 174, 57–64.

17. Kondo, S., Kubota, S., Shimo, T., et al. (2002) Connective tissue growth factor increased by hypoxia may initiate angiogenesis in collaboration with matrix metalloproteinases. *Carcinogenesis* 23, 769–776.

18. Ambrosini, G., Nath, A. K., Sierra-Honigmann, M. R., Flores-Riveros, J. (2002) Transcriptional activation of the human leptin gene in response to hypoxia. Involvement of hypoxia-inducible factor 1. *J Biol Chem* 277, 34601–34609.

19. Hitchon, C., Wong, K., Ma, G., Reed, J., Lyttle, D., El-Gabalawy, H. (2002) Hypoxia-induced production of stromal cell-derived factor 1 (CXCL12) and vascular endothelial growth factor by synovial fibroblasts. *Arthritis Rheum* 46, 2587–2597.

20. Bacher, M., Schrader, J., Thompson, N., et al. (2003) Up-regulation of macrophage migration inhibitory factor gene and protein expression in glial tumor cells during hypoxic and hypoglycemic stress indicates a critical role for angiogenesis in glioblastoma multiforme. *Am J Pathol* 162, 11–17.

21. Green, C. J., Lichtlen, P., Huynh, N. T., et al. (2001) Placenta growth factor gene expression is induced by hypoxia in fibroblasts: a central role for metal transcription factor-1. *Cancer Res* 61, 2696–2703.

22. Murray, B., Wilson, D. J. (2001) A study of metabolites as intermediate effectors in angiogenesis. *Angiogenesis* 4, 71–77.

23. Senger, D. R., Connolly, D. T., Van de Water, L., Feder, J., Dvorak, H. F. (1990) Purification and NH2-terminal amino acid sequence of guinea pig tumor-secreted vascular permeability factor. *Cancer Res* 50, 1774–1778.

24. Shweiki, D., Itin, A., Soffer, D., Keshet, E. (1992) Vascular endothelial growth factor induced by hypoxia may mediate hypoxia-initiated angiogenesis. *Nature* 359, 843–845.

25. Olsson, A. K., Dimberg, A., Kreuger, J., Claesson-Welsh, L. (2006) VEGF receptor signalling - in control of vascular function. *Nat Rev Mol Cell Biol* 7, 359–371.

26. Fong, G. H., Rossant, J., Gertsenstein, M., Breitman, M. L. (1995) Role of the Flt-1 receptor tyrosine kinase in regulating the assembly of vascular endothelium. *Nature* 376, 66–70.

27. Kendall, R. L., Thomas, K. A. (1993) Inhibition of vascular endothelial cell growth factor activity by an endogenously encoded soluble receptor. *Proc Natl Acad Sci U S A* 90, 10705–10709.

28. Goldman, C. K., Kendall, R. L., Cabrera, G., et al. (1998) Paracrine expression of a native soluble vascular endothelial growth factor receptor inhibits tumor growth, metastasis, mortality rate. *Proc Natl Acad Sci U S A* 95, 8795–8800.

29. Autiero, M., Waltenberger, J., Communi, D., et al. (2003) Role of PlGF in the intra- and intermolecular cross talk between the VEGF receptors Flt1 and Flk1. *Nat Med* 9, 936–943.

30. Hiratsuka, S., Maru, Y., Okada, A., Seiki, M., Noda, T., Shibuya, M. (2001) Involvement of Flt-1 tyrosine kinase (vascular endothelial growth factor receptor-1) in pathological angiogenesis. *Cancer Res* 61, 1207–1213.

31. Dougher, M., Terman, B. I. (1999) Autophosphorylation of KDR in the kinase

domain is required for maximal VEGF-stimulated kinase activity and receptor internalization. *Oncogene* 18, 1619–1627.

32. Takahashi, T., Yamaguchi, S., Chida, K., Shibuya, M. (2001) A single autophosphorylation site on KDR/Flk-1 is essential for VEGF-A-dependent activation of PLC-gamma and DNA synthesis in vascular endothelial cells. *EMBO J* 20, 2768–2778.

33. Sakurai, Y., Ohgimoto, K., Kataoka, Y., Yoshida, N., Shibuya, M. (2005) Essential role of Flk-1 (VEGF receptor 2) tyrosine residue 1173 in vasculogenesis in mice. *Proc Natl Acad Sci U S A* 102, 1076–1081.

34. Matsumoto, T., Bohman, S., Dixelius, J., et al. (2005) VEGF receptor-2 Y951 signaling and a role for the adapter molecule TSAd in tumor angiogenesis. *EMBO J* 24, 2342–2353.

35. Zeng, H., Sanyal, S., Mukhopadhyay, D. (2001) Tyrosine residues 951 and 1059 of vascular endothelial growth factor receptor-2 (KDR) are essential for vascular permeability factor/vascular endothelial growth factor-induced endothelium migration and proliferation, respectively. *J Biol Chem* 276, 32714–32719.

36. Kaipainen, A., Korhonen, J., Mustonen, T., et al. (1995) Expression of the fms-like tyrosine kinase 4 gene becomes restricted to lymphatic endothelium during development. *Proc Natl Acad Sci U S A* 92, 3566–3570.

37. Jussila, L., Alitalo, K. (2002) Vascular growth factors and lymphangiogenesis. *Physiol Rev* 82, 673–700.

38. Dumont, D. J., Jussila, L., Taipale, J., et al. (1998) Cardiovascular failure in mouse embryos deficient in VEGF receptor-3. *Science* 282, 946–949.

39. Valtola, R., Salven, P., Heikkila, P., et al. (1999) VEGFR-3 and its ligand VEGF-C are associated with angiogenesis in breast cancer. *Am J Pathol* 154, 1381–1390.

40. Partanen, T. A., Alitalo, K., Miettinen, M. (1999) Lack of lymphatic vascular specificity of vascular endothelial growth factor receptor 3 in 185 vascular tumors. *Cancer* 86, 2406–2412.

41. Wang, J. F., Zhang, X., Groopman, J. E. (2004) Activation of vascular endothelial growth factor receptor-3 and its downstream signaling promote cell survival under oxidative stress. *J Biol Chem* 279, 27088–27097.

42. Makinen, T., Veikkola, T., Mustjoki, S., (2001) Isolated lymphatic endothelial cells transduce growth, survival and migratory signals via the VEGF-C/D receptor VEGFR-3. *EMBO J* 20, 4762–4773.

43. Walter, J. W., North, P. E., Waner, M., et al. (2002) Somatic mutation of vascular endothelial growth factor receptors in juvenile hemangioma. *Genes Chromosomes Cancer* 33, 295–303.

44. Stevens, A., Soden, J., Brenchley, P. E., Ralph, S., Ray, D. W. (2003) Haplotype analysis of the polymorphic human vascular endothelial growth factor gene promoter. *Cancer Res* 63, 812–816.

45. Bajou, K., Masson, V., Gerard, R. D., et al. (2001) The plasminogen activator inhibitor PAI-1 controls in vivo tumor vascularization by interaction with proteases, not vitronectin. Implications for antiangiogenic strategies. *J Cell Biol* 152, 777–784.

46. Luttun, A., Tjwa, M., Moons, L., et al. (2002) Revascularization of ischemic tissues by PlGF treatment, inhibition of tumor angiogenesis, arthritis and atherosclerosis by anti-Flt1. *Nat Med* 8, 831–840.

47. Pipp, F., Heil, M., Issbrucker, K., et al. (2003) VEGFR-1-selective VEGF homologue PlGF is arteriogenic: evidence for a monocyte-mediated mechanism. *Circ Res* 92, 378–385.

48. Adini, A., Kornaga, T., Firoozbakht, F., Benjamin, L. E. (2002) Placental growth factor is a survival factor for tumor endothelial cells and macrophages. *Cancer Res* 62, 2749–2752.

49. Eriksson, A., Cao, R., Pawliuk, R., et al. (2002) Placenta growth factor-1 antagonizes VEGF-induced angiogenesis and tumor growth by the formation of functionally inactive PlGF-1/VEGF heterodimers. *Cancer Cell* 1, 99–108.

50. Barbera-Guillem, E., Nyhus, J. K., Wolford, C. C., Friece, C. R., Sampsel, J. W. (2002) Vascular endothelial growth factor secretion by tumor-infiltrating macrophages essentially supports tumor angiogenesis, and IgG immune complexes potentiate the process. *Cancer Res* 62, 7042–7049.

51. Pasterkamp, R. J., Kolodkin, A. L. (2003) Semaphorin junction: making tracks toward neural connectivity. *Curr Opin Neurobiol* 13, 79–89.

52. Kruger, R. P., Aurandt, J., Guan, K. L. (2005) Semaphorins command cells to move. *Nat Rev Mol Cell Biol* 6, 789–800.

53. Guttmann-Raviv, N., Kessler, O., Shraga-Heled, N., Lange, T., Herzog, Y., Neufeld, G. (2006) The neuropilins and their role in tumorigenesis and tumor progression. *Cancer Lett* 231, 1–11.

54. Bagri, A., Tessier-Lavigne, M. (2002) Neuropilins as Semaphorin receptors: in vivo functions in neuronal cell migration

and axon guidance. *Adv Exp Med Biol* 515, 13–31.

55. Song, H., Ming, G., He, Z., et al. (1998) Conversion of neuronal growth cone responses from repulsion to attraction by cyclic nucleotides. *Science* 281, 1515–1518.

56. de Castro, F., Hu, L., Drabkin, H., Sotelo, C., Chedotal, A. (1999) Chemoattraction and chemorepulsion of olfactory bulb axons by different secreted semaphorins. *J Neurosci* 19, 4428–4436.

57. Wong, J. T., Wong, S. T., O'Connor, T. P. (1999) Ectopic semaphorin-1a functions as an attractive guidance cue for developing peripheral neurons. *Nat Neurosci* 2, 798–803.

58. Bagnard, D., Lohrum, M., Uziel, D., Puschel, A. W., Bolz, J. (1998) Semaphorins act as attractive and repulsive guidance signals during the development of cortical projections. *Development* 125, 5043–5053.

59. Miao, H. Q., Soker, S., Feiner, L., Alonso, J. L., Raper, J. A., Klagsbrun, M. (1999) Neuropilin-1 mediates collapsin-1/semaphorin III inhibition of endothelial cell motility: functional competition of collapsin-1 and vascular endothelial growth factor-165. *J Cell Biol* 146, 233–242.

60. Miao, H. Q., Lee, P., Lin, H., Soker, S., Klagsbrun, M. (2000) Neuropilin-1 expression by tumor cells promotes tumor angiogenesis and progression. *FASEB J* 14, 2532–2539.

61. Tse, C., Xiang, R. H., Bracht, T., Naylor, S. L. (2002) Human Semaphorin 3B (SEMA3B) located at chromosome 3p21.3 suppresses tumor formation in an adenocarcinoma cell line. *Cancer Res* 62, 542–546.

62. Xiang, R., Davalos, A. R., Hensel, C. H., Zhou, X. J., Tse, C., Naylor, S. L. (2002) Semaphorin 3F gene from human 3p21.3 suppresses tumor formation in nude mice. *Cancer Res* 62, 2637–2643.

63. Roche, J., Boldog, F., Robinson, M., et al. (1996) Distinct 3p21.3 deletions in lung cancer and identification of a new human semaphorin. *Oncogene* 12, 1289–1297.

64. Brambilla, E., Constantin, B., Drabkin, H., Roche, J. (2000) Semaphorin SEMA3F localization in malignant human lung and cell lines: a suggested role in cell adhesion and cell migration. *Am J Pathol* 156, 939–950.

65. Bielenberg, D. R., Hida, Y., Shimizu, A., et al. (2004) Semaphorin 3F, a chemorepulsant for endothelial cells, induces a poorly vascularized, encapsulated, nonmetastatic tumor phenotype. *J Clin Invest* 114, 1260–1271.

66. Neufeld, G., Shraga-Heled, N., Lange, T., Guttmann-Raviv, N., Herzog, Y., Kessler, O. (2005) Semaphorins in cancer. *Front Biosci* 10, 751–760.

67. Kessler, O., Shraga-Heled, N., Lange, T., et al. (2004) Semaphorin-3F is an inhibitor of tumor angiogenesis. *Cancer Res* 64, 1008–1015.

68. Ochi, K., Mori, T., Toyama, Y., Nakamura, Y., Arakawa, H. (2002) Identification of semaphorin3B as a direct target of p53. *Neoplasia* 4, 82–87.

69. Xu, X. M., Fisher, D. A., Zhou, L., et al. (2000) The transmembrane protein semaphorin 6A repels embryonic sympathetic axons. *J Neurosci* 20, 2638–2648.

70. Dhanabal, M., Wu, F., Alvarez, E., et al. (2005) Recombinant semaphorin 6A-1 ectodomain inhibits in vivo growth factor and tumor cell line-induced angiogenesis. *Cancer Biol Ther* 4, 659–668.

71. Kolodkin, A. L., Levengood, D. V., Rowe, E. G., Tai, Y. T., Giger, R. J., Ginty, D. D. et al. (1997) Neuropilin is a semaphorin III receptor. *Cell* 90, 753–762.

72. Kawasaki, T., Kitsukawa, T., Bekku, Y., et al. (1999) A requirement for neuropilin-1 in embryonic vessel formation. *Development* 126, 4895–4902.

73. Soker, S., Takashima, S., Miao, H. Q., Neufeld, G., Klagsbrun, M. (1998) Neuropilin-1 is expressed by endothelial and tumor cells as an isoform-specific receptor for vascular endothelial growth factor. *Cell* 92, 735–745.

74. Rossignol, M., Gagnon, M. L., Klagsbrun, M. (2000) Genomic organization of human neuropilin-1 and neuropilin-2 genes: identification and distribution of splice variants and soluble isoforms. *Genomics* 70, 211–222.

75. Gagnon, M. L., Bielenberg, D. R., Gechtman, Z., et al. (2000) Identification of a natural soluble neuropilin-1 that binds vascular endothelial growth factor: in vivo expression and antitumor activity. *Proc Natl Acad Sci U S A* 97, 2573–2578.

76. Yamada, Y., Takakura, N., Yasue, H., Ogawa, H., Fujisawa, H., Suda, T. (2001) Exogenous clustered neuropilin 1 enhances vasculogenesis and angiogenesis. *Blood* 97, 1671–1678.

77. Gluzman-Poltorak, Z., Cohen, T., Herzog, Y., Neufeld, G. (2000) Neuropilin-2 is a receptor for the vascular endothelial growth factor (VEGF) forms VEGF-145 and VEGF-165 [corrected]. *J Biol Chem* 275, 18040–18045.

78. Herzog, Y., Kalcheim, C., Kahane, N., Reshef, R., Neufeld, G. (2001) Differential

expression of neuropilin-1 and neuropilin-2 in arteries and veins. *Mech Dev* 109, 115–119.

79. Yuan, L., Moyon, D., Pardanaud, L., et al. (2002) Abnormal lymphatic vessel development in neuropilin 2 mutant mice. *Development* 129, 4797–4806.

80. Karkkainen, M. J., Saaristo, A., Jussila, L., et al. (2001) A model for gene therapy of human hereditary lymphedema. *Proc Natl Acad Sci U S A* 98, 12677–12682.

81. Takashima, S., Kitakaze, M., Asakura, M., et al. (2002) Targeting of both mouse neuropilin-1 and neuropilin-2 genes severely impairs developmental yolk sac and embryonic angiogenesis. *Proc Natl Acad Sci U S A* 99, 3657–3662.

82. Shen, J., Samul, R., Zimmer, J., et al. (2004) Deficiency of neuropilin 2 suppresses VEGF-induced retinal neovascularization. *Mol Med* 10, 12–18.

83. Comoglio, P. M., Trusolino, L. (2002) Invasive growth: from development to metastasis. *J Clin Invest* 109, 857–862.

84. Giordano, S., Corso, S., Conrotto, P., et al. (2002) The semaphorin 4D receptor controls invasive growth by coupling with Met. *Nat Cell Biol* 4, 720–724.

85. Gu, C., Yoshida, Y., Livet, J., et al. (2005) Semaphorin 3E and plexin-D1 control vascular pattern independently of neuropilins. *Science* 307, 265–268.

86. Toyofuku, T., Zhang, H., Kumanogoh, A., et al. (2004) Dual roles of Sema6D in cardiac morphogenesis through region-specific association of its receptor, Plexin-A1, with off-track and vascular endothelial growth factor receptor type 2. *Genes Dev* 18, 435–447.

87. Takahashi, T., Fournier, A., Nakamura, F., et al. (1999) Plexin-neuropilin-1 complexes form functional semaphorin-3A receptors. *Cell* 99, 59–69.

88. Tamagnone, L., Artigiani, S., Chen, H., et al (1999) Plexins are a large family of receptors for transmembrane, secreted, and GPI-anchored semaphorins in vertebrates. *Cell* 99, 71–80.

89. Suto, F., Murakami, Y., Nakamura, F., Goshima, Y., Fujisawa, H. (2003) Identification and characterization of a novel mouse plexin, plexin-A4. *Mech Dev* 120, 385–396.

90. Takahashi, T., Strittmatter, S. M. (2001) Plexin1 autoinhibition by the plexin sema domain. *Neuron* 29, 429–439.

91. Iso, T., Hamamori, Y., Kedes, L. (2003) Notch signaling in vascular development. *Arterioscler Thromb Vasc Biol* 23, 543–553.

92. Shawber, C. J., Das, I., Francisco, E., Kitajewski, J. (2003) Notch signaling in primary endothelial cells. *Ann N Y Acad Sci* 995, 162–170.

93. Uyttendaele, H., Ho, J., Rossant, J., Kitajewski, J. (2001) Vascular patterning defects associated with expression of activated Notch4 in embryonic endothelium. *Proc Natl Acad Sci U S A* 98, 5643–5648.

94. Mailhos, C., Modlich, U., Lewis, J., Harris, A., Bicknell, R., Ish-Horowicz, D. (2001) Delta4, an endothelial specific notch ligand expressed at sites of physiological and tumor angiogenesis. *Differentiation* 69, 135–144.

95. Shutter, J. R., Scully, S., Fan, W., et al. (2000) Dll4, a novel Notch ligand expressed in arterial endothelium. *Genes Dev* 14, 1313–1318.

96. Villa, N., Walker, L., Lindsell, C. E., et al. (2001) Vascular expression of Notch pathway receptors and ligands is restricted to arterial vessels. *Mech Dev* 108, 161–164.

97. Kalaria, R. N., Low, W. C., Oakley, A. E., et al. (2002) CADASIL and genetics of cerebral ischaemia. *J Neural Transm Suppl*, 75–90.

98. Xue, Y., Gao, X., Lindsell, C. E., et al. (1999) Embryonic lethality and vascular defects in mice lacking the Notch ligand Jagged1. *Hum Mol Genet* 8, 723–730.

99. Hrabe de Angelis, M., McIntyre, J., 2nd, Gossler, A. (1997) Maintenance of somite borders in mice requires the Delta homologue Dll1. *Nature* 386, 717–721.

100. Krebs, L. T., Xue, Y., Norton, C. R., (2000) Notch signaling is essential for vascular morphogenesis in mice. *Genes Dev* 14, 1343–1352.

101. Huppert, S. S., Le, A., Schroeter, E. H., et al. (2000) Embryonic lethality in mice homozygous for a processing-deficient allele of Notch1. *Nature* 405, 966–970.

102. Limbourg, F. P., Takeshita, K., Radtke, F., Bronson, R. T., Chin, M. T., Liao, J. K. (2005) Essential role of endothelial Notch1 in angiogenesis. *Circulation* 111, 1826–1832.

103. Fischer, A., Schumacher, N., Maier, M., Sendtner, M., Gessler, M. (2004) The Notch target genes Hey1 and Hey2 are required for embryonic vascular development. *Genes Dev* 18, 901–911.

104. Carlson, T. R., Yan, Y., Wu, X., et al. (2005) Endothelial expression of constitutively active Notch4 elicits reversible arteriovenous malformations in adult mice. *Proc Natl Acad Sci U S A* 102, 9884–9889.

105. Holder, N., Klein, R. (1999) Eph receptors and ephrins: effectors of morphogenesis. *Development* 126, 2033–2044.

106. Adams, R. H., Klein, R. (2000) Eph receptors and ephrin ligands, essential mediators of vascular development. *Trends Cardiovasc Med* 10, 183–188.

107. Zhang, J., Hughes, S. (2006) Role of the ephrin and Eph receptor tyrosine kinase families in angiogenesis and development of the cardiovascular system. *J Pathol* 208, 453–461.

108. Wilkinson, D. G. (2001) Multiple roles of EPH receptors and ephrins in neural development. *Nat Rev Neurosci* 2, 155–164.

109. Kullander, K., Klein, R. (2002) Mechanisms and functions of Eph and ephrin signalling. *Nat Rev Mol Cell Biol* 3, 475–486.

110. Palmer, A., Klein, R. (2003) Multiple roles of ephrins in morphogenesis, neuronal networking, and brain function. *Genes Dev* 17, 1429–1450.

111. Gale, N. W., Holland, S. J., Valenzuela, D. M., et al. (1996) Eph receptors and ligands comprise two major specificity subclasses and are reciprocally compartmentalized during embryogenesis. *Neuron* 17, 9–19.

112. Adams, R. H., Diella, F., Hennig, S., Helmbacher, F., Deutsch, U., Klein, R. (2001) The cytoplasmic domain of the ligand ephrinB2 is required for vascular morphogenesis but not cranial neural crest migration. *Cell* 104, 57–69.

113. Wang, H. U., Chen, Z. F., Anderson, D. J. (1998) Molecular distinction and angiogenic interaction between embryonic arteries and veins revealed by ephrin-B2 and its receptor Eph-B4. *Cell* 93, 741–753.

114. Gerety, S. S., Wang, H. U., Chen, Z. F., Anderson, D. J. (1999) Symmetrical mutant phenotypes of the receptor EphB4 and its specific transmembrane ligand ephrin-B2 in cardiovascular development. *Mol Cell* 4, 403–414.

115. Adams, R. H., Wilkinson, G. A., Weiss, C., et al. (1999) Roles of ephrinB ligands and EphB receptors in cardiovascular development: demarcation of arterial/venous domains, vascular morphogenesis, sprouting angiogenesis. *Genes Dev* 13, 295–306.

116. Pandey, A., Shao, H., Marks, R. M., Polverini, P. J., Dixit, V. M. (1995) Role of B61, the ligand for the Eck receptor tyrosine kinase, in TNF-alpha-induced angiogenesis. *Science* 268, 567–569.

117. Stein, E., Lane, A. A., Cerretti, D. P., et al. (1998) Eph receptors discriminate specific ligand oligomers to determine alternative signaling complexes, attachment, assembly responses. *Genes Dev* 12, 667–678.

118. Ogawa, K., Pasqualini, R., Lindberg, R. A., Kain, R., Freeman, A. L., Pasquale, E. B. (2000) The ephrin-A1 ligand and its receptor, EphA2, are expressed during tumor neovascularization. *Oncogene* 19, 6043–6052.

119. Brantley, D. M., Cheng, N., Thompson, E. J., et al. (2002) Soluble Eph A receptors inhibit tumor angiogenesis and progression in vivo. *Oncogene* 21, 7011–7026.

120. Cheng, N., Brantley, D. M., Chen, J. (2002) The ephrins and Eph receptors in angiogenesis. *Cytokine Growth Factor Rev* 13, 75–85.

121. Cheng, N., Brantley, D. M., Liu, H., et al. (2002) Blockade of EphA receptor tyrosine kinase activation inhibits vascular endothelial cell growth factor-induced angiogenesis. *Mol Cancer Res* 1, 2–11.

122. Wilson, B. D., Ii, M., Park, K. W., et al. (2006) Netrins promote developmental and therapeutic angiogenesis. *Science* 313, 640–644.

123. Lu, X., Le Noble, F., Yuan, L., et al. (2004) The netrin receptor UNC5B mediates guidance events controlling morphogenesis of the vascular system. *Nature* 432, 179–186.

124. Klagsbrun, M., Eichmann, A. (2005) A role for axon guidance receptors and ligands in blood vessel development and tumor angiogenesis. *Cytokine Growth Factor Rev* 16, 535–548.

125. Park, K. W., Crouse, D., Lee, M., et al. (2004) The axonal attractant Netrin-1 is an angiogenic factor. *Proc Natl Acad Sci U S A* 101, 16210–16215.

126. Huminiecki, L., Bicknell, R. (2000) In silico cloning of novel endothelial-specific genes. *Genome Res* 10, 1796–1806.

127. Suchting, S., Heal, P., Tahtis, K., Stewart, L. M., Bicknell, R. (2005) Soluble Robo4 receptor inhibits in vivo angiogenesis and endothelial cell migration. *FASEB J* 19, 121–123.

128. Wang, B., Xiao, Y., Ding, B. B., et al. (2003) Induction of tumor angiogenesis by Slit-Robo signaling and inhibition of cancer growth by blocking Robo activity. *Cancer Cell* 4, 19–29.

129. Park, K. W., Morrison, C. M., Sorensen, L. K., et al. (2003) Robo4 is a vascular-specific receptor that inhibits endothelial migration. *Dev Biol* 261, 251–267.

130. Bedell, V. M., Yeo, S. Y., Park, K. W., et al. (2005) Roundabout4 is essential for angiogen-

esis in vivo. *Proc Natl Acad Sci U S A* 102, 6373–6378.

131. Pola, R., Ling, L. E., Silver, M., et al. (2001) The morphogen Sonic hedgehog is an indirect angiogenic agent upregulating two families of angiogenic growth factors. *Nat Med* 7, 706–711.

132. Lawson, N. D., Vogel, A. M., Weinstein, B. M. (2002) sonic hedgehog and vascular endothelial growth factor act upstream of the Notch pathway during arterial endothelial differentiation. *Dev Cell* 3, 127–136.

133. Hacohen, N., Kramer, S., Sutherland, D., Hiromi, Y., Krasnow, M. A. (1998) sprouty encodes a novel antagonist of FGF signaling that patterns apical branching of the *Drosophila* airways. *Cell* 92, 253–263.

134. Tefft, J. D., Lee, M., Smith, S., et al. (1999) Conserved function of mSpry-2, a murine homolog of *Drosophila sprouty*, which negatively modulates respiratory organogenesis. *Curr Biol* 9, 219–222.

135. de Maximy, A. A., Nakatake, Y., Moncada, S., Itoh, N., Thiery, J. P., Bellusci, S. (1999) Cloning and expression pattern of a mouse homologue of *Drosophila sprouty* in the mouse embryo. *Mech Dev* 81, 213–216.

136. Minowada, G., Jarvis, L. A., Chi, C. L., et al. (1999) Vertebrate *Sprouty* genes are induced by FGF signaling and can cause chondrodysplasia when overexpressed. *Development* 126, 4465–4475.

137. Sutherland, D., Samakovlis, C., Krasnow, M. A. (1996) branchless encodes a *Drosophila* FGF homolog that controls tracheal cell migration and the pattern of branching. *Cell* 87, 1091–1101.

138. Metzger, R. J., Krasnow, M. A. (1999) Genetic control of branching morphogenesis. *Science* 284, 1635–1639.

139. Klambt, C., Glazer, L., Shilo, B. Z. (1992) breathless, a *Drosophila* FGF receptor homolog, is essential for migration of tracheal and specific midline glial cells. *Genes Dev* 6, 1668–1678.

140. Lee, T., Hacohen, N., Krasnow, M., Montell, D. J. (1996) Regulated Breathless receptor tyrosine kinase activity required to pattern cell migration and branching in the *Drosophila* tracheal system. *Genes Dev* 10, 2912–2921.

141. Lee, S. H., Schloss, D. J., Jarvis, L., Krasnow, M. A., Swain, J. L. (2001) Inhibition of angiogenesis by a mouse sprouty protein. *J Biol Chem* 276, 4128–4133.</bh>

Chapter 2

Imaging Angiogenesis

Natalie Charnley, Stephanie Donaldson, and Pat Price

Abstract

There is a need for direct imaging of effects on tumor vasculature in assessment of response to anti-angiogenic drugs and vascular disrupting agents. Imaging tumor vasculature depends on differences in permeability of vasculature of tumor and normal tissue, which cause changes in penetration of contrast agents. Angiogenesis imaging may be defined in terms of measurement of tumor perfusion and direct imaging of the molecules involved in angiogenesis. In addition, assessment of tumor hypoxia will give an indication of tumor vasculature. The range of imaging techniques available for these processes includes positron emission tomography (PET), dynamic contrast-enhanced magnetic resonance imaging (DCE-MRI), perfusion computed tomography (CT), and ultrasound (US).

Key words: Dynamic contrast-enhanced MRI (DCE-MRI), hypoxia, perfusion, perfusion computed tomography (CT), positron emission tomography (PET), ultrasound, volume of distribution of water (V_d).

1. Introduction

Angiogenesis is the formation of new blood vessels essential for tumor growth, which is required when tumors reach up to 1 mm³ *(1)*. Imaging angiogenesis in tumors is of tantamount importance for early assessment of the efficacy of anti-cancer treatment. There is a need for direct imaging of effects on tumor vasculature in assessment of response to anti-angiogenic drugs and vascular-disrupting agents (VDAs), which impair angiogenesis and destroy existing vasculature, respectively. Conventional imaging that examines tumor size is less useful for assessing efficacy of these agents as they are cytostatic rather than tumor shrinking. Perfusion, the nutritive flow of blood through tissues, is governed by

S. Martin and C. Murray (eds.), *Methods in Molecular Biology, Angiogenesis Protocols, Second edition, Vol. 467*
© Humana Press, a part of Springer Science+ Business Media, LLC 2009
DOI: 10.1007/978-1-59745-241-0_2

the tumor vasculature. Assessment of tumor angiogenesis in terms of perfusion can also be a measure of pharmacodynamic response to many chemotherapeutic agents *(2)*, monitoring response to radiotherapy, and drug delivery *(3)*.

In vivo imaging of angiogenesis or perfusion has the potential to characterise indeterminate lung lesions and identify occult metastases *(4)*. Imaging perfusion may also correlate with grade in some tumors *(4)* and provide prognostic information in patients with cerebral glioma *(5)*, lymphoma *(6)*, and head and neck cancer *(7,8)*. However, this must be approached cautiously as there are conflicting reports on the relationship between perfusion and outcome. Perfusion can be both a poor and a good prognostic factor, depending on the imaging technique employed *(7, 8)*.

Tumor vasculature is chaotic. Tumor blood vessels have structural and functional abnormalities. This includes irregular branching, increased permeability, and independence from normal flow control mechanisms. Such abnormalities lead to variable and inadequate perfusion. Differences in permeability of vasculature of tumor and normal tissue cause changes in penetration of contrast agents, and this is exploited in imaging tumor vasculature. There may be differences in tumor perfusion depending on where the region of interest is defined. For example, there may be high perfusion in the angiogenic periphery of a tumor but low perfusion in a necrotic centre (**Fig. 2.1**).

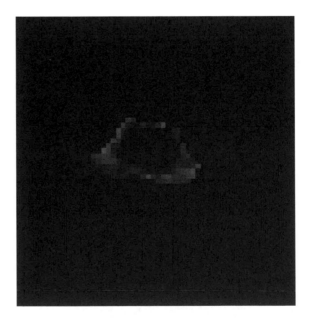

Fig. 2.1. Map of K^{trans} for a prostate tumor showing "rim-and-core" characteristics. Higher values of K^{trans} are seen in the rim of the tumor compared to the centre of the tumor, indicating higher enhancement and possibly vascularity in the rim. (Courtesy of Gio Buonocorssi, ISBE, University of Manchester.).

Angiogenesis is triggered when tumors reach a certain size. This may be determined by hypoxia, acidosis, and hypoglycaemia *(9)*. Angiogenesis is coordinated by a plethora of cytokines, including vascular endothelial growth factor (VEGF), which also has effects on vascular permeability. Other cytokines are not vascular endothelium specific but do promote angiogenesis. These include fibroblast growth factor (FGF), platelet-derived growth factor (PDGF), and transforming growth factor-β (TGF-β). Expression of FGF and VEGF is up-regulated in tumors, and levels correlate with vascularity *(10)*. Angiogenesis is also potentiated by matrix metalloproteases produced by activated endothelial cells, which degrade the basement membrane, and integrins, which enable endothelial cell migration toward the tumor.

Angiogenesis imaging may be defined in terms of measurement of tumor perfusion, direct imaging of the molecules involved in angiogenesis; in addition, assessment of tumor hypoxia will give an indication of tumor vasculature. The range of imaging techniques available for these processes include positron emission tomography (PET), dynamic contrast-enhanced magnetic resonance imaging (DCE-MRI), perfusion computed tomography (CT), and ultrasound (US).

2. Positron Emission Tomographic Imaging of Blood Flow

Imaging of blood flow by PET is a well-established and well-validated technique *(11)* and gives truly quantitative data (**Fig. 2.2**). This method may be considered the gold standard. PET imaging is based on positron emission from short-lived radioisotopes produced by a cyclotron. The radiotracer or probe is produced by replacing a molecule of interest with a radio-labelled molecule, and this is subsequently administered to the patient by injection or inhalation. Positrons are emitted by nuclear decay from the tracer and collide with electrons in the tissues in an annihilation reaction. During this process, two 511-keV gamma (γ) rays are produced at 180° to each other. Pairs of scintillation ring detectors transmit a coincident signal when both are stimulated simultaneously by the γ-rays at 180°. Information is corrected for photon attenuation, and three-dimensional tomographic images of tissue concentration are reconstructed.

Regional tumor blood flow is the usual parameter measured in nuclear medicine imaging. Regional flow is flow standardized to a unit quantity of tissue and has the units per millilitre blood/minute/decilitre tissue. In the case of arteriovenous shunting, for which the capillary bed is bypassed, flow may occur without perfusion.

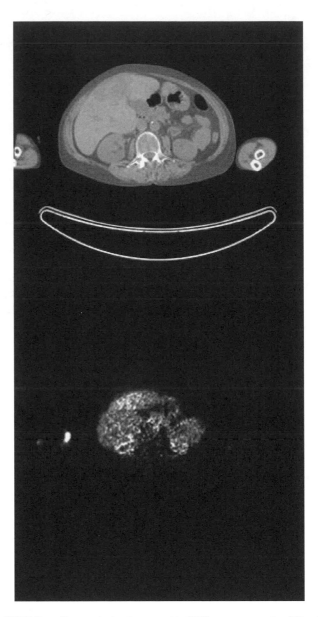

Fig. 2.2. [15][O]H$_2$O positron emission tomographic (PET) measurements of blood flow in liver metastases. Computed tomographic (CT; *superior*) and [15][O]H$_2$O PET (*inferior*) images of metastases from a neuroendocrine tumor. Note the high blood flow in liver metastases and normal kidneys.

Volume of distribution (V_d) can also be measured. This is the steady-state ratio of the activity in tissue to activity in blood/plasma or a regional volume of distribution, that is, the fractional volume of blood/plasma to account for the activity within a region.

The classical measurement of tumor perfusion in nuclear medicine depends on the injection of a tracer into the bloodstream and subsequent measurement in the tumor. These methods are based on the Fick principle *(12)* and produce quantitative measurements of blood flow.

$$P = \frac{C_t}{\int (C_a(t) - C_v(t))dt} \tag{1}$$

where- $C_t(t)$ is the concentration of tracer in tissue; $C_a(t)$ is the concentration of tracer in arterial blood; and $C_v(t)$ is the concentration of tracer in venous blood.

This method is not clinically practical as it requires both arterial and venous dynamic measurements and assumes homogeneity of venous blood. For clinical practice, the freely diffusible tracer method is used in which the rate of tracer uptake is measured. This does not require venous data.

Either the steady-state or dynamic method can be utilised to obtain values for perfusion. Both these methods necessitate the use of a freely diffusible tracer, which is commonly the isotope ^{15}O. This isotope is often used in the forms of $^{15}[O]\,CO_2$, or $^{15}[O]H_2O$ and has a half-life of 123 s. An on-site cyclotron is essential for its generation.

2.1. Steady-State Method

The steady-state method was used extensively in early studies of PET and perfusion. This method involves inhalation of $^{15}[O]$ CO_2, which is subsequently converted to $^{15}[O]H_2O$ in the lungs. Tracer diffuses not only from arterial blood to tissue but also from tissue to venous blood. Equilibrium is reached after 10 min so that the movement of tracer from arterial blood to tissue is matched by that from tissue to venous blood and radioactive decay.

Regional flow is given as

$$F = \frac{\lambda}{\dfrac{C_a(t)}{C_t(t)} - 1/v_d} \tag{2}$$

where λ is the decay constant for ^{15}O (0.338/min) *(11)*; $C_a(t)$ is the arterial concentration of $H_2{}^{15}O$ (Bq/mL); $C_t(t)$ is the tissue concentration of $H_2{}^{15}O$ (Bq/mL); and v_d is the volume of distribution of water in millilitre (blood)/millilitre (tissue).

A region of interest is defined on the PET scan, and from this tissue radioactivity concentration is quantitatively measured. The arterial radioactivity concentration is assessed by arterial sampling using a cross-calibrated well counter or from a large arterial pool from the PET image, such as the abdominal aorta.

Advantages of the steady-state method compared to the dynamic method are that it requires only a simple calculation and minimal blood sampling. However, it is particularly sensitive to effects from tissue heterogeneity. There are inaccuracies in the predefined partition coefficient, which is often not known in diseased tissues such as tumor, and this causes particular problems with this method *(13)*. In addition, perfusion in heterogeneous tissues is often underestimated *(13, 14)*.

2.2. Dynamic Method

Inhalation of $^{15}[O]CO_2$ or injection of a bolus of $^{15}[O]H_2O$ can be used for both steady-state and dynamic methods. However, modern dynamic methods usually involve injection of a bolus of $^{15}[O]H_2O$. Compared to steady-state inhaled $^{15}[O]CO_2$, a lower dose is needed, and the scan time is shorter. For example, a steady-state acquisition time would typically be approximately 30 min, compared with 10 min for dynamic scanning.

For dynamic scanning, kinetics of activity are measured in tissue within individual time frames. Continuous or rapid discrete arterial data are obtained from the image or arterial blood sampling. This enables construction of arterial input corrected time activity curves of changing ^{15}O activity in a region of interest (**Fig. 2.2**). Data must be corrected for delay and dispersion of the arterial curve due to blood passage through sampling tubing when using blood samples.

The Kety-Schmidt model based on Kety's original model *(15)* is employed to model kinetic data, and estimates of blood flow and V_d are put into the model. This model incorporates a term for radioactive decay. Estimates of flow and V_d are perturbed until a good fit is obtained.

Change in tissue concentration with time is given by

$$\frac{dC_t(t)}{dt} = F.C_a(t) - \left(\frac{F}{V_d} + \lambda\right).C_t(t) \qquad (3)$$

The solution is given by the equation

$$C_t(t) = F_a.C_a(t) * exp\left(-\left(\frac{F}{V_d} + \lambda.t\right)\right) \qquad (4)$$

where the asterisk (*) denotes the operation of convolution.

There are several advantages to dynamic scanning. Data acquisition is faster, so a lower radiation dose is delivered to the patient, and scans can be repeated every 10 min. The model is less sensitive to heterogeneity than the steady-state method. Flow and V_d are estimated separately. However, long reconstruction times and large data sets are needed, and data are noisier. In addition, data modelling is highly complex, and extensive blood sampling is required (50 mL per scan). As for the steady-state method, it makes the assumption that water is freely diffusible and fully extracted from circulation and is poor for high blood flow regions.

2.3. Imaging Blood Volume

Blood volume imaging involves inhalation of a tracer dose of $^{15}[O]CO$ in oxygen, then breathing room air. This continues until arteriovenous $C^{15}O$-Hb carboxyhaemoglobin equilibrates within blood. This typically only takes 1–2 min. Tissue radioactivity and arterial $C^{15}O$-Hb are recorded. As carboxyhaemoglobin remains in vessels, the ratio of activity in tissue to blood is the blood volume within the imaged tissues. Problems associated with this technique include the low signal-to-noise ratio *(16)*, particularly compared with MRI.

PET imaging of blood flow using labelled water is well validated and gives reproducible data *(3)*. As the technique involves following kinetics of labelled water, there are no issues associated with contrast agent toxicity. However, there are limitations. Imaging with $^{15}[O]H_2O$ PET requires an on-site cyclotron and multidisciplinary expertise and is thus very limited in availability. Spatial resolution is limited, which can lead to partial volume effects: spillover of radioactivity into neighbouring structures and underlying tissue inhomogeneity. In addition, there is also spillover of counts from adjacent structures with high blood flow. Multiple tracers need to be utilised for multiple parameters (e.g., $^{15}[O]H_2O$ and $^{15}[O]CO$ for perfusion and blood volume, respectively) *(17)*.

2.4. Imaging Blood Flow with DCE-MRI

The use of DCE-MRI is increasingly popular for diagnosis. A contrast agent is injected intravenously, which is taken up more readily by tumor tissue than the surrounding normal tissue. This contrast agent produces changes in $1/T_1$ values on magnetic resonance (MR) images in proportion to its concentration in tissue. Signal enhancement can therefore be seen on T_1-weighted images. By analysing signal enhancement over time, tracer kinetic parameters can be estimated that give information on the physiology of tissues and tumors.

DCE-MRI is now used routinely in the diagnosis of various cancers, particularly of the breast, cervix, and prostate. Different types of signal–time curves have been shown in breast cancer to be capable of distinguishing benign from "suspicious" and malignant tumors *(18)*. DCE-MRI is also being investigated as a method for monitoring the effects of anti-angiogenic and anti-vascular drugs, which aim to halt the development of tumors by stopping the process of angiogenesis or destroying existing tumor vasculature, respectively *(19)*.

Angiogenesis in tumors produces leaky blood vessels with high blood flow, which results in the high uptake of the contrast agent by tumors in comparison to normal tissues. Analysis of signal–time curves in DCE-MRI may be able to provide a non-invasive assessment of angiogenesis.

The contrast agents used in MRI of angiogenesis contain either paramagnetic or superparamagnetic ions, which enhance both the T_1 and T_2 relaxation rates of neighbouring water protons in direct proportion to the concentration of the contrast agent. The effects seen in T_1- and T_2-weighted DCE-MRI are very different and are exploited using different MR sequences to acquire either T_1- or T_2-weighted data. This chapter discusses mostly T_1-weighted DCE-MRI as this method is more popular than T_2-weighted DCE-MRI. T_2-weighted DCE-MRI is discussed briefly at the end of this section.

Gadolinium chelates, such as Gd-DTPA (Magnevist), are the most commonly used contrast agents in MR. Gd-DTPA has a high T_1 relaxivity and an excellent safety record *(20)*. It is a low molecular weight agent, does not enter the intracellular space, and has a high extraction fraction E in tumors (E ~ 45%) (the fraction of contrast agent that diffuses from the plasma to the extracellular, extravascular space [EES] in each circulatory pass) *(21)*. Macromolecular contrast media (MMCM), which have lower extraction fractions, are in the early stages of clinical trials and may provide exciting opportunities for imaging angiogenesis in the future. For an overview, see Choyke *(22)*.

Requirements for an MR protocol in T_1-weighted DCE-MRI are the ability to image a volume rapidly. Pulse sequences are mostly spoiled two- or three-dimensional (2D or 3D) gradient echo (GE) sequences capable of rapidly scanning a volume while providing high T_1-weighted contrast. Parallel imaging is now standard on most scanners and may be used to further speed image acquisition *(23, 24)*.

The equation relating signal intensity to T_1 for a spoiled echo GE sequence such as (fast low angle shot) FLASH is shown below:

$$S = M_0 . \frac{\left(1 - \exp\left(-\frac{TR}{T_1}\right)\right).\sin(\alpha)}{\left(1 - \cos(\alpha).\exp\left(-\frac{TR}{T_1}\right)\right)} . \exp\left(\frac{-TE}{T_2{}^*}\right) \qquad (5)$$

where S is the signal produced by a particular flip angle α; M_0 is a term that includes proton density and machine-dependent gain; TR is the pulse repetition time; α is the flip angle used; TE is the echo time; and $T_2{}^*$ is the $T_2{}^*$ relaxation time (this term is generally ignored as short TE values are chosen to minimise the influence of $T_2{}^*$ on the MR signal).

At a time t following injection of the contrast agent, the tissue T_1 will be given by

$$\frac{1}{T_1(t)} = \frac{1}{T_1(0)} + r_1 C(t) \qquad (6)$$

where r_1 is the T_1 relaxivity of the contrast agent; $T_1(0)$ is the pre-contrast T_1 value of the tissue; and $C_t(t)$ is the concentration of the contrast agent in the tissue at time t.

Equations 5 and **6** can be combined to relate signal intensity to contrast agent concentration. It follows that a non-linear relationship exists between contrast agent concentration and signal intensity, particularly at the high concentrations of contrast agent *(25)* that may occur in blood vessels. It is therefore necessary to estimate pre-contrast T_1 values, from which a calculation of the contrast agent concentration can be made.

Pre-contrast T_1 values are obtained using either inversion *(26)* or saturation recovery *(27)* pulse sequences with varying delays between the preparation and normal imaging pulses or variable saturation techniques, such as multiple spin echo (SE) or GE sequences using multiple flip angles *(28)*. T_1 can be estimated in a number of ways, the most popular of which involves a non-linear least-squares fit *(29)* of the multiple signals obtained for the signal equation.

Most tracer kinetic models include a plasma concentration term or arterial input function (AIF), which provides an estimate of the arterial blood supplying the tissue of interest. Studies have assumed various forms of the AIF *(30, 31)*; however, it is becoming accepted practice to obtain an AIF on a patient-by-patient basis by rapidly imaging a supplying artery in parallel with imaging of the tumor. This has been shown to improve the accuracy of tracer kinetic parameters *(32)* as blood supply has been shown to vary between patients *(33)*. The fields of view acquired in DCE-MRI must therefore include both the tissue of interest and a suitable artery.

Most models in common use are based on a situation in which the tumor/tissue of interest is separated into two compartments: the plasma and the EES. The contrast agent (tracer) is carried to the tissue of interest via the arteries and removed via the veins. It diffuses into the EES at a rate dependent on the capillary permeability–surface area product PS, a measure of "leakiness," and the blood flow F. The main model in use is the generalised kinetic model *(15)*. The main drawback of this model is its inability to separate out perfusion P and flow F. The St. Lawrence and Lee model *(34)* is able to do this, so these two models are discussed. Parameter estimation for most models is commonly carried out by non-linear least-squares fitting to the concentration–time equation *(35)*.

The generalised kinetic model is based on the Kety model *(15)* and adapted by Tofts et al. *(36)* for use in DCE-MRI. The rate of transfer equation is shown next, where $C_p(t)$ and $C_e(t)$ are the concentrations at time t in the plasma and EES, respectively. This equation is equivalent to **Equation 3** used in dynamic PET modelling.

$$v_e \frac{dC_e(t)}{dt} = E.F_p.(C_p(t) - C_e(t)) \qquad (7)$$

where v_e is the volume of the EES; F_p is plasma flow; and E is known as the extraction fraction of the tracer, the fraction of tracer that is removed from the plasma to the EES in a single capillary transit *(32, 37)*:

$$E = 1 - \exp\left(-\frac{PS}{F_p}\right) \qquad (8)$$

where PS is the permeability-surface area product.

The "extended" kinetic model includes the addition of a vascular term $C_p(t)$:

$$C_t(t) = v_p C_p(t) + v_e C_e(t) \qquad (9)$$

where v_p is the volume of the plasma.

The vascular concentration $C_p(t)$ is provided directly via the AIF. The solution to the rate-of-transfer equation is

$$C_t(t) = v_p C_p(t) + E.F_p.\int_0^t C_p(t') \exp\left(\frac{-E.F_p.(t-t')}{v_e}\right) dt' \qquad (10)$$

Using this model, it is impossible to separate out permeability PS (see **Equation 8**) and plasma flow F_p. On fitting to this model, it is therefore common to derive a general transfer constant K^{trans}, a rate constant governing transfer from the EES to the plasma. The physiological meaning of K^{trans} varies depending on the permeability and flow of the tracer in the tissue of interest. When blood flow is fast, the transfer of the tracer between the plasma and EES is limited by permeability, and K^{trans} = PS. When blood flow is slow or permeability is high, the transfer of tracer is flow limited, and K^{trans} = F_p. Most situations will be of mixed permeability— and flow-limited regimes. In this situation, K^{trans} = E.F_p with E defined as in **Equation 8**.

A further parameter k_{ep} relating this transfer constant and the volume of the EES v_e can be derived using

$$k_{ep} = \frac{K^{trans}}{v_e} \qquad (11)$$

The St. Lawrence and Lee model *(34)* attempts to separate out the effects of flow F_p and permeability PS. It is an adiabatic approximation to the tissue homogeneity model of transport across a permeable blood–brain barrier developed by Johnson and Wilson *(38)*. The tissue homogeneity model defines tracer concentration within the plasma as a function of both time and distance along the length of the capillary.

The tissue concentration $C_p(t)$ is related to plasma concentration thus:

$$C_t(t) = F_p.C_p(t)*R(t) \qquad (12)$$

where the asterisk (*) denotes the convolution operation, and $R(t)$ is the tissue residue function:

$$R(t) = \left\{ E.exp\left(-\frac{E.F_p^1}{v_e}(t-\tau) \right) \begin{array}{l} 0 < t \le \frac{v_p}{F_p} \\[2ex] t \ge \frac{v_p}{F_p} \end{array} \right. \qquad (13)$$

where τ is the capillary transit time.

The solution to this equation is given by

$$C_t(t) = F_p.\int_0^\tau C_p(t-t').dt' + E.F_p.\int_\tau^t C_p(t').exp\left(\frac{-E.F_p.(t-\tau-t')}{v_e} \right) dt' \qquad (14)$$

Standard (model-free) parameters can be obtained by analysing the signal–time curve and include relative signal increase and time-to-peak enhancement. These parameters are computationally less intensive to calculate than tracer kinetic parameters; however, they are difficult to relate to physiological properties, such as blood flow and permeability *(32)*, and depend largely on the scanner and imaging protocol used, so a comparison of results between MR centres is nearly impossible. Providing an intermediate step between tracer kinetic parameters and standard parameters is the initial area under the (concentration–time) curve IAUC, *(39)* up to, for example, 60 s following the onset of signal enhancement, the $IAUC_{60s}$:

$$IAUC_t = \int_0^t C_t(t')dt' \qquad (15)$$

Calculation of the IAUC again relies on knowledge of pre-contrast T_1 values to enable conversion from signal intensity to concentration of the contrast agent, although calculation of an AIF is not necessary. Values are not indicative of any physiological parameter but are generally though to reflect the amount of contrast agent delivered and retained in the tumor within the defined time period *(32)*. MRI has the advantage of high spatial resolution compared to imaging modalities such as PET and ultrasound. Data, however, are semi-quantitative. It has been shown to be sensitive to even small changes in flow *(40)*. There is no ionising radiation dose associated with MRI, making it more suitable for repeated examinations in, for example, clinical trials of anti-angiogenic drugs. In addition, Gd-DTPA has an excellent safety record, and there is a vast amount of clinical experience in its use. As an imaging modality, MR is capable of measuring many different things; therefore, in one examination it may be possible to obtain T_1-, T_2-weighted images, DCE-MRI measurements, MR angiography, and diffusion-weighted images of the same anatomy.

Reproducibility of DCE-MRI results between centres has been difficult largely due to differences in the following:

- Data acquisition: The contrast agent media and dose administered, the injection rate, and the MR protocol employed may differ.

- Data analysis: The tracer kinetic model used, the parameters obtained, whether data are analysed on a pixel-by-pixel basis or average values are obtained over a whole region of interest (ROI) may be different.

- The definition of the ROI: Is the whole tumor to be looked at or the most-enhancing part of the tumor ("hot spot"?).

As yet, there is no consensus on how best to carry out DCE-MRI examinations; however, recommendations for carrying out DCE-MRI studies in oncology have been published in an effort to make comparison between studies easier *(19, 39, 41)*.

Measurement of an AIF requires high temporal resolution imaging sequences as the concentration of the contrast agent in a supplying artery changes rapidly, and undersampling of the signal–time course can result in the peak enhancement being underestimated. This often means that spatial resolution has to be sacrificed. There are also concerns over the best way to obtain the AIF: Manually defining a region of interest in an artery can be subject to user bias, and so automated methods *(42, 43)* may be preferable. Finding a suitable artery may be difficult, and the resulting AIF may be subject to partial volume and inflow effects. To avoid the problems associated with obtaining an AIF, many centres rely on AIF data published in literature *(30, 44)*.

The usual MR contraindications (such as pacemakers and metal implants) apply to the suitability of patients for DCE-MRI examinations, and MR-related artefacts can occur when imaging certain anatomy (e.g., ghosting artefacts due to respiration or cardiac motion, flow in arteries and veins, and susceptibility effects at air–tissue interfaces).

Dynamic T_2- and T_2*-weighted MR techniques, also known as dynamic susceptibility contrast MRI (see **ref.** *45* for a review), are also used for imaging of angiogenesis. A decrease in T_2 occurs due to variations in the local magnetic field introduced by the presence of the contrast agent. This results in a decrease in T_2 and a signal loss on T_2-weighted MR images. Modelling of this signal loss is more difficult in T_2-weighted DCE-MRI than the modelling in T_1-weighted DCE-MRI due to a number of effects: susceptibility effects occurring at air–tissue boundaries, such as the sinuses, resulting in image distortion; and T_1-weighted enhancement occurring in tissues where normal leakage of the contrast agent into the EES has taken place. T_2-weighted techniques are therefore most commonly used in the brain, where the contrast agent leaks into the EES only if the blood–brain barrier has been disrupted, for example, in brain tumors. Modelling can be applied to the signal changes to estimate relative blood flow, relative blood volume, and mean transit time (MTT). Studies suggested that these parameters may reflect angiogenesis levels in tumors (46–48).

2.5. Imaging Flow with Perfusion CT

Perfusion CT methodology has been developed by Miles's group (4, 49). Images are acquired over time, before and after administration of iodinated contrast, leading to generation of time–attenuation curves. Enhancement in tissues is compared to that in a regional artery. For first-pass imaging, contrast remains intravascular, enabling measurement of perfusion and blood volume. Subsequent delayed enhancement assesses movement of contrast into the extravascular space. In contrast to MRI, it is possible to obtain truly quantitative parameters as there is a linear relationship between enhancement change and iodine concentration. As perfusion may be affected by patients' cardiac output, the term standardised perfusion value (SPV) has been described, which is tumor perfusion divided by whole-body perfusion. Cardiac output, determined from the arterial time attenuation curve, is divided by patient weight to give whole-body perfusion (50).

CT protocols vary in terms of timing of imaging following administration of contrast, spatial and temporal resolution, noise, and radiation dose. Choice of protocol is determined by the relative importance of these factors. For example, increasing the number of CT images results in better-quality perfusion data, and larger tube current reduces noise, but both these measures increase radiation dose (4).

Kinetic analysis of the data may involve a compartmental model (slope method, Mullani-Gould formulation), deconvolution (linear systems approach), or Patlak analysis (two-compartmental model). Choice of model is determined by the particular perfusion parameter studied.

Compartmental analysis is based on the Fick principle. As for PET, the Fick principle necessitates the additional analysis of venous data. The Mullani-Gould formulation and slope methods obviate the need for venous data by making the assumption that there is no washout of contrast material before maximum tissue enhancement (49). The slope method provides perfusion data earlier than with the Mullani-Gould formulation, so ensures compliance with the assumption of no venous washout (49). Perfusion ($\frac{E}{V}$) is defined as the maximum slope of the tissue enhancement curve divided by the maximum arterial enhancement. Thus (51),

$$\frac{F}{V} = \frac{\frac{d}{dt}\left[C_t(t)\right]_{max}}{\left[C_a(t)\right]_{max}} \tag{16}$$

Compartmental analysis enables calculation of several parameters. Blood volume is calculated from the ratio of peak tissue enhancement to peak arterial enhancement. Time to peak tissue enhancement may also be obtained. MTT may also be determined from the time attenuation curve. Parameters derived by this method may be affected by bolus volume and rate and patient cardiac output *(49)*.

The deconvolution method involves the use of arterial and tissue time attenuation curves to calculate the tissue impulse response function (IRF). This is the time attenuation curve from an idealised instantaneous injection of one unit of tracer and is analogous to the range of tissue transit times. *(49)*. The arterial time attenuation curve may be convolved with the impulse response function to give the tissue time attenuation curve. This is given by the convolution

$$(t) = C_a(t) * \frac{F}{V} \bullet R(t) \tag{17}$$

where R(t) is the impulse response function *(52)*. The perfusion-scaled impulse response curve enables calculation of the parameters perfusion, MTT, and blood volume *(49)*. Assessment of delayed enhancement enables calculation of quantitative permeability data using Cenic's alteration *(53)*.

Choice of CT protocol must account for the method of data analysis used. The advantage of the deconvolution method over compartmental analysis is reduced sensitivity to noise *(4)*, but at the expense of a longer acquisition time. Rate of injection of bolus is also determined by the data analysis used. A rapid bolus is usually needed for compartmental analysis compared to deconvolution.

Perfusion CT is potentially widely available and relatively cheap. It may be particularly good for assessing flow at certain sites. For example, CT is better than MRI for assessment of flow at the pulmonary hilum due to MRI-related artefacts *(54)*. The technique has the advantage that commercial analysis software approved by the U.S. Food and Drug Administration (FDA) is now available. However, there are some limitations. Contrast agent may be contraindicated in some patients due to renal toxicity. Patient respiration can lead to attenuation and motion artefacts. CT is associated with a radiation dose, and the field of view is smaller than for PET or MRI. The maximum detector width for current CT systems is 4 cm, although with a table-toggling technique, this can give 8-cm coverage in the Z-direction.

2.6. Ultrasound Measurements of Flow

Ultrasound is able to image tumor vascularity and blood flow. Both pulsed and continuous-wave Doppler are used to image flow *(55)*. Standard clinical Doppler ultrasound (DU) in the range 2–10 MHz can detect blood flow in vessels greater than 200 μm; however, most of these vessels will be larger than those involved in angiogenesis *(55)*. DU has been used clinically to image flow in vessels of patients with cervical cancer *(56)* and, by establishing variation in vascularisation, has been found to differentiate low-risk prostate tumors from high-risk, hypervascular tumors *(57)*. High-frequency DU in the range 20 to 100 MHz can scan with a resolution of 15 to 100 μm and so has the ability to assess flow velocity in specific microvessels *(55)*.

Ultrasound derives the velocity of blood-borne particles, including erythrocytes. Velocity v has been related to perfusion, ρ by the equation given by Dymling *(3)*. This model does not necessitate the use of an AIF.

$$\rho = \frac{F_c}{V_t} \alpha \, n_c \, \bar{v}_c \qquad (18)$$

where F_c is total capillary blood flow; v_t is the sample volume of tissue; n_c is the number of capillaries in the volume; and $\bar{v}_c$ is the mean capillary blood velocity.

For contrast-enhanced ultrasound imaging of flow, the contrast agents in question are microbubbles measuring 1–10 μm. These gas-filled bubbles are coated with a shell, which may be lipid or albumin based *(55)*. They are retained in the intravascular compartment, including capillaries, and their vascular path may be traced by enhancement of the ultrasound signal. Microbubbles can be infused at a constant rate, resulting in a steady-state concentration. When destroyed by a high-power ultrasound pulse, their rate of reaccumulation and signal intensity can be detected using intermittent imaging with low-amplitude ultrasound. This enables derivation of vascular volume as well as flow *(58)*. Quantitative data can be obtained using time intensity curves. This may be achieved using a simple one-compartment model, in contrast to PET, CT, and MRI, as the contrast remains intravascular *(55)*. Parameters such as area under the curve and time to peak intensity can be generated *(59)*.

Ultrasound has the advantage of being cheap and widely available, but it has several drawbacks. It is very operator dependent and has a small field of view. Certain regions may be inaccessible to the ultrasound probe or poorly imaged (e.g., lung). Furthermore, ultrasound has poor resolution in comparison to other imaging techniques and is subject to unpredictable attenuation by tissues such as body fat.

3. Results of Imaging in Clinical Trials

The incorporation of imaging techniques in recent clinical studies of anti-angiogenic agents and VDAs has been discussed by Laking et al. *(3)*. In contrast to the experience with anti-angiogenic agents, all the studies using VDAs could demonstrate physiological responses to the treatment given. **Tables 2.1 and 2.2** summarise the use of PET, DCE-MRI, CT, or Doppler in assessing change in vascular function following administration of antiangiogenic agents or VDAs.

DCE-MRI has been used to study the effects of a number of anti-vascular and anti-angiogenic drugs. In spite of recommendations made when acquiring DCE-MRI data in oncology *(19, 39, 41)*, comparisons between the results of drug trials are difficult due to differences in the doses of drugs given, the timing of MR examinations, the MR protocols used to acquire the data, and subsequent analysis of the data.

DCE-MRI parameters reflecting the rate of enhancement were shown to decrease within 24 h of treatment in studies on the effects of the VDA CA4P *(60–62)* and the anti-angiogenic drugs PTK787 and ZK222584 *(63, 64)*, suggesting that they are capable of disrupting the vascular supply to the tumor. A study into another anti-angiogenic drug, bevacizumab, demonstrated that decreases in tracer kinetic parameters occurred with each cycle of treatment. DCE-MRI studies into the effects of other anti-angiogenic drugs—SU5416 *(65)*, CDP860 *(66)*, and recombinant human endostatin *(67)*—showed no change in DCE-MRI parameters obtained pre- and post-treatment. Some studies have gone further by investigating whether changes in DCE-MRI parameters obtained early in the course of treatment can predict the response to treatment, with those demonstrating better response having larger decreases in DCE-MRI parameters *(63, 64)*.

3.1. Histological Correlates of Blood Flow Measurements

Validation of angiogenesis with histology is complicated by the plethora of surrogates of angiogenesis, such as assessment of microvessel density (MVD), microarray analysis of angiogenic genes, and expression of proteins, VEGF, CD34, and factor VIII. There is no gold standard. In addition, presence of microvessels does not necessarily equate with perfused tissue.

There are few data published relating ^{15}[O]H$_2$O PET to angiogenesis. However, CT and MRI have been validated against angiogenic markers in a range of tumor sites *(68, 69)*. A summary of the DCE-MRI measures that correlate with histopathology and survival were given by Brasch *(69)*. Significant positive correlations between MR parameters and MVD have been shown in a number of studies in breast *(70–72)* and cervix carcinomas

Table 2.1
Monitoring antiangiogenic agents by imaging vascular changes

Study	n	Treatment	Tumor	Test	Follow-up	Findings
Herbst, 2002 (103)	24	Endostatin[a]	Various	PET	28, 56 d	Complex BF dose-response relationship
Eder, 2002 (67)	10	Endostatin[a]	Various	MRI$_2$	3 mo	No consistent change in MRI parameters
Thomas, 2003 (104)	21	Endostatin	Various	CT MRI PET	4,8 wk	No consistent effect on tumor vasculature
Anderson, 2003 (105)	6	Razoxane	Renal cell	PET	4–8 wk	No change in individual BF or VD; no correlation with razoxane outcomes
Kurdziel, 2003 (106)	6	Thalidomide	Prostate	PET	9 wk	Negligible correlation ΔPSA, ΔBF
Lassau, 2004 (107)	40	Thalidomide	Renal cell	DU	6 weekly	↓ BV → S.D.; ↑ BV → PD
Jayson, 2002 (87)	4	HUMV833	Various	MRI	2, 35 d	K^{trans} in all 4 pts at 48 h
Willett, 2004 (108)	5	Bevacizumab	Various	CT	12 d	↓ Tumor BF, BV, MVD, IFP
Morgan, 2003 (64)	22	PTK/ZK[a]	Various	MRI	2, 30, 58 d	Evidence for a dose-response effect; negative correlation AUC_{drug}, $ΔK_i$
Jayson, 2005 (66)	8	CDP860	Abdominal	MRI	1, 7, 27, 45 d	Fluid retention 7/8, ascites 3/8 pts; no change in MRI parameters
Liu, 2005 /(109)	17	AG-0173736	Various	MRI	2 d	Evidence for a dose-response effect
McNeel, 2005 (110)	25	MEDI-522	Various	CT	8 wk	↑ in contrast mean transit time with ↑ doses of drug
Xiong, 2004 (111)	6	SU6668	Various	CT MRI	4 wk, 12 wk	5/6 ↓ flow on CT 2/4 ↓ AUC and/or maximum slope on MRI

Source: Adapted from ref. *3.BF* blood flow, *BV* blood volume, *IFP* interstitial fluid pressure, *pts* patients, ↑ increase, ↓ decrease, → "is associated with." *See text* for additional abbreviations.

[a]Phase I dose-ranging trial. Endostatin promotes endothelial cell apoptosis. Razoxane has mixed cytotoxic/antiangiogenic activity. The precise mechanism of action of thalidomide is unclear. HUMV833 and bevacizumab are anti-VEGF antibodies. PTK/ZK inhibits the VEGF receptor. CDP860 blocks the PDGF receptor β-subunit. AG-0173736 inhibits both VEGF and PDGF receptors. MEDI-522 is an anti-alphavbeta3 integrin antibody. SU6668 is an inhibitor of angiogenic receptor tyrosine kinase.

Table 2.2
Monitoring vascular disruptive agents by imaging vascular changes

Study	n	Treatment	Tumor	Test	Follow-up	Findings
Falk, 1994 (112)	6	BW12C	Various	CT	1h	Slight ↓ tumor BF in 5/6 pts
Logan, 2002 (113)	8	IL-1	Lung mets	PET	2, 4, 24h	Arterial BP, ↓ tumor BF, VD, ↑ BV at 2h
Galbraith, 2002 (114)	16	DMXAA[a]	Various	MRI_1	4, 24h	Tumor-specific ↓ G, E, and AUC_{MRI} at 24h
Anderson, 2003 (115)	6	CA4P[a]	Various	PET	0.5, 24h	Dose-dependent ↓ ρ, BV at 30 min
Dowlati, 2002 (60)	7	CA4P[a]	Various	MRI_1	4–6h	↓ G_{peak} in 6/7 pts at 60 mg/m² dose
Stevenson, 2003 (62)	10	CA4P[a]	Various	MRI_2	6–8h	↓ K^{trans} in 8/10 pts; negative correlation AUC_{drug}, ΔK^{trans}
Galbraith, 2003 (61)	18	CA4P[a]	Various	MRI_2	4, 24h	↓ K^{trans} in 8/16 pts with evidence of a dose effect; negative correlation AUC_{drug}, ΔK^{trans}

Source: Adapted from ref. 3. *Mets* metastases, *pts* patients, ↑ increase, ↓ decrease, → "is associated with." *See text* for additional abbreviations. Phase I dose-ranging trial. BW12C stabilizes oxyhaemoglobin, causing ↓ BF, hypoxia. IL-1 (interleukin-1) causes haemorrhagic necrosis in animals, and DMXAA causes TNF-α-mediated vascular shutdown. CA4P binds tubulin, with selective toxicity to the tumor vasculature.

(73–75). Other studies have shown positive correlations between MR parameters and VEGF in breast (76) and rectal carcinomas (77). There have been several studies correlating CT flow parameters with histological measures of angiogenesis. Some of these studies, however, have given conflicting results. In studies of pancreatic cancer, renal cancer, and oral cancer, CT enhancement was positively correlated with MVD (78–80). In lung cancer, one study showed no relationship between MVD and flow but a significant relationship between enhancement and VEGF expression (81), whereas a second study demonstrated a positive correlation between CT enhancement and MVD as determined by factor VIII staining (82).

Using DU, the vascularity index has been positively correlated with MVD in patients with cervical cancer (56), and in prostate cancer, a positive relationship has been demonstrated between Doppler imaging of flow and MVD (83).

Measurements of flow can be affected by tumor factors such as presence of shunts, necrosis, and interstitial pressure and by imaging factors such as parameter measured; permeability of contrast agent can also affect flow. Perhaps different imaging tools will reflect different aspects of tumor biology.

3.2. Molecular Imaging of Angiogenesis

Angiogenesis has a significant role in determining tumor perfusion. Non-invasive molecular imaging can detect molecules that are selectively expressed on tumor vasculature and involved in angiogenesis. Imaging involves production of a labelled ligand or a labelled antibody to the ligand. Many of these studies are still in an early phase of clinical development or indeed preclinical.

The integrin alphavbeta3 is expressed selectively on activated tumor vasculature. This has been imaged in a phase II study using a labelled antibody specific for this integrin, and retention occurred only in the patient who had an alphavbeta3-positive tumor (84). The TGF-β receptor endoglin is a proliferation-associated endothelial marker. Labelled anti-endoglin bound specifically to neovasculature in an animal model (85). VEGF expression has been imaged in patients with pancreatic cancer. Iodine-123-labelled VEGF enabled imaging of the majority of primary tumors and their metastases (86). A phase I proof-of-principle study of iodine-124-labelled anti-VEGF antibody, however, showed heterogeneous antibody distribution and clearance between and within patients (87).

There have been some developments in imaging matrix metalloproteinases (MPs) and FGF expression. An inhibitor of MP labelled with indium-111 was administered safely to patients with Kaposi's sarcoma, but there was no tumor uptake (88). Preclinical studies of labelled FGF antibody showed retention of antibody only in tumor that expressed FGF (89).

Developments in MRI technology have enabled imaging of angiogenic pathways and involve linking of the contrast agent to an angiogenic marker. Expression of alphavbeta3 has been demonstrated in an animal model. MRI-detectable liposomes were constructed carrying Arg-Gly-Asp peptides (RGD), which accumulated specifically on activated tumor endothelium *(90)*. Expression of the endothelial inflammatory marker e-selectin in human umbilical cells has been detected using MRI of constructs of iron oxide nanoparticles and anti-human e-selectin *(91)*. Angiogenic markers are also detectable by ultrasound microbubbles linked to angiogenic ligands. Microbubbles were constructed with avidin incorporated into an albumin shell and linked to monoclonal antibodies specific for endoglin. The microbubbles specifically bound to endoglin-expressing endothelial cells *(92)*. In a pre-clinical study, microbubbles were conjugated to peptides containing tripeptide arginine-arginine-leucine (RRL), which specifically binds angiogenic endothelium. Ultrasound showed enhancement specifically in tumor *(93)*.

This is an expanding area, and choice of the imaging technique is determined by local availability and expertise. PET had the best sensitivity and is able to detect molecules in picomolar concentrations. PET-CT combines sensitivity with improved resolution. However, considerable radiopharmaceutical skills are necessary for development of ligands, as is an on-site cyclotron. Progress in this field depends on development of new labelled ligands or antibodies involved in tumor angiogenesis. Candidate tracers should demonstrate not only a relatively high level of specific signal bound to the site but also kinetics that are amenable to correction for background/non-specific binding and quantitation.

3.3. Imaging of Hypoxia

Hypoxia stimulates both angiogenesis and blood flow in tumors and occurs as a result of the aberrant tumor vasculature that develops. Tumor hypoxia adversely affects prognosis and is associated with reduced survival following surgery or radiotherapy *(94, 95)*. There is good evidence for a benefit of hypoxic modification therapy *(96)*. A number of PET probes are being developed for measuring hypoxia, including ^{64}Cu-(ATSM) diacetyl-bis (N^4-methylthiosemicarbazone) and ^{18}F-fluoromisonidazole (^{18}F-FMISO), and of these ^{18}F-FMISO is the most widely studied. This is a member of the group of nitroimidazoles that diffuses into a cell and undergoes electron reduction to form reactive species. In the presence of oxygen, the molecule is reoxidised. In hypoxic conditions, it forms covalent bonds with intracellular macromolecules and is trapped in the cell. There are studies of hypoxic imaging with ^{18}F-FMISO in human head and neck cancer, soft tissue sarcoma, breast cancer, and glioblastoma multiforme *(97)*, and these have shown inter-tumor heterogeneity of ^{18}F-FMISO uptake *(97, 98)*. ^{18}F-FMISO has recently been tested against a range of hypoxic markers, including oxygen electrodes *(99)*.

The early delivery of ^{18}F-FMISO into tumors correlates with blood flow. Flow-independent measurements of tissue hypoxia are possible around 90–120 min following injection. Tumor uptake is generally quantified relative to plasma or muscle levels. However, a kinetic model for analysis of dynamic ^{18}F-FMISO PET data has been described *(100)*. This was used to derive parameters that reflected well-oxygenated/perfused tissue, diffusion-limited hypoxia, and strong hypoxia/necrosis and was able to predict prognosis of patients undergoing radiotherapy. New tracers are being developed (e.g., etanidazole) that may have better imaging characteristics than ^{18}F-FMISO.

Blood oxygenation level-dependent (BOLD) imaging using MRI can monitor oxygen status in tissues. Deoxygenated blood is paramagnetic, thereby causing changes in the local magnetic field, which can be visualised as low signal intensity on T_2^*-weighted MR images. Oxygenated blood is diamagnetic, resulting in little or no change in signal intensity. Areas with a "hypoxic blood volume" are therefore demonstrated by areas of low signal intensity on T_2^*-maps. MR sequences used are either echo planar imaging (EPI) based or multi-echo gradient echo sequences. Studies have shown improvements in the BOLD signal in a variety of tumors during carbogen breathing *(101, 102)* as well as correlations between BOLD signal intensity and MVD.

4. Conclusions

Imaging angiogenesis is an important area for development of anti-angiogenic drugs and VDAs. Angiogenesis itself may be imaged by novel molecular markers, but also important are measurement of perfusion and tumor hypoxia resulting from inadequate perfusion. Future challenges in this field include production of new specific ligands for molecular imaging of angiogenesis and improved understanding of the complex interaction among angiogenesis, perfusion, and hypoxia in tumors.

Acknowledgements

We acknowledge Professor K. Miles, Dr. D. Buckley and Dr. J. Matthews for their advice.

References

1. Folkman, J. (1971) Tumor angiogenesis: therapeutic implications. *N Engl J Med* 285, 1182–1186.

2. Mankoff, D. A., Dunnwald, L. K., Gralow, J. R., et al. (2003) Changes in blood flow and metabolism in locally advanced breast cancer treated with neoadjuvant chemotherapy. *J Nucl Med* 44, 1806–1814.

3. Laking, G. R., West, C., Buckley, D. L., et al. (2006) Imaging vascular physiology to monitor cancer treatment. *Crit Rev Oncol Hematol* 58, 95–113.

4. Miles, K. A. (2003) Perfusion CT for the assessment of tumour vascularity: which protocol? *Br J Radiol* 76 (Spec No 1), S36–S42.

5. Leggett, D. A., Miles, K. A., Kelley, B.B. (1998) Blood–brain barrier and blood volume imaging of cerebral glioma using functional CT: a pictorial review. *Australas Radiol* 42, 335–340.

6. Dugdale, P. E., Miles, K. A., Bunce, I., et al. (1999) CT measurement of perfusion and permeability within lymphoma masses and its ability to assess grade, activity, and chemotherapeutic response. *J Comput Assist Tomogr* 23, 540–547.

7. Hermans, R., Meijerink, M., Van den Bogaert, W., et al. (2003) Tumor perfusion rate determined noninvasively by dynamic computed tomography predicts outcome in head-and-neck cancer after radiotherapy. *Int J Radiat Oncol Biol Phys* 57, 1351–1356.

8. Lehtio, K., Eskola, O., Viljanen, T., et al. (2004) Imaging perfusion and hypoxia with PET to predict radiotherapy response in head-and-neck cancer. *Int J Radiat Oncol Biol Phys* 59, 971–982.

9. Carmeliet, P., Jain, R. K. (2000) Angiogenesis in cancer and other diseases. *Nature* 407, 249–257.

10. Shemirani, B., Crowe, D. L. (2000) Head and neck squamous cell carcinoma lines produce biologically active angiogenic factors. *Oral Oncol* 36, 61–66.

11. Laking, G. R., Price, P. M. (2003) Positron emission tomographic imaging of angiogenesis and vascular function. *Br J Radiol* 76 (Spec No 1), S50–S59.

12. Acierno, L. J. (2000) Adolph Fick, mathematician, physicist, physiologist. *Clin Cardiol* 23, 390–391.

13. Frackowiak, R. S., Jones, T., Lenzi, G. L., et al.(1980) Regional cerebral oxygen utilization and blood flow in normal man using oxygen-15 and positron emission tomography. *Acta Neurol Scand* 62, 336–344.

14. Lammertsma, A. A., Jones, T. (1992). Low oxygen extraction fraction in tumours measured with the oxygen-15 steady state technique: effect of tissue heterogeneity. *Br J Radiol* 65, 697–700.

15. Kety, S. S. (1951) The theory and applications of the exchange of inert gas at the lungs and tissues. *Pharm Rev* 3, 1–41.

16. Ito, H., Kanno, I., Kato, C., et al.(2004) Database of normal human cerebral blood flow, cerebral blood volume, cerebral oxygen extraction fraction and cerebral metabolic rate of oxygen measured by positron emission tomography with ^{15}O-labelled carbon dioxide or water, carbon monoxide and oxygen: a multicentre study in Japan. *Eur J Nucl Med Mol Imaging* 31, 635–643.

17. Bacharach, S. L., Libutti, S. K., Carrasquillo, J. A. (2000) Measuring tumor blood flow with H(2)(15)O: practical considerations. *Nucl Med Biol* 27, 671–676.

18. Kuhl, C. K., Mielcareck, P., Klaschik, S., et al. (1999) Dynamic breast MR imaging: are signal intensity time course data useful for differential diagnosis of enhancing lesions? *Radiology* 211, 101–110.

19. Leach, M. O., Brindle, K. M., Evelhoch, J. L., et al. (2003) Assessment of antiangiogenic and antivascular therapeutics using MRI: recommendations for appropriate methodology for clinical trials. *Br J Radiol* 76 (Spec No 1), S87–S91.

20. Runge, V. M. (2000) Safety of approved MR contrast media for intravenous injection. *J Magn Reson Imaging* 12, 205–213.

21. Preda, A., van Vliet, M., Krestin, G. P., et al. (2006) Magnetic resonance macromolecular agents for monitoring tumor microvessels and angiogenesis inhibition. *Invest Radiol* 41, 325–331.

22. Choyke, P. L. (2005) Contrast agents for imaging tumor angiogenesis: is bigger better? *Radiology* 235, 1–2.

23. Pruessmann, K. P., Weiger, M., Scheidegger, M. B., et al. (1999) SENSE: sensitivity encoding for fast MRI. *Magn Reson Med* 42, 952–962.

24. Sodickson, D. K., Manning, W. J. (1997) Simultaneous acquisition of spatial harmonics (SMASH): fast imaging with radiofrequency coil arrays. *Magn Reson Med* 38, 591–603.

25. Armitage, P., Behrenbruch, C., Brady, M., et al. (2005) Extracting and visualizing physiological parameters using dynamic contrast-enhanced magnetic resonance imaging of the breast. *Med Image Anal* 9, 315–329.

26. Bluml, S., Schad, L. R., Stepanow, B. et al. (1993) Spin-lattice relaxation time measurement by means of a TurboFLASH technique. *Magn Reson Med* 30, 289–295.

27. Parker, G. J., Baustert, I., Tanner, S. F., et al. (2000) Improving image quality and T(1) measurements using saturation recovery turboFLASH with an approximate K-space normalisation filter. *Magn Reson Imaging* 18, 157–167.

28. Brookes, J. A., Redpath, T. W., Gilbert, F. J., et al. (1999) Accuracy of T1 measurement in dynamic contrast-enhanced breast MRI using two- and three-dimensional variable flip angle fast low-angle shot. *J Magn Reson Imaging* 9, 163–171.

29. Press, W. H. (1992) Modelling of data, in *Numerical Recipes in C: The Art of Scientific Computing*, pp. 656–706.

30. Tofts, P. S., Kermode, A. G. (1991) Measurement of the blood-brain barrier permeability and leakage space using dynamic MR imaging. 1. Fundamental concepts. *Magn Reson Med* 17, 357–367.

31. Weinmann, H. J. (1984). Pharmacokinetics of Gd-DTPA/dimeglumine after intravenous injection into healthy volunteers. *Physiol Chem Phys Med NMR* 16, 167–172.

32. Parker, G. J. (2005) Tracer kinetic modelling for T1-weighted DCE-MRI, in *Dynamic Contrast-Enhanced Magnetic Resonance Imaging in Oncology*. Springer, New York.

33. Port, R. E., Knopp, M. V., Brix, G. (2001) Dynamic contrast-enhanced MRI using Gd-DTPA: interindividual variability of the arterial input function and consequences for the assessment of kinetics in tumors. *Magn Reson Med* 45, 1030–1038.

34. St. Lawrence, K. S., Lee, T. Y. (1998) An adiabatic approximation to the tissue homogeneity model for water exchange in the brain: II. Experimental validation. *J Cereb Blood Flow Metab* 18, 1378–1385.

35. Buckley, D. L. (2002) Uncertainty in the analysis of tracer kinetics using dynamic contrast-enhanced T1-weighted MRI. *Magn Reson Med* 47, 601–606.

36. Tofts, P. S., Brix, G., Buckley, D. L., et al. (1999) Estimating kinetic parameters from dynamic contrast-enhanced T(1)-weighted MRI of a diffusable tracer: standardized quantities and symbols. *J Magn Reson Imaging* 10, 223–232.

37. Crone, C. (1963) The permeability of capillaries in various organs as determined by use of the "indicator diffusion" method. *Acta Physiol Scand* 58, 292–305.

38. Johnson, J. A., Wilson, T. A. (1966). A model for capillary exchange. *Am J Physiol* 210, 1299–1303.

39. Evelhoch, J. L. (1999) Key factors in the acquisition of contrast kinetic data for oncology. *J Magn Reson Imaging* 10, 254–259.

40. Simons, M. et al. (2005) Angiogenesis: where do we stand now? *Circulation* 111, 1556–1566.

41. Leach, M. O., Brindle, K. M., Evelhoch, J. L., et al. (2005) The assessment of antiangiogenic and antivascular therapies in early-stage clinical trials using magnetic resonance imaging: issues and recommendations. *Br J Cancer* 92, 1599–1610.

42. Parker, G. J. (2003) Automated arterial input function extraction for T1-weighted DCE-MRI. *Proc. ISMRM*.

43. Ripjkema, M. (2001) Method for quantitative mapping of dynamic MRI contrast agent uptake in human tumors. *J Magn Reson Imaging* 14, 457–463.

44. Weinmann, H. J., Laniado, M., Mutzel, W. (1984) Pharmacokinetics of GdDTPA/dimeglumine after intravenous injection into healthy volunteers. *Physiol Chem Phys Med NMR* 16, 167–172.

45. Calamante, F., Thomas, D. L., Pell, G. S., et al. (1999) Measuring cerebral blood flow using magnetic resonance imaging techniques. *J Cereb Blood Flow Metab* 19, 701–735.

46. Pathak, A. P., Schmainda, K. M., Ward, B. D., et al. (2001) MR-derived cerebral blood volume maps: issues regarding histological validation and assessment of tumor angiogenesis. *Magn Reson Med* 46, 735–747.

47. Fuss, M., Wenz, F., Essig, M., et al. (2001) Tumor angiogenesis of low-grade astrocytomas measured by dynamic susceptibility contrast-enhanced MRI (DSC-MRI) is predictive of local tumor control after radiation therapy. *Int J Radiat Oncol Biol Phys* 51, 478–482.

48. Jackson, A., Kassner, A., Annesley-Williams, D., et al (2002). Abnormalities in the recirculation phase of contrast agent bolus passage in cerebral gliomas: comparison with relative blood volume and tumor grade. *AJNR Am J Neuroradiol* 23, 7–14.

49. Miles, K. A., Griffiths, M. R. (2003) Perfusion CT: a worthwhile enhancement? *Br J Radiol* 76, 220–231.

50. Miles, K. A., Griffiths, M. R., Fuentes, M. A. (2001) Standardized perfusion value: universal CT contrast enhancement scale that correlates with FDG PET in lung nodules. *Radiology* 220, 548–553.

51. Mullani, N. A., Gould, K. L. (1983). First-pass measurements of regional blood flow with external detectors. *J Nucl Med* 24, 577–581.

52. Yeung, W. T., Lee, T. Y., Del Maestro, R. F., et al. (1992) An absorptiometry method for the determination of arterial blood concentration of injected iodinated contrast agent. *Phys Med Biol* 37, 1741–1758.

53. Cenic, A., Nabavi, D. G., Craen, R. A., et al. (2000) A CT method to measure hemodynamics in brain tumors: validation and application of cerebral blood flow maps. *AJNR Am J Neuroradiol* 21, 462–470.

54. Goh, V., Padhani, A. R. (2006) Imaging tumor angiogenesis: functional assessment using MDCT or MRI? *Abdom Imaging* 31, 194–199.

55. Ferrara, K. W., Merritt, C. R., Burns, P. N., et al. (2000) Evaluation of tumor angiogenesis with US: imaging, Doppler, and contrast agents. *Acad Radiol* 7, 824–839.

56. Cheng, W.F., Lee, C. N., Chu, J. S., et al. (1999) Vascularity index as a novel parameter for the in vivo assessment of angiogenesis in patients with cervical carcinoma. *Cancer* 85, 651–657.

57. Cornud, F., Hamida, K., Flam, T., et al. (2000) Endorectal color Doppler sonography and endorectal MR imaging features of nonpalpable prostate cancer: correlation with radical prostatectomy findings. *AJR Am J Roentgenol* 175, 1161–1168.

58. Padhani, A. R., Harvey, C. J., Cosgrove, D. O. (2005) Angiogenesis imaging in the management of prostate cancer. *Nat Clin Pract Urol* 2, 596–607.

59. Harvey, C. J., Pilcher, J. M., Eckersley, R. J., et al. (2002) Advances in ultrasound. *Clin Radiol* 57, 157–177.

60. Dowlati, A., Robertson, K., Cooney, M., et al. (2002) A phase I pharmacokinetic and translational study of the novel vascular targeting agent combretastatin a-4 phosphate on a single-dose intravenous schedule in patients with advanced cancer. *Cancer Res* 62, 3408–3416.

61. Galbraith, S. M., Maxwell, R. J., Lodge, M. A., et al. (2003) Combretastatin A4 phosphate has tumor antivascular activity in rat and man as demonstrated by dynamic magnetic resonance imaging. *J Clin Oncol* 21, 2831–2842.

62. Stevenson, J. P., Rosen, M., Sun, W., et al. (2003) Phase I trial of the antivascular agent combretastatin A4 phosphate on a 5-day schedule to patients with cancer: magnetic resonance imaging evidence for altered tumor blood flow. *J Clin Oncol* 21, 4428–4438.

63. Thomas, A. L., Morgan, B., Horsfield, M. A., et al. (2005) Phase I study of the safety, tolerability, pharmacokinetics, and pharmacodynamics of PTK787/ZK 222584 administered twice daily in patients with advanced cancer. *J Clin Oncol* 23, 4162–4171.

64. Morgan, B., Thomas, A. L., Drevs, J., et al. (2003) Dynamic contrast-enhanced magnetic resonance imaging as a biomarker for the pharmacological response of PTK787/ZK 222584, an inhibitor of the vascular endothelial growth factor receptor tyrosine kinases, in patients with advanced colorectal cancer and liver metastases: results from two phase I studies. *J Clin Oncol* 21, 3955–3964.

65. O'Donnell, A., Padhani, A., Hayes, C., et al. (2005) A Phase I study of the angiogenesis inhibitor SU5416 (semaxanib) in solid tumours, incorporating dynamic contrast MR pharmacodynamic end points. *Br J Cancer* 93, 876–883.

66. Jayson, G. C., Parker, G. J., Mullamitha, S., et al. (2005) Blockade of platelet-derived growth factor receptor-beta by CDP860, a humanized, PEGylated di-Fab, leads to fluid accumulation and is associated with increased tumor vascularized volume. *J Clin Oncol* 23, 973–981.

67. Eder, J. P., Jr., Supko, J. G., Clark, J. W., et al. (2002) Phase I clinical trial of recombinant human endostatin administered as a short intravenous infusion repeated daily. *J Clin Oncol* 20, 3772–3784.

68. Wang, J. H., Min, P. Q., Wang, P. J., et al. (2006) Dynamic CT evaluation of tumor vascularity in renal cell carcinoma. *AJR Am J Roentgenol* 186, 1423–1430.

69. Brasch, R. C., Li, K. C., Husband, J. E., et al. (2000) *In vivo* monitoring of tumor angiogenesis with MR imaging. *Acad Radiol* 7, 812–823.

70. Ikeda, O., Yamashita, Y., Takahashi, M. (1999) Gd-enhanced dynamic magnetic resonance imaging of breast masses. *Top Magn Reson Imaging* 10, 143–151.

71. Buckley, D. L., Drew, P. J., Mussurakis, S., et al. (1997) Microvessel density of invasive

breast cancer assessed by dynamic Gd-DTPA enhanced MRI. *J Magn Reson Imaging* 7, 461–464.

72. Buadu, L. D., Murakami, J., Murayama, S., et al. (1996) Breast lesions: correlation of contrast medium enhancement patterns on MR images with histopathologic findings and tumor angiogenesis. *Radiology* 200, 639–649.

73. Hawighorst, H., Knapstein, P. G., Weikel, W., et al. (1997) Angiogenesis of uterine cervical carcinoma: characterization by pharmacokinetic magnetic resonance parameters and histological microvessel density with correlation to lymphatic involvement. *Cancer Res* 57, 4777–4786.

74. Hawighorst, H., Knapstein, P. G., Knopp, M. V., et al. (1998) Uterine cervical carcinoma: comparison of standard and pharmacokinetic analysis of time-intensity curves for assessment of tumor angiogenesis and patient survival. *Cancer Res* 58, 3598–3602.

75. Hawighorst, H., Weikel, W., Knapstein, P. G., et al. (1998) Angiogenic activity of cervical carcinoma: assessment by functional magnetic resonance imaging-based parameters and a histomorphological approach in correlation with disease outcome. *Clin Cancer Res* 4, 2305–2312.

76. Knopp, M. V., Weiss, E., Sinn, H. P., et al. (1999) Pathophysiologic basis of contrast enhancement in breast tumors. *J Magn Reson Imaging* 10, 260–266.

77. Dzik-Jurasz, A. (2000) Is there an association between systemic VEGF and permeability in locally advanced rectal adenocarcinoma? Initial observations. *Eighth scientific meeting of the International Society of Magnetic Resonance in Medicine (ISMRM)* , Denver, CO.

78. Wang, Z. Q., Li, J. S., Lu, G. M., et al. (2003) Correlation of CT enhancement, tumor angiogenesis and pathologic grading of pancreatic carcinoma. *World J Gastroenterol* 9, 2100–2104.

79. Jinzaki, M., Tanimoto, A., Mukai, M., et al. (2000) Double-phase helical CT of small renal parenchymal neoplasms: correlation with pathologic findings and tumor angiogenesis. *J Comput Assist Tomogr* 24, 835–842.

80. Hayashi, K., Tozaki, M., Sugisaki, M., et al. (2002) Dynamic multislice helical CT of ameloblastoma and odontogenic keratocyst: correlation between contrast enhancement and angiogenesis. *J Comput Assist Tomogr* 26, 922–926.

81. Tateishi, U., Kusumoto, M., Nishihara, H., et al. (2002) Contrast-enhanced dynamic computed tomography for the evaluation of tumor angiogenesis in patients with lung carcinoma. *Cancer* 95, 835–842.

82. Swensen, S. J., Brown, L. R., Colby, T. V., et al. (1996) Lung nodule enhancement at CT: prospective findings. *Radiology* 201, 447–455.

83. Strohmeyer, D., Frauscher, F., Klauser, A., et al. (2001) Contrast-enhanced transrectal color Doppler ultrasonography (TRCDUS) for assessment of angiogenesis in prostate cancer. *AntiCancer Res* 21, 2907–2913.

84. Posey, J. A., Khazaeli, M. B., DelGrosso, A., et al. (2001) A pilot trial of Vitaxin, a humanized anti-vitronectin receptor (anti alpha v beta 3) antibody in patients with metastatic cancer. *Cancer Biother Radiopharm* 16, 125–132.

85. Bredow, S., Lewin, M., Hofmann, B., et al. (2000) Imaging of tumour neovasculature by targeting the TGF-beta binding receptor endoglin. *Eur J Cancer* 36, 675–681.

86. Li, S., Peck-Radosavljevic, M., Kienast, O., et al. (2004) Iodine-123-vascular endothelial growth factor-165 (123I-VEGF165). Biodistribution, safety and radiation dosimetry in patients with pancreatic carcinoma. *Q J Nucl Med Mol Imaging* 48, 198–206.

87. Jayson, G. C., Zweit, J., Jackson, A., et al. (2002) Molecular imaging and biological evaluation of HuMV833 anti-VEGF antibody: implications for trial design of antiangiogenic antibodies. *J Natl Cancer Inst* 94, 1484–1493.

88. Kulasegaram, R., Giersing, B., Page, C. J., et al. (2001) In vivo evaluation of 111In-DTPA-N-TIMP-2 in Kaposi sarcoma associated with HIV infection. *Eur J Nucl Med* 28, 756–761.

89. Kobayashi, H., Sakahara, H., Hosono, M., et al. (1993) Scintigraphic detection of xenografted tumors producing human basic fibroblast growth factor. *Cancer Immunol Immunother* 37, 281–285.

90. Mulder, W. J., Strijkers, G. J., Habets, J. W., et al. (2005) MR molecular imaging and fluorescence microscopy for identification of activated tumor endothelium using a bimodal lipidic nanoparticle. *FASEB J* 19, 2008–2010.

91. Kang, H. W., Josephson, L., Petrovsky, A., et al. (2002) Magnetic resonance imaging of inducible E-selectin expression in human endothelial cell culture. *Bioconjug Chem* 13, 122–127.

92. Korpanty, G., Grayburn, P. A., Shohet, R. V., et al. (2005) Targeting vascular endothelium with avidin microbubbles. *Ultrasound Med Biol* 31, 1279–1283.

93. Weller, G. E., Wong, M. K., Modzelewski, R. A., et al. (2005) Ultrasonic imaging of tumor angiogenesis using contrast microbubbles targeted via the tumor-binding peptide arginine-arginine-leucine. *Cancer Res* 65, 533–539.

94. Nordsmark, M., Overgaard, M., Overgaard, J. (1996) Pretreatment oxygenation predicts radiation response in advanced squamous cell carcinoma of the head and neck. *Radiother Oncol* 41, 31–39.

95. Hockel, M., Schlenger, K., Mitze, M., et al. (1996) Hypoxia and radiation response in human tumors. *Semin Radiat Oncol* 6, 3–9.

96. Overgaard, J., Hansen, H. S., Overgaard, M., et al. (1998) A randomized double-blind phase III study of nimorazole as a hypoxic radiosensitizer of primary radiotherapy in supraglottic larynx and pharynx carcinoma. Results of the Danish Head and Neck Cancer Study (DAHANCA) Protocol 5-85. *Radiother Oncol* 46, 135–146.

97. Rajendran, J. G., Mankoff, D. A., O'Sullivan, F., et al. (2004) Hypoxia and glucose metabolism in malignant tumors: evaluation by [18F]fluoromisonidazole and [18F]fluorodeoxyglucose positron emission tomography imaging. *Clin Cancer Res* 10, 2245–2252.

98. Bruehlmeier, M., Roelcke, U., Schubiger, P. A., et al. (2004) Assessment of hypoxia and perfusion in human brain tumors using PET with 18F-fluoromisonidazole and 15O-H2O. *J Nucl Med* 45, 1851–1859.

99. Gagel, B., Reinartz, P., Dimartino, E., et al. (2004) pO(2) Polarography versus positron emission tomography ([(18)F]fluoromisonidazole, [(18)F]-2-fluoro-2′-deoxyglucose). An appraisal of radiotherapeutically relevant hypoxia. *Strahlenther Onkol* 180, 616–622.

100. Thorwarth, D., Eschmann, S. M., Scheiderbauer, J., et al. (2005) Kinetic analysis of dynamic 18F-fluoromisonidazole PET correlates with radiation treatment outcome in head-and-neck cancer. *BMC Cancer* 5, 152.

101. Griffiths, J. R., Taylor, N. J., Howe, F. A., et al. (1997) The response of human tumors to carbogen breathing, monitored by gradient-recalled echo magnetic resonance imaging. *Int J Radiat Oncol Biol Phys* 39, 697–701.

102. Taylor, N. J., Baddeley, H., Goodchild, K. A., et al. (2001) BOLD MRI of human tumor oxygenation during carbogen breathing. *J Magn Reson Imaging* 14, 156–163.

103. Herbst, R. S., Mullani, N. A., Davis, D. W., et al. (2002) Development of biologic markers of response and assessment of antiangiogenic activity in a clinical trial of human recombinant endostatin. *J Clin Oncol* 20, 3804–3814.

104. Thomas, J. P., Arzoomanian, R. Z., Alberti, D., et al. (2003) Phase I pharmacokinetic and pharmacodynamic study of recombinant human endostatin in patients with advanced solid tumors. *J Clin Oncol* 21, 223–231.

105. Anderson, H., Yap, J. T., Wells, P., et al. (2003) Measurement of renal tumour and normal tissue perfusion using positron emission tomography in a phase II clinical trial of razoxane. *Br J Cancer* 89, 262–267.

106. Kurdziel, K. A., Figg, W. D., Carrasquillo, J. A., et al. (2003) Using positron emission tomography 2-deoxy-2-[18F]fluoro-D-glucose, 11CO, and 15O-water for monitoring androgen independent prostate cancer. *Mol Imaging Biol* 5, 86–93.

107. Lassau, N., Chawi, I., Rouffiac, V., et al. (2004) [Interest of color Doppler ultrasonography to evaluate a new anti-angiogenic treatment with thalidomide in metastatic renal cell carcinoma]. *Bull Cancer* 91, 629–635.

108. Willett, C. G., Boucher, Y., di Tomaso, E., et al. (2004) Direct evidence that the VEGF-specific antibody bevacizumab has antivascular effects in human rectal cancer. *Nat Med* 10, 145–147.

109. Liu, G., Rugo, H. S., Wilding, G., et al. (2005) Dynamic contrast-enhanced magnetic resonance imaging as a pharmacodynamic measure of response after acute dosing of AG-013736, an oral angiogenesis inhibitor, in patients with advanced solid tumors: results from a phase I study. *J Clin Oncol* 23, 5464–5473.

110. McNeel, D. G., Eickhoff, J., Lee, F. T., et al. (2005) Phase I trial of a monoclonal antibody specific for alphavbeta3 integrin (MEDI-522) in patients with advanced malignancies, including an assessment of effect on tumor perfusion. *Clin Cancer Res* 11, 7851–7860.

111. Xiong, H. Q., Herbst, R., Faria, S. C., et al. (2004) A phase I surrogate endpoint study of SU6668 in patients with solid tumors. *Invest New Drugs* 22, 459–466.

112. Falk, S. J., Ramsay, J. R., Ward, R., et al. (1994) BW12C perturbs normal and tumour tissue oxygenation and blood flow in man. *Radiother Oncol* 32, 210–217.

113. Logan, T. F., Jadali, F., Egorin, M. J., et al. (2002) Decreased tumor blood flow as measured by positron emission tomography in cancer patients treated with interleukin-1 and carboplatin on a phase I trial. *Cancer Chemother Pharmacol* 50, 433–444.

114. Galbraith, S. M., Rustin, G. J., Lodge, M. A., et al. (2002) Effects of 5,6-dimethylxanthenone-4-acetic acid on human tumor microcirculation assessed by dynamic contrast-enhanced magnetic resonance imaging. *J Clin Oncol* 20, 3826–3840.

115. Anderson, H. L., Yap, J. T., Miller, M. P., et al. (2003) Assessment of pharmacodynamic vascular response in a phase I trial of combretastatin A4 phosphate. *J Clin Oncol* 21, 2823–2830.

Section II

Microscopic Assessment

Chapter 3

Assessing Tumor Angiogenesis in Histological Samples

Stephen B. Fox

Abstract

Tumor neovascularization acquires vessels through a number of processes, including angiogenesis, vasculogenesis, vascular remodelling, intussusception, and possibly vascular mimicry in certain tumors. The end result of the tumor vasculature has been quantified by counting the number of immunohistochemically identified microvessels in areas of maximal vascularity so-called hot spots. Other techniques have been developed, such as Chalkley counting and the use of image analysis systems that are robust and reproducible as well as more objective. Many of the molecular pathways that govern tumor neovascularization have been identified, and many reagents are now available to study these tissue sections. These include angiogenic growth factors and their receptors, cell adhesion molecules, proteases, and markers of activated, proliferating, cytokine-stimulated, or angiogenic vessels, such as CD105. It is also possible to differentiate quiescent from active vessels. Other reagents that can identify proteins involved in microenvironmental influences such as hypoxia have also been generated. Although the histological assessment of tumor vascularity is used mostly in the research context, it may also have clinical applications if appropriate methodology and trained observers perform the studies.

Key words: Angiogenic factors, Chalkley counts, hypoxia, microvessel density, tumor angiogenesis, vascular grading.

1. Introduction

Although it has been recognised for many centuries that tumors are more vascular than their normal counterpart, it is only since Folkman's hypothesis on anti-angiogenesis *(1)* that a more quantitative method for measuring the blood vasculature in tissue sections has been pursued. Folkman and colleagues recognised that quantitation of the tumor vasculature might play an important a role in predicting tumor behaviour and patient management

S. Martin and C. Murray (eds.), *Methods in Molecular Biology, Angiogenesis Protocols, Second edition, Vol. 467*
© Humana Press, a part of Springer Science+ Business Media, LLC 2009
DOI: 10.1007/978-1-59745-241-0_3

and therefore developed a microscopic angiogenesis grading system designated the MAGS score. The score was calculated by measuring vessel number, endothelial cell hyperplasia, and cytology in tinctorially stained tissue sections (2). It was designed to be an objective method for quantifying tumor angiogenesis that would yield important information on the relationship to other clinicopathological tumor characteristics and aid the testing of anti-angiogenic therapies. However, although it was possible to classify tumors into "endothelial poor" or "rich," the technical limitations of sample selection, inter- and intra-observer variation, and conceptual biological problems were such that the technique could not be easily applied. Interest in grading tumor angiogenesis was rekindled in the 1980s and 1990s with the advent of non-specific endothelial markers (3–5), but it has been only more recently with the advent of more specific endothelial markers that quantitation studies have been performed. Most investigators have highlighted endothelium using immunohistochemistry and based quantification on a method developed by Weidner et al (6). This method highlights the tumor vasculature with immunohistochemistry and counts individual vessels in the most vascular areas (so-called hot spots) of the tumor. These studies have shown that an increased microvessel density is a powerful prognostic tool in many human tumor types (reviewed in ref. 7). Nevertheless, due to limitations in capillary identification and quantitation, not all investigators have been able to confirm a relationship between tumor vascularity and prognosis (reviewed in ref. 8).

This chapter briefly discusses the considerations in quantifying tumor angiogenesis by microvessel quantitation in tissue sections and gives the current optimal protocol for assessment. Since an increased understanding of how tumors acquire a neovasculature has emerged over the last few years along with the identification of the molecules involved in these processes, I also discuss pathways involved in tumor neovascularization and the microenvironmental influences (e.g., hypoxia) that profoundly affect the vascular program.

1.1. Mechanisms of Tumor Neovascularization

The hypothesis presented by Judah Folkman that tumors are angiogenesis dependent (9) is likely only to be partly the case, with the tumors using a variety of mechanisms to establish a blood supply. A non-angiogenic mechanism of tumor growth that occurs in lung carcinomas (10) and in secondary breast cancer metastasis to the lung (11) is where tumor cells fill the existing structure within the lung (i.e., the alveolar spaces), whether primary or metastatic, without destroying the underlying architecture, with the tumor using the established blood vessels rather than generating new vessels as occurs during angiogenesis. Thus, there is no associated stromal desmoplasia, and unlike many solid tumors that are more vascular than their normal tissue counterpart, these tumors have a similar vascularity as normal lung. A similar co-option model has also been described in a brain tumor model by which early in development the tumor uses the existing blood vessels without eliciting an angiogenic response (12, 13). The co-option model is also likely to occur in other tumor types, such as colorectal hepatic metastases (14). Thus, in some circumstances tumors are able to "parasitise" the normal stroma and sinusoidal vasculature for their metabolic needs.

Vasculogenesis, the *de novo* generation of blood vessels from endothelial progenitors, is another method exploited by tumors to generate neovessels. This has been reported in animal models and in some human tumors, such as inflammatory sub-type breast tumors (15–19). The importance of this method of tumor vascularisation may be more relevant in early tumor development since inhibition of stem cells or endothelial cell precursor mobilisation prevents xenografts from inducing the initial angiogenic response (20).

Although currently there is a paucity of evidence, it has also been suggested that intussusception contributes to the establishment of a tumor blood supply. This is the process by which larger vessels within a tumor are divided by columns of tumor cells splitting an individual large vessel into two or more channels. Unlike conventional sprouting angiogenesis, endothelial cell proliferation is not a feature (21, 22). This may be part of vascular remodelling, which may be the dominant mechanism in the establishment of the tumor vascular bed (23–25).

Another somewhat contentious mechanism of neovascularisation that has been reported for aggressive ocular melanomas and ovarian tumors (26) is vascular mimicry. The neoplastic cells rather than endothelial cells line the blood vessels and conduct the blood (27, 28). The tumor cells acquire both the morphology and phenotype of endothelium and co-express some vascular markers. Partial lining of the blood vessels by tumor cells has been known for many years (29) and more recently was reported in animal models using advanced techniques (30), but in our experience using morphology and double immunohistochemistry, this is not frequently observed in the common solid tumors (unpublished data).

2. Materials

1. Silane-coated or charged microscope slides (e.g., Superfrost Plus®).
2. Dry incubator/oven at 37°C for 30 minutes or overnight.
3. Xylene or xylene substitute (e.g., Citroclear, Histolene).

4. Graded alcohols (100%, 90%, and 70% ethanol).

5. 5% H_2O_2 in PBS or methanol (methanol and 30% hydrogen peroxide).

6. IHC running buffer: Phosphate-buffered saline (PBS) or Tris-buffered saline (TBS).

7. Tris EDTA buffer, pH 9.0.

8. Microwave oven or pressure cooker.

9. Antibodies to CD31 (JC70; Dako M0823) or CD34 (QBEND10; Dako M7165).

10. Detection system: Streptavidin biotin kit or Chain polymer detection systems.

11. Chromogens: Diaminobenzidine, 3-amino-9-ethylcarbazole (AEC), and or 5-Bromo-4-Chloro-3-Indolyl Phosphate/Nitro Blue Tetrazolium (BCIP/NBT).

12. Permanent mountant for DAB or aqueous mountant (for alcohol-soluble chromogens, e.g., new fuchsin, BCIP/NBT).

13. Microscope coverslips.

14. Chalkley graticule (25 dot).

3. Methods

3.1. Highlighting the Vasculature in Tissue Sections

It should be emphasised that time should be devoted to optimisation and validation of the immunohistochemical staining procedure since quality staining with little background greatly facilitates assessment. Consideration should be given to antibody clone choice, antigen retrieval methods, primary antibody titration, and the choice of detection system. Some chromogens form alcohol-soluble end products (e.g., AEC and BCIP/NBT), and care must be taken to use an aqueous mountant with these products. Many histopathology laboratories are well versed in immunohistochemistry, necessitating only minor adjustments to the in-house protocol.

The preferred immunohistochemical protocols that are employed in our laboratory are outlined next.

3.1.1. Staining Procedure

1. Cut 4-μ formalin-fixed paraffin-embedded sections of the representative tumor block onto silane-coated or -charged slides.

2. Dry at 37°C overnight in an incubator.

3. Dewax using citroclear (HD Supplies) or xylene for 15 min before passing through graded alcohols (100%, 90%, and 70% ethanol) into water.

4. Place in IHC running buffer for 5 min.

5. Perform antigen retrieval as required and apply primary antibody at room temperature as outlined in **Table 3.1**.

Table 3.1
Antibodies used for highlighting endothelium in tumors

Antibody	Antigen retrieval	Clone/source	Dilution	Incubation
CD31 (monoclonal)	2 mins pressure cooker 10 mM citrate Tris EDTA buffer, PH 9.0	JC70a; Dako (M0823)	1/300	30 min
CD34 (monoclonal)	2 mins pressure cooker Tris EDTA buffer, PH 9.0	QBEND10: Dako (M7165)	1/200	30 min

6. Block endogenous peroxidase if using a horseradish peroxidase (HRP) detection system. Incubate the slides in 3% to 5% hydrogen peroxide in PBS or methanol for 10 min (methanol is gentler on sections that show poor adhesion to the slides).

7. Rinse three times in IHC running buffer for 5 min.

8. Apply the appropriate detection system as outlined next. Ensure that the detection system takes into account the type of primary antibody (i.e., whether mouse monoclonal or rabbit polyclonal or rabbit monoclonal).

3.1.2. Detection Systems

The choice of detection system will depend on the type of primary antibody and the sensitivity required for the assay.

3.1.3. Chain Polymer-Conjugated Technology

Chain polymer-conjugated technology is a new type of detection system that uses an inert dextran backbone that has been labelled with multiple copies of HRP to enhance the detection system. Most major antibody suppliers now offer these detection systems (e.g., Dako Envision or Envision+, SuperPicTure Zymed, Biocare, BioGenex). The main advantage these systems over older biotin-based detection systems is that there is no possibility of binding with endogenous tissue biotin. These detection kits give good sensitivity and are more rapid than the alkaline phosphatase-anti-alkaline phosphatase (APAAP) technique as they only require a single step for detection. Although some antibodies come already attached to the polymer, allowing a single-step staining method, many antibodies are not available and thus utilize a two-step method in which a primary antibody is followed by a mouse- or rabbit-labelled dextran–HRP complex. The sensitivity results from the conjugate having hundreds of enzyme molecules and multiple secondary anti-mouse (or -rabbit) antibody molecules per backbone. The chromogen 3,3-diaminobenzidine (DAB) forms a brown, insoluble end product when it is oxidised in the presence of hydrogen peroxide. Other chromogens may also be used with these kits, such as AEC, which forms a red alcohol-soluble precipitate. Some kits contain all the necessary reagents for each step, including blocking if required. If double immunohistochemical staining is considered (see below 4.3 and 4.4) the technology allows for discrimination using different secondary reagents and substrates.

1. Wash the slides in IHC running buffer after the primary antibody incubation.
2. Incubate the sections with the polymer solution for 30 min.
3. Rinse the sections three times with IHC running buffer.
4. Apply the chromogen substrate (e.g., DAB) for 10 min.
5. Rinse in tap water for 2–3 min.
6. Counterstain, then wash off excess in running tap water.
7. Mount in aqueous mountant if using an alcohol-soluble chromogen. Otherwise, dehydrate in graded alcohols, then clear in xylene and mount with permanent mountant.

3.1.4. StreptABC (Avidin–Biotin Complex)

The StreptABC (avidin–biotin complex) method uses the high affinity of streptavidin for biotin. It requires sequential application of a biotinylated secondary antibody followed by a tertiary antibody complex of streptavidin–biotin–HRP. The open sites on the streptavidin complex bind to the biotin on the secondary antibody. StreptABC kits may come prediluted, ready-to-use reagents or may require titration of the secondary antibody. Ensure that the secondary antibody is appropriate for the primary antibody. A blocking step for endogenous biotin may be required if background binding is observed. Commercial kits are available for this purpose.

1. Apply the secondary antibody for 15 to 30 min.
2. Wash the slides in IHC running buffer.
3. Apply the streptavidin complex for 15 to 30 min.
4. Rinse the sections three times with IHC running buffer.
5. Apply the chromogen substrate for 5 to 10 min.
6. Rinse the slides in tap water and counterstain if desired.
7. Mount in aqueous mountant if using an alcohol-soluble chromogen.
    Otherwise, dehydrate in graded alcohols, then clear in xylene and mount with permanent mountant.

3.1.5. Alkaline Phosphatase-Anti-alkaline Phosphatase

The APAAP method uses a soluble enzyme anti-enzyme antibody complex (calf intestinal APAAP) to act on new fuschin substrate. The primary and final antibody complex is bridged by excess rabbit anti-mouse antibody that binds to the primary mouse antibody with one Fab, leaving a Fab site free to bind the tertiary complex. Repeated rounds of application with secondary and tertiary antibodies amplify the staining intensity. The enzyme hydrolyses the naphthol esters in the substrate to phenols, which couple to colourless diazonium salts in the chromogen to produce a red colour. Endogenous alkaline phosphatase is inhibited by the addition of 5 mM levamisole, which does not inhibit calf intestinal alkaline phosphatase.

1. Apply the secondary antibody for 30 min.

2. Wash the slides in IHC running buffer.

3. Apply the tertiary complex for 30 min.

4. Wash the slides in IHC running buffer and apply chromogen for 20 mins.

5. Wash the slides in running tap water and counterstain if required.

6. Mount in aqueous mountant if using an alcohol-soluble chromogen. Otherwise, dehydrate in graded alcohols, then clear in xylene and mount with permanent mountant.

3.2. Assessment of Tumor Blood Vascularity

Confirm satisfactory staining using normal entrapped vasculature as internal positive control. An optional parallel negative control section using an immunoglobulin G_1 (IgG_1) isotype antibody may be used. The three hot spot areas containing the maximum number of *discrete* microvessels should be identified by scanning the entire tumor at low power (×40 and ×100) (**Fig. 3.1**). This is the most subjective step of the procedure. It has been demonstrated that the experience of the observer determines the success of identifying the relevant hot spots *(31)*. Poor selection will in turn lead to an inability to classify patients into different prognostic groups. Therefore, it is recommended that inexperienced

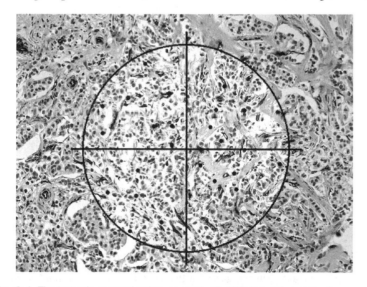

Fig. 3.1. The tumor is scanned at low power (×40–100) (*centre*), and the three areas that contain the highest number of discrete microvessels are selected.

observers spend time in a laboratory where a period of training can be undertaken. Ideally, comparisons between hot spots chosen by an experienced investigator and trainee should be performed and continued on different series until there is more than 90% agreement. Training can be completed by assessing sections from a series that already contains prognostic information *(32)*.

Inexperienced observers tend to be drawn to areas with dilated vascular channels, often within the sclerotic body of the tumor. These central areas together with necrotic tumor should be ignored. Vascular lumina or erythrocytes are not a requirement to be considered a countable vessel; indeed, many of the microvessels have a collapsed configuration. Although the hot spot areas can occur anywhere within the tumor, they are generally at the tumor periphery, making it important to include the normal tumor interface in the representative area to be assessed. Vessels outside the tumor margin by one ×200–250 field diameter and immediately adjacent to benign tissues should not be counted. The procedure takes 2–5 min.

3.3. Chalkley Counting

Once selected, a 25-point Chalkley point eyepiece graticule *(33)* at ×200–250 should then be oriented over each hot spot region so that the maximum number of graticule points are on or within areas of highlighted vessels (**Fig. 3.2**). Particular care should be

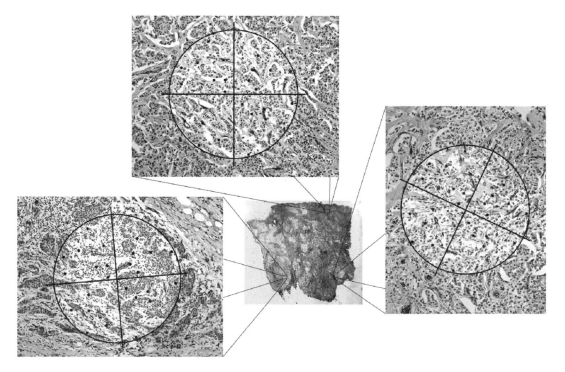

Fig. 3.2. The areas of tumor containing the highest number of discrete microvessels are examined at high power (×200–250), and the Chalkley graticule is then rotated in the eyepiece so that the maximum number of graticule dots coincide with the vessels or their lumens. This number is then recorded.

taken in the occasional case (<1% breast cancers) for which an intense plasma cell infiltrate can mimic a hot spot and obscure the underlying tumor vasculature. Plasma cells can otherwise be disregarded on morphological grounds. The mean of the three Chalkley counts is then generated for each tumor and used for statistical analysis. The procedure takes 2–3 min.

3.4. Intratumoral Microvessel Density

For the intratumoral microvessel density index, any endothelial cell or endothelial cell cluster separate from adjacent microvessel, tumor cells, or matrix elements is considered a countable vessel. Those that appear to be derived from the same vessel if distinct should also be counted. Again, vessel lumens and erythrocytes are not included in the criteria defining a microvessel. There is no cutoff for vessel calibre. The procedure takes 3–6 min.

3.5. Vascular Grading

To facilitate assessing angiogenesis in tissue sections, akin to semi-quantitative tumor grading, vascular grading based on subjective appraisal by trained observers over a conference microscope has been assessed (6, 34). Significant correlations were demonstrated between vascular grade and both microvessel density ($p = 0.002$) and Chalkley count ($p = 0.0001$). Although the method is reproducible (35), delineating criteria are difficult due to the subjective nature of the system, and a considerable investment in time would be required to align the cutoffs required for multicentre studies. However, although there is some loss of power associated with translation of numerical to categorical data, the overall time savings engendered by this make it an attractive proposition. Nevertheless, further validation in a large series of randomised patients is warranted to determine its prognostic utility before its application can be considered in such studies.

3.6. Novel Angiogenic Antigens

Instead of highlighting all the tumor-associated endothelium, an alternative approach would be to selectively identify only the vasculature that is undergoing active neovascularisation. This not only might be valuable in more accurately quantifying tumor angiogenesis but also might have important implications for anti-vascular targeting (36). A number of antibodies have been identified that recognise antigens that have been reported to be up-regulated in tumor-associated endothelium compared with normal tissues and include EN7/44, endoglin, and endosialin (32, 37). However, there are few data assessing their utility in human tissues (38).

3.7. Tumor Vascular Architecture

The vascular morphology of tumors is different within tumors of similar and different histological types (29). It has been suggested that particular vascular patterns both might help distinguish benign from malignant lesions (39, 40) and be a prognostic marker; in ocular melanomas, a closed back-to-back loop vascular pattern was associated with death from metastasis *(41)*, and in lung carcinomas distinct patterns of neovascularisation might potentially respond differently to anti-cancer treatments *(42)*.

3.8. Computer-Aided Image Analysis

We and others have attempted to automate the counting procedure by using computer image analysis systems *(4, 31, 43–53)*. These systems have several drawbacks, not including the capital and running costs over those shared with manual methods. An endothelial marker that gives sensitive and specific capillary staining is essential to reduce background signal. Currently, there is no software available that can identify hot spots, but when developed, the requirement for motorised stages will add further expense. Although partially automated systems with area and shape filters using defined colour tolerances are available, most systems are not fully automated, require a high degree of operator interaction, and like manual counting suffer from observer bias. Currently, computer image analysis is more costly, time consuming, and no more accurate than a trained observer; these factors make it unsuitable to routine diagnostic practice. The recent advances in whole slide scanner technology and image analysis software packages designed to complement these systems look promising. Thus if validated although still expensive, fully automated systems may be able to be used to calculate many vascular parameters that may be of research and diagnostic use.

Data from these such studies have demonstrated that most vascular indices, including microvessel density, vessel perimeter, and vascular area, are significantly correlated, suggesting that they are equivalent indices of angiogenesis *(45)*. Previously, it had been hypothesised that microvessel density might not be the most important vascular parameter since a large vascular perimeter or area might be a better measure of angiogenesis because these may reflect the functional aspects of endothelial surface and volume of blood available for interaction with the tumor *(45)*.

3.9. Additional Information

3.9.1. Choice of Antibody

Since endothelium is highly heterogeneous *(54)*, the choice of antibody profoundly influences the number of microvessels available for assessment. Many, such as those directed to vimentin *(43)*, lectin *(4, 55)*, alkaline phosphatase *(3)*, and type IV collagen *(44, 56)*, suffer from low specificity and are present on many non-endothelial elements. Others, including antibodies to factor VIII-related antigen, the marker used in most studies *(6, 34, 57–62)*, identify only a proportion of capillaries and detect lymphatic endothelium. The most specific and sensitive endothelial marker currently available is CD31, which is present on most capillaries and is a reliable epitope for immunostaining in routinely handled formalin-fixed paraffin-embedded tissues *(63)*. CD34 has been recommended by the Second International Consensus on the Methodology and Criteria of Evaluation of Angiogenesis

Quantification in Solid Human Tumors. However, care is required when interpreting staining since some stromal cells also express this antigen, particularly in breast *(32)*.

3.9.2. Selection of Hot-spots

Once the tumor vasculature has been immunohistochemically highlighted, the tumor is scanned at low magnification (×40–100) to identify angiogenic hot spots *(6, 34)*. The number of vessels is then quantified at high magnification (×200–400; field area 0.15–0.74 mm²) in these regions. These areas of high vascularity are chosen on the basis that these are the tumor regions likely to be biologically important. A high magnification (which will identify more microvessels by virtue of increased resolution; *[64]*) used over too small an area will always give a high vessel index, whereas a low magnification over too large an area will dilute out the hot spot. Further, tumors naturally have a limited number of hot spots, which would be diluted if too many were counted. Thus, although the number of hot spots assessed varies from 1 to 5 *(6, 34, 57, 58, 65–67)*, most studies have examined three from a single representative tissue block. Nevertheless, both the magnification used and its corresponding tumor field area will determine the vessel number derived from each hot spot. It is thus recommended that three regions are examined using a microscope magnification of field area between ×200 and ×400 (corresponds to areas of approximately 0.15–0.74 mm², depending on the microscope type) *(32, 45)*.

3.9.3. Chalkley Counting

Although less subjective than identifying angiogenic hot spots *(32)*, the process of counting vessels has also resulted in significant variation in published series. This was emphasised in the study of Axelsson et al. *(68)*; the authors, after an initial training period with Weidner, who defined the criteria regarding what constituted individual microvessels (see protocols section), did not observe a correlation between microvessel density and patient survival. Even experienced observers occasionally disagree about what constitutes a microvessel. To overcome these problems, after selection of each hot spot, a 25-dot Chalkley microscope eyepiece graticule *(33)* has been used to quantify tumor angiogenesis (see protocols section). This method not only is objective, since no decision is required regarding whether adjacent stained structures are separate, but also is rapid (2–3 min per section) and reproducible and gives independent prognostic information in breast *(45, 69, 70)* and bladder *(71)* cancers. Thus, it is currently the preferred method of a recent multi-centre discussion article *(32)*.

3.9.4. Analysis

The final consideration in quantifying angiogenesis is the difference in the value used for stratification into different study groups. This alone will result in different conclusions being drawn from the same data set. Studies have used the highest, the mean, the median *(64)*, tertiles *(45)*, mean count in node-negative patients with recurrence *(60)*, or variable cutoffs given as a function of tumor area *(6, 34)* or microscope magnification *(67)*. The use of the median and tertile groups avoids strong assumptions about the relationship between tumor vascularity and other variables, including survival, and is therefore useful clinically. However, there is some loss of information, making it optimal to use continuous data if possible.

4. Other Measures of Tumor Vascularity

4.1. Other Measures of Angiogenesis

The microvessels highlighted by immunohistochemistry in a tissue section are the conclusion of a dynamic multistep process. The evolving neovasculature is the result of a complex interplay between extracellular matrix remodelling, endothelial cell migration and proliferation, capillary differentiation, and anastomosis *(72, 73)*. Although it may soon be possible to measure these continuous processes *in vivo*, for human tissues measurement of molecules involved in these events might be surrogate endpoints of angiogenesis. Thus, partly due to many of the inherent and methodological difficulties of vascular counts, these alternative strategies for quantifying tumor angiogenesis have also been pursued. Since many studies now use tissue microarrays (TMAs) I have suggested a scoring system that can also be applied to whole tissue sections (*see* **Section 5**).

4.2. Angiogenic Factors and Receptors

Angiogenesis is the result of the net change in the balance of angiogenic stimulators and inhibitors (i.e., gain of promoters or loss of inhibitors). There are now numerous reports documenting upregulation of several angiogenic factors and their receptors at the messenger RNA (mRNA) and protein level using a variety of techniques in a range of histological tumor types, including breast *(74–81)*.

The prototypical factor is vascular endothelial growth factor (VEGF). A significant relationship between tumor VEGF levels and microvessel density has been shown in breast tumors *(82–84)*. Furthermore, some studies demonstrated that tumor VEGF expression levels gave prognostic information in breast carcinomas *(85–90)*, the power of which may be improved by combining it in a ratio with soluble flt-1 *(91)*.

However, this is a complex pathway with four direct members: VEGF-A, VEGF-B, VEGF-C, and VEGF-D, with the VEGF-A and VEGF-C genes generating additional isoforms through alternative splicing. These variably bind to three receptors: VEGF receptor (VEGFR) 1, VEGFR2, and VEGFR3, with the latter receptor largely restricted to lymphatic endothelium in normal tissues. VEGF-C and VEGF-D are expressed in areas of lymphatic sprouting during embryonic development, and some studies have shown a correlation between VEGF-C levels *(92, 93)* or VEGFR3 *(92)* and lymph node metastases. Other family ligand members and co receptors such as placenta growth factor and neuropilins, respectively, further complicate the pathway. However, not all studies have shown associations with clinicopathological factors *(92–94)*. It is also unknown for human breast tumors what is the dominant factor or isoforms at different stages of neoplastic progression. Similarly, some studies of VEGF-D expression in breast, colon, and endometrium have suggested that this may also be an independent prognostic marker *(95–97)*. It has been further suggested that VEGF-D may be particularly important in inflammatory breast cancer *(98)*.

Thymidine phosphorylase (TP) is another angiogenic factor that appears to be important in human cancer. This migratory rather than mitogenic angiogenic factor is expressed in *in situ* *(99)*, and invasive breast cancer *(100)* has also been associated in some studies with microvessel density *(100, 101)* and patient survival *(102–105)*.

Different tumors use different angiogenic factors during the various phases of their development; for example, breast carcinomas co-express VEGF and TP *(106)*, whereas they are reciprocally expressed in bladder cancers *(107, 108))*. It is likely that determining specific profiles for individual tumor types might assume greater importance in quantitative tumor angiogenesis. Nevertheless, some sera investigations may be compromised by the high VEGF levels present in platelets and not reflect tumor-derived VEGF, suggesting plasma measurement may be more accurate *(109)*.

4.3. Endothelial Cell Proliferation

It is now possible to measure endothelial cell proliferation in tissue sections using double immunohistochemistry with antibodies to endothelial markers to discriminate endothelial cells from other tissue elements in conjunction with antibodies to proliferation markers. We have used combinations of CD31 or CD34 (for endothelium) and BrdU and Ki67 (MIB-1) (as proliferation markers) with good results, but some of the newer cell cycle markers that can also be used on archival tissue, such as minichromosome maintenance proteins 2 and 5, may also be of use *(110, 111)*. This technique allows simultaneous assessment of tumor and endothelial cell proliferation and may aid the stratification of patients for novel therapies. Studies of lymphatic endothelial cell proliferation using lymphatic markers and proliferation markers can also be undertaken. Similarly the use of apoptotic markers such as caspases or annexins can also be applied to study endothelial cell apoptosis.

4.4. Vessel Maturation Index

A late event in the establishment of a tumor blood supply, which accompanies downregulation of endothelial cell proliferation, is pericyte recruitment and secretion of a basement membrane. This basement membrane is irregular and is composed of abnormal ratios of fibronectin, laminin, and collagen, depending on the maturation state of the capillary. Although many studies have documented the heterogeneity, few studies have assessed the significance. Nevertheless, some studies have examined the ratio of endothelial cells with a pericyte *(112)* or basement membrane *(25)* cover as a surrogate of vessel maturation. There is great variation between tumor types *(112)*, but studies in breast carcinomas have shown that the vessel maturation index (VMI) gives a different measure compared with that of microvessel density. There is continual remodelling of vessels in normal breast, and a subset of patients can be identified who have an elevated risk of node recurrence; the VMI potentially gives more functional information *(25)*.

4.5. Cell Adhesion Molecules

Increasing evidence suggests that many of the endothelial cell adhesion molecules (CAMs) of the immunoglobulin, selectin, and integrin superfamilies, which have physiological roles in immune trafficking and tumor metastasis, also play a major role in angiogenesis. Some clinical studies indicated that melanoma patients with upregulated CAMs on endothelium have a significantly worse prognosis, and these studies validated the interest in CAMs and their cognate ligands in tumor angiogenesis *(113, 114)*. Indeed, soluble CAMs are readily identified in sera of cancer-bearing patients, although their relationship to tumor angiogenesis is yet unknown *(115, 116)*. Similarly, integrins, including $\beta_3\alpha_v$, also appear to be upregulated in human breast carcinomas compared to normal or benign breast and might also be a potential surrogate marker for angiogenesis *(117, 118)*.

4.6. Proteolytic Enzymes

Several studies have demonstrated that proteolytic enzymes, including the plasminogen activators and the matrix metalloproteinases, that are important in tumor cell invasion and migration are also important in angiogenesis *(119–124)*. Although no correlation was observed between microvessel density and both uPA and PAI-1 *(125)*, the poor prognosis in tumors *(126–132)* associated with elevated levels of the uPA system are likely to be partly due to the angiogenic activity of these tumors. Thus, measurement of proteases, particularly components of the urokinase system, might give some indication of the angiogenic activity of a tumor.

4.7. Hypoxic Markers

Once a tumor vasculature has been established, there is still continued remodelling of vessels. The remodelling process is likely to be related to an exaggerated stress response and therefore is profoundly influenced by the tumor microenvironment. Hypoxia has been frequently reported despite the increased microvessel density in tumors. The hypoxia may be due to a reduction in blood flow from structural differences in blood vessels *(29, 133)* or from the influence of permeability factors such as VEGF-A *(134)* or from shunting of blood across the tumor vascular bed, as reported in several tumor types *(135)*. In any of these scenarios, tissue hypoxia results in stabilisation of the hypoxic-inducible factors that mediate transcription of angiogenic pathways, thereby enhancing tumor angiogenesis. A pivotal pathway is the regulation by hypoxia of VEGF through the transcription factor hypoxia-inducible factor (HIF).

Hypoxia-inducible factor 1α (and 2α) bind to the aryl hydrocarbon nuclear translocator (ARNT) (HIF-1β), which then binds a specific DNA hypoxia response element (HRE), increasing mRNA transcription. In normoxia, the HIF-α units are unstable and are rapidly degraded by the proteasome pathway through ubiquitin E3 ligase complex and the recognition component of von Hippel-Lindau (VHL) disease. This is regulated through enzymatic hydroxylation of either two critical prolyl residues within the oxygen-dependent degradation domains of HIF-α subunits by prolyl hydroxylases (PHDs) 1, 2, and 3 and dioxygen as a co-substrate. This results in one oxygen incorporated into the prolyl residue of HIF-α, the other into 2-oxoglutarate to yield succinate and carbon dioxide. However, in hypoxic conditions, since no oxygen is available for hydroxylation, the HIF-α subunit is stabilized and translocated to the nucleus, where it binds to HIF-β. The complex then recruits co-activators and binds a specific DNA HRE, resulting in increased mRNA transcription (reviewed in **ref.***136)*. A further level of control is mediated by factor inhibitor of HIF (FIH), which interferes with coactivator binding.

Many HIF target genes are beneficial to the tumor, including those involved in iron metabolism (transferrin and ceruloplasmin); angiogenesis (VEGF, VEGFR1, TP, Ang2); glucose metabolism (glucose transporters); proliferation (IGF II); endothelial adhesion and pH regulation (carbonic anhydrase IX). Antibodies to both HIFs *(137)* and carbonic anhydrase IX *(138)* have been shown to be surrogates of hypoxia *(139)* and associated with a poor prognosis *(140, 141)*. Thus, measurement of the factors involved in the regulation of the hypoxic response may be clinically important in patient management *(142, 143)*.

Nevertheless, these techniques require a degree of quantitation and are currently unsuitable for a general diagnostic pathology laboratory. The presence of a fibrotic focus (FF) as a surrogate marker of hypoxia is being examined. This is defined as a scar-like area consisting of fibroblasts and collagen fibres in the centre of an invasive ductal carcinoma of the breast. It was first proposed in 1996 by Hasebe et al. as an indicator of tumor aggressiveness *(144, 145)*, and the presence of an FF is associated with higher microvessel density *(146)* and more lattery as a marker of intratumoral hypoxia. Thus, an FF may be a useful surrogate marker of hypoxia-driven ongoing angiogenesis *(147)*.

4.8. Additional Information

1. The tumor block should be selected by examining haematoxylin- and eosin-stained slides.

2. If using TMAs, the periphery of the tumor should be selected since this is the area where angiogenesis is most active; in addition, it is likely that at least three tissue cores should be examined.

3. Also, 8-μ cryostat sections can be used, but the area of tumor assessed is less representative.

4. If sections continually float off after antigen retrieval, drying at 56°C overnight or using distilled water when cutting sections will increase tissue adherence.

5. Conjugated polymer or APAAP methodology is preferred in tissues such as liver and kidney that contain high endogenous biotin.

6. PBS works well for chain polymer-conjugated or StreptABC methodologies, and TBS works well for APAAP.

7. The Chalkley graticule can be obtained from Graticules (Morely Road, Botany Trading Estate, Tonbridge Wells, Kent, TN9 1ZN, UK.) The size of the graticule required will depend on the eyepiece diameter of the microscope lens.

8. Studies using a magnification of ×200–400 and field areas between 0.74 and 0.15 mm^2 have derived prognostic information.

9. One block is justified by the prevailing evidence that shows a high concordance in vessel number between different blocks *(57, 148, 149)*.

10. Stromal cell immunoreactivity may interfere with microvessel counts if using CD34 as a vascular marker.

5. Tissue Microarrays and Their Assessment of Staining

Tissue microarrays (TMAs) have become an important tool for the high-throughput assessment of candidate markers. They have also helped standardise testing across samples for which it was previously difficult to ensure reproducibility when using cohorts of individual slides. They have also enabled more accurate correlation analyses since the same part (core) of the tumor is examined across markers, in contrast to the use of whole tissue sections when different parts of so-called positive tumors are compared. Interestingly, for several markers such as oestrogen receptor in breast tumors, the use of a single core is representative of the whole tumor. Most TMAs use the core system pioneered by Kononen et al. *(150)* by which cores of tissue from donor blocks are arrayed at high density into a recipient block using a dedicated instrument. These may hold many hundreds of tumor cores and give 150 serial sections. When constructing TMAs, we usually take four cores of tumor and two cores of normal tissue, all arrayed in separate recipient blocks to give six TMAs that are representative of normal tissue and tumor, taking into account any heterogeneity that may be apparent for some novel markers. More recently, this technology has been extended to cutting edge matrix assembly arrays that maximise array density *(151)*.

There is no single correct method for the scoring of immunohistochemistry in tissue sections. Although some authorities do not use intensity in their system, we use a combination of both intensity and proportion of cells staining when assessing TMAs. However, the range of intensity scores is dependent on the dynamic range of the antibody. Thus, for most epitopes we use the standard 1, no staining; 1, weak staining; 2, moderate staining; and 3, strong staining, but for the HIFs, we use 0, 1, and 2 since these demonstrate a narrower range of staining in our hands. For the proportion of cells staining, we use the broad categories of 0, 1–10%, 11–50%, 51–80%, and 81–100%, which are reproducible in our hands. For analysis, we have used the product of intensity and proportion of cells staining with variable cutoff depending on the distribution of cases. We have tended to use medians and tertiles to avoid assumptions on what level of staining is significant.

The analysis of data derived from TMAs has proved challenging. Akin to the bioinformatic analysis of complementary DNA microarrays, similar algorithms of interrogation are required. We have used several methods, including multivariate logistic regression and the elastic net method *(152)*, but others are also available *(153, 154)*.

6. Summary

Continuing research into angiogenesis using quantitative data not only will broaden our understanding of the angiogenic process but also will have several potential clinical applications beyond its use for prognosis *(37)*. It might help in stratifying patients for cytotoxic therapy *(155)*, aid monitoring and prediction of their response

(156), and with the advent of anti-angiogenesis and vascular targeting, stratifying and altering treatment based on these angiogenic measurements. The next few years will provide the data regarding the reliability of quantitation of angiogenesis in tissue sections. During this time, it is also probable that basic research will describe several candidate molecules that might become objective, sensitive, and specific enough to supersede the presently used assays.

References

1. Folkman, J. (1971) Tumour angiogenesis: Therapeutic implications. *N Engl J Med* 285, 82–86.

2. Brem, S., Cotran, R., Folkman, J. (1972) Tumor angiogenesis: a quantitative method for histological grading. *J Natl Cancer Inst* 48, 347–356.

3. Mlynek, M., van Beunigen, D., Leder, L.-D., Streffer, C. (1985) Measurement of the grade of vascularisation in histological tumour tissue sections. *Br J Cancer* 52, 945–948.

4. Svrivastava, A., Laidler, P., Davies, R., Horgan, K., Hughes, L. (1988) The prognostic significance of tumor vascularity in intermediate-thickness (0.76–4.0 mm thick) skin melanoma. *Am J Pathol* 133, 419–423.

5. Porschen, R., Classen, S., Piontek, M., Borchard, F. (1994) Vascularization of carcinomas of the esophagus and its correlation with tumor proliferation. *Cancer Res* 54, 587–591.

6. Weidner, N., Semple, J. P., Welch, W. R., Folkman, J. (1991) Tumor angiogenesis and metastasis—correlation in invasive breast carcinoma. *N Engl J Med* 324, 1–8.

7. Fox, S. B. (1997) Tumour angiogenesis and prognosis. *Histopathology* 30, 294–301,

8. Fox, S., Harris, A. (2004) The biology of breast tumor angiogenesis, in (Harris, J., Lippman, M. E., Morrow, M., Osborne, C. K., eds.), *Diseases of the Breast*, 3rd ed., pp. 441–458. Lippincott, Williams & Wilkins, Philadelphia.

9. Folkman, J. (1990) What is the evidence that tumours are angiogenesis dependent. *J Natl Cancer Inst* 82, 4–6.

10. Pezzella, F., Pastorin, O. U., Tagliabue, E., et al. (1996) Non-small-cell lung carcinoma tumor growth without morphological evidence of neo-angiogenesis. *Am J Pathol* 151(5), 1417–1423.

11. Pezzella, F. (2000) Evidence for novel non-angiogenic pathway in breast-cancer metastasis. Breast Cancer Progression Working Party. *Lancet* 355, 1787–1788.

12. Holash, J., Maisonpierre, P. C., Compton, D., et al. (1999) Vessel cooption, regression, and growth in tumors mediated by angiopoietins and VEGF. *Science* 284, 1994–1998.

13. Holash, J., Wiegand, S. J., Yancopoulos, G. D. (1999) New model of tumor angiogenesis: dynamic balance between vessel regression and growth mediated by angiopoietins and VEGF. *Oncogene* 18, 5356–5362.

14. Vermeulen, P. B., Colpaert, C., Salgado, R., et al. (2001) Liver metastases from colorectal adenocarcinomas grow in three patterns with different angiogenesis and desmoplasia. *J Pathol* 195, 336–342.

15. Shirakawa, K., Wakasugi, H., Heike, Y., et al. (2002) Vasculogenic mimicry and pseudo-comedo formation in breast cancer. *Int J Cancer* 99, 821–828.

16. Asahara, T., Masuda, H., Takahashi, T., et al. (1999) Bone marrow origin of endothelial progenitor cells responsible for postnatal vasculogenesis in physiological and pathological neovascularization. *Circ Res* 85, 221–228.

17. Gunsilius, E., Duba, H. C., Petzer, A. L., et al. (2000) Evidence from a leukaemia model for maintenance of vascular endothelium by bone-marrow-derived endothelial cells. *Lancet* 355, 1688–1691.

18. Rafii, S. Circulating endothelial precursors: mystery, reality, and promise [comment]. *J Clin Invest* 105, 17–19, 2000.

19. Asahara, T., Takahashi, T., Masuda, H., et al. (1999) VEGF contributes to postnatal neovascularization by mobilizing bone marrow-derived endothelial progenitor cells. *EMBO J,* 18, 3964–3972.

20. Lyden, D., Hattori, K., Dias, S., et al. (2001) Impaired recruitment of bone-marrow-derived endothelial and hematopoietic precursor cells blocks tumor angiogenesis and growth. *Nat Med* 7, 1194–1201.

21. Patan, S., Munn, L. L., Jain, R. K. (1996) Intussusceptive microvascular growth in a human colon adenocarcinoma xenograft: a novel mechanism of tumor angiogenesis. *Microvasc Res* 51, 260–272.

22. Patan, S. (2000) Vasculogenesis and angiogenesis as mechanisms of vascular network formation, growth and remodeling. *J Neurooncol* 50, 1–15.

23. Fox, S., Gatter, K., Bicknell, R., et al. (1993) Relationship of endothelial cell proliferation to tumor vascularity in human breast cancer. *Cancer Res* 53, 9161–9163.

24. Kakolyris, S., Giatromanolaki, A., Koukourakis, M., et al. (1999) Assessment of vascular maturation in non-small cell lung cancer using a novel basement membrane component, LH39: correlation with p53 and angiogenic factor expression [in process citation]. *Cancer Res* 59, 5602–5607.

25. Kakolyris, S., Fox, S. B., Koukourakis, M., et al. (2000) Relationship of vascular maturation in breast cancer blood vessels to vascular density and metastasis, assessed by expression of a novel basement membrane component, LH39. *Br J Cancer* 82, 844–851.

26. Sood, A. K., Seftor, E. A., Fletcher, M. S., et al. (2001) Molecular determinants of ovarian cancer plasticity. *Am J Pathol* 158, 1279–1288.

27. Folberg, R., Hendrix, M. J., Maniotis, A. J. (2000) Vasculogenic mimicry and tumor angiogenesis. *Am J Pathol* 156, 361–381.

28. McDonald, D. M., Munn, L., Jain, R. K. (2000) Vasculogenic mimicry: how convincing, how novel, how significant? *Am J Pathol* 156, 383–388.

29. Warren, B. (1979) The vascular morphology of tumors, in (Peterson, H., ed.), *Tumor Blood Circulation*, pp. 1–47. CRC Press, Boca Raton, FL.

30. Chang, Y. S., di Tomaso, E., McDonald, D. M., Jones, R., Jain, R. K., Munn, L. L. (2000) Mosaic blood vessels in tumors: frequency of cancer cells in contact with flowing blood. *Proc Natl Acad Sci U S A* 97, 14608–14613.

31. Barbareschi, M., Weidner, N., Gasparini, G., et al. (1995) Microvessel quantitation in breast carcinomas. *Appl Immunochem* 3, 75–84.

32. Vermeulen, P. B., Gasparini, G., Fox, S. B., et al. (2002) Second international consensus on the methodology and criteria of evaluation of angiogenesis quantification in solid human tumours. *Eur J Cancer* 38, 1564–1579.

33. Chalkley, H. (1943) Method for the quantative morphological analysis of tissues. *J Natl Cancer Inst* 4, 47–53.

34. Weidner, N., Folkman, J., Pozza, F., et al. (1992) Tumor angiogenesis: a new significant and independent prognostic indicator in early-stage breast carcinoma. *J Natl Cancer Inst* 84, 1875–1887.

35. Fox, S. B., Leek, R. D., Bliss, J., et al. (1997) Association of tumor angiogenesis with bone marrow micrometastases in breast cancer patients. *J Natl Cancer Inst* 89, 1044–1049.

36. Burrows, F. J., Thorpe, P. E. (1994) Vascular targeting—a new approach to the therapy of solid tumors. *Pharmacol Ther* 64, 155–174.

37. Fox, S., Harris, A. (1997) Markers of tumor angiogenesis: clinical applications in prognosis and anti-angiogenic therapy. *Invest New Drugs* 15, 15–28.

38. Kumar, S., Ghellal, A., Li, C., et al. (1999) Breast carcinoma: vascular density determined using CD105 antibody correlates with tumor prognosis. *Cancer Res* 59, 856–861.

39. Smolle, J., Soyer, H. P., Hofmann-Wellenhof, R., Smolle-Juettner, F. M., Kerl, H. (1989) Vascular architecture of melanocytic skin tumors. *Pathol Res Pract* 185, 740–745.

40. Cockerell, C. J., Sonnier, G., Kelly, L., Patel, S. (1994) Comparative analysis of neovascularisation in primary cutaneous melanoma and Spitz nevus. *Am J Dermatopathol* 16, 9–13.

41. Folberg, R., Rummelt, V., Ginderdeuren, R.-V., et al. (1993) The prognostic value of tumor blood vessel morphology in primary uveal melanoma. *Ophthalmology* 100, 1389–1398.

42. Pezzella, F., Dibacco, A., Andreola, S., Nicholson, A. G., Pastorino, U., Harris, A. L. (1996) Angiogenesis in primary lung cancer and lung secondaries. *Eur J Cancer*, 32A, 2494–2500.

43. Wakui, S. (1992) Epidermal growth factor receptor at endothelial cell and pericyte interdigitation in human granulation tissue. *Microvasc Res* 44, 255–262.

44. Visscher, D., Smilanetz, S., Drozdowicz, S., Wykes, S. (1993) Prognostic significance of image morphometric microvessel enumeration in breast carcinoma. *Anal Quant Cytol* 15, 88–92.

45. Fox, S. B., Leek, R. D., Weekes, M. P., Whitehouse, R. M., Gatter, K. C., Harris,

A. L. (1995) Quantitation and prognostic value of breast cancer angiogenesis: comparison of microvessel density, Chalkley count, and computer image analysis. *J Pathol* 177, 275–283.

46. Simpson, J., Ahn, C., Battifora, H., Esteban, J. (1994) Vascular surface area as a prognostic indicator in invasive breast carcinoma. *Lab Invest* 70, 22A.

47. Brawer, M. K., Deering, R. E., Brown, M., Preston, S. D., Bigler, S. A. (1994) Predictors of pathologic stage in prostatic carcinoma. The role of neovascularity. *Cancer* 73, 678–687.

48. Furusato, M., Wakui, S., Sasaki, H., Ito, K., Ushigome, S. (1994) Tumour angiogenesis in latent prostatic carcinoma. *Br J Cancer* 70, 1244–1246.

49. Bigler, S., Deering, R., Brawer, M. Comparisons of microscopic vascularity in benign and malignant prostate tissue. *Human Pathol* 24, 220–226.

50. Williams, J. K., Carlson, G. W., Cohen, C., Derose, P. B., Hunter, S., Jurkiewicz, M. J. (1994) Tumor angiogenesis as a prognostic factor in oral cavity tumors. *Am J Surg* 168, 373–380,

51. Wesseling, P., van der Laak, J. A., Link, M., Teepen, H. L., Ruiter, D. J. (1998) Quantitative analysis of microvascular changes in diffuse astrocytic neoplasms with increasing grade of malignancy. *Hum Pathol* 29, 352–358.

52. Charpin, C., Devictor, B., Bergeret, D., (1995) CD31 quantitative immunocytochemical assays in breast carcinomas. Correlation with current prognostic factors. *Am J Clin Pathol* 103, 443–448.

53. Van der Laak, J., Westphal, J., Schalkwijk, L., (1998) An improved procedure to quantify tumour vascularity using true colour image analysis: comparison with the manual hot-spot procedure in a human melanoma xenograft model. *J Pathol* 184, 136–143.

54. McCarthy, S. A., Kuzu, I., Gatter, K. C., Bicknell, R. (1991) Heterogeneity of the endothelial cell and its role in organ preference of tumour metastasis. *Trends Pharmacol Sci* 12, 462–467.

55. Carnochan, P., Briggs, J. C., Westbury, G., Davies, A. J. (1991) The vascularity of cutaneous melanoma: a quantitative histological study of lesions 0.85–1.25 mm in thickness. *Br J Cancer* 64, 102–107.

56. Vesalainen, S., Lipponen, P., Talja, M., Alhava, E., Syrjanen, K. (1994) Tumor vascularity and basement membrane structure as prognostic factors in T1–2M0 pro-

static adenocarcinoma. *AntiCancer Res* 14, 709–714.

57. Van Hoef, M. E., Knox, W. F., Dhesi, S. S., Howell, A., Schor, A. M. (1993) Assessment of tumour vascularity as a prognostic factor in lymph node negative invasive breast cancer. *Eur J Cancer* 29A, 1141–1145.

58. Hall, N. R., Fish, D. E., Hunt, N., Goldin, R. D., Guillou, P. J., Monson, J.R. (1992) Is the relationship between angiogenesis and metastasis in breast cancer real? *Surg Oncol* 1, 223–229.

59. Ottinetti, A., Sapino, A. (1988) Morphometric evaluation of microvessels surrounding hyperplastic and neoplastic mammary lesions. *Breast Cancer Res Treat* 11, 241–248.

60. Bosari, S., Lee, A. K., DeLellis, R. A., Wiley, B. D., Heatley, G. J., Silverman, M. L. (1992) Microvessel quantitation and prognosis in invasive breast carcinoma. *Hum Pathol* 23, 755–761.

61. Bundred, N., Bowcott, M., Walls, J., Faragher, E., Knox, F. (1994) Angiogenesis in breast cancer predicts node metastasis and survival [abstract]. *Br J Surgery* 81, 768.

62. Li, V. W., Folkerth, R. D., Watanabe, H., (1994) Microvessel count and cerebrospinal fluid basic fibroblast growth factor in children with brain tumours. *Lancet* 344, 82–86.

63. Parums, D., Cordell, J., Micklem, K., Heryet, A., Gatter, K., Mason, D. (1990) JC70: a new monoclonal antibody that detects vascular endothelium associated antigen on routinely processed tissue sections. *J Clin Pathol* 43, 752–757.

64. Horak, E. R., Harris, A. L., Stuart, N., Bicknell, R. (1993) Angiogenesis in breast cancer. Regulation, prognostic aspects, and implications for novel treatment strategies. *Ann N Y Acad Sci* 698, 71–84.

65. Sightler, H., Borowsky, A., Dupont, W., Page, D., Jensen, R. Evaluation (1994) of tumor angiogenesis as a prognostic marker in breast cancer [abstract]. *Lab Invest* 70, 22A.

66. Barnhill, R. L., Fandrey, K., Levy, M. A., Mihm, M. J., Hyman, B. (1992) Angiogenesis and tumor progression of melanoma. Quantification of vascularity in melanocytic nevi and cutaneous malignant melanoma. *Lab Invest* 67, 331–337.

67. Sahin, A., Sneige, N., Singletary, E., Ayala, A. (1992) Tumor angiogenesis detected by factor-VIII immunostaining in node-negative breast carcinoma (NNBC): a possible

predictor of distant metastasis [abstract]. *Mod Pathol* 5, 17A.

68. Axelsson, K., Ljung, B. M., Moore, D. H., 2nd, (1995) Tumor angiogenesis as a prognostic assay for invasive ductal breast carcinoma [see comments]. *J Natl Cancer Inst* 87, 997–1008.

69. Fox, S. B., Leek, R. D., Smith, K., Hollyer, J., Greenall, M., Harris, A. L. (1994) Tumor angiogenesis in node-negative breast carcinomas—relationship with epidermal growth factor receptor, estrogen receptor, and survival. *Breast Cancer Res Treat* 29, 109–116.

70. Hansen, S., Grabau, D. A., Sorensen, F. B., Bak, M., Vach, W., Rose, C. (2000) The prognostic value of angiogenesis by Chalkley counting in a confirmatory study design on 836 breast cancer patients. *Clin Cancer Res* 6, 139–146.

71. Dickinson, A. J., Fox, S. B., Persad, R. A., Hollyer, J., Sibley, G. N., Harris, A. L. (1994) Quantification of angiogenesis as an independent predictor of prognosis in invasive bladder carcinomas. *Br J Urol* 74, 762–766.

72. Paweletz, N., Knierim, M. (1989) Tumor-related angiogenesis. *Crit Rev Oncol Hematol* 9, 197–242.

73. Blood, C. H., Zetter, B. R. (1990) Tumor interactions with the vasculature: angiogenesis and tumor metastasis. *Biochim Biophys Acta* 1032, 89–118.

74. Brown, L. F., Berse, B., Jackman, R. W., et al. (1995) Expression of vascular permeability factor (vascular endothelial growth factor) and its receptors in breast cancer. *Hum Pathol* 26, 86–91.

75. Moghaddam, A., Bicknell, R. (1992) Expression of platelet-derived endothelial cell growth factor in *Escherichia coli* and confirmation of its thymidine phosphorylase activity. *Biochemistry* 31, 12141–12146.

76. Anandappa, S. Y., Winstanley, J. H., Leinster, S., Green, B., Rudland, P. S., Barraclough, R. (1994) Comparative expression of fibroblast growth factor mRNAs in benign and malignant breast disease. *Br J Cancer* 69, 772–776.

77. Relf, M., LeJeune, S., Scott, P. A., et al. (1997) Expression of the angiogenic factors vascular endothelial cell growth factor, acidic and basic fibroblast growth factor, tumor growth factor beta-1, platelet-derived endothelial cell growth factor, placenta growth factor, and pleiotrophin in human primary breast cancer and its relation to angiogenesis. *Cancer Res* 57, 963–969.

78. Garver, R. J., Radford, D. M., Donis, K. H., Wick, M. R., Milner, P. G. (1994) Midkine and pleiotrophin expression in normal and malignant breast tissue. *Cancer* 74, 1584–1590.

79. Smith, K., Fox, S. B., Whitehouse, R., et al. (1999) Upregulation of basic fibroblast growth factor in breast carcinoma and its relationship to vascular density, oestrogen receptor, epidermal growth factor receptor and survival. *Ann Oncol* 10, 707–713.

80. Wong, S. Y., Purdie, A. T., Han, P. (1992) Thrombospondin and other possible related matrix proteins in malignant and benign breast disease. An immunohistochemical study. *Am J Pathol* 140, 1473–1482.

81. Visscher, D. W., DeMattia, F., Ottosen, S., Sarkar, F. H., Crissman, J. D. (1995) Biologic and clinical significance of basic fibroblast growth factor immunostaining in breast carcinoma. *Mod Pathol* 8, 665–670.

82. Toi, M., Kondo, S., Suzuki, H., et al. (1996) Quantitative analysis of vascular endothelial growth factor in primary breast cancer. *Cancer* 77, 1101–1106.

83. Lantzsch, T., Hefler, L., Krause, U., et al. (2002) The correlation between immunohistochemically-detected markers of angiogenesis and serum vascular endothelial growth factor in patients with breast cancer. *AntiCancer Res* 22, 1925–1928.

84. Valkovic, T., Dobrila, F., Melato, M., Sasso, F., Rizzardi, C., Jonjic, N. (2002) Correlation between vascular endothelial growth factor, angiogenesis, and tumor-associated macrophages in invasive ductal breast carcinoma. *Virchows Arch* 440, 583–588.

85. Linderholm, B., Tavelin, B., Grankvist, K., Henriksson, R. (1998) Vascular endothelial growth factor is of high prognostic value in node- negative breast carcinoma. *J Clin Oncol* 16, 3121–3128.

86. Gasparini, G., Toi, M., Gion, M., et al. (1997) Prognostic-significance of vascular endothelial growth-factor protein in node negative breast-carcinoma. *J Natl Cancer Inst* 89, 139–147.

87. Obermair, A., Bancher-Todesca, D., Bilgi, S., (1997) Correlation of vascular endothelial growth factor expression and microvessel density in cervical intraepithelial neoplasia. *J Natl Cancer Inst* 89, 1212–1217.

88. Manders, P., Beex, L. V., Tjan-Heijnen, V. C., et al. (2002) The prognostic value of vascular endothelial growth factor in 574 node-negative breast cancer patients who did not receive adjuvant systemic therapy. *Br J Cancer* 87, 772–778.

89. Eppenberger, U., Kueng, W., Schlaeppi, J. M., et al. (1998) Markers of tumor angiogenesis and proteolysis independently define high- and low-risk subsets of node-negative breast cancer patients [in process citation]. *J Clin Oncol* 16, 3129–3136.

90. Coradini, D., Boracchi, P., Daidone, M. G., et al. (2001) Contribution of vascular endothelial growth factor to the Nottingham prognostic index in node-negative breast cancer. *Br J Cancer* 85, 795–797.

91. Toi, M., Bando, H., Ogawa, T., Muta, M., Hornig, C., Weich, H. A. (2002) Significance of vascular endothelial growth factor (VEGF)/soluble VEGF receptor-1 relationship in breast cancer. *Int J Cancer* 98, 14–18.

92. Gunningham, S., Currie, M., Cheng, H., et al. (2000) The short form of the alternatively spliced flt-4 but not its ligand VEGF-C is related to lymph node metastasis in human breast cancers. *Clin Cancer Res* 6,4278–4286.

93. Kinoshita, J., Kitamura, K., Kabashima, A., Saeki, H., Tanaka, S., Sugimachi, K. (2001) Clinical significance of vascular endothelial growth factor-C (VEGF-C) in breast cancer. *Breast Cancer Res Treat* 66, 159–164.

94. Gunningham, S., Currie, M., Cheng, H., et al. (2000) VEGF-B expression in human breast cancers is associated with positive lymph node status. *J Pathol* 193, 325–332.

95. Onogawa, S., Kitadai, Y., Tanaka, S., Kuwai, T., Kimura, S., Chayama, K. (2004) Expression of VEGF-C and VEGF-D at the invasive edge correlates with lymph node metastasis and prognosis of patients with colorectal carcinoma. *Cancer Sci* 95, 32–39.

96. Yokoyama, Y., Charnock-Jones, D. S., Licence, D., et al. (2003) Expression of vascular endothelial growth factor (VEGF)-D and its receptor, VEGF receptor 3, as a prognostic factor in endometrial carcinoma. *Clin Cancer Res* 9, 1361–1369.

97. Nakamura, Y., Yasuoka, H., Tsujimoto, M., et al. (2003) Prognostic significance of vascular endothelial growth factor D in breast carcinoma with long-term follow-up. *Clin Cancer Res* 9, 716–721.

98. Kurebayashi, J., Otsuki, T., Kunisue, H., et al. (1999) Expression of vascular endothelial growth factor (VEGF) family members in breast cancer. *Jpn J Cancer Res* 90, 977–981.

99. Engels, K., Fox, S. B., Whitehouse, R. M., Gatter, K. C., Harris, A. L. (1997) Up-regulation of thymidine phosphorylase expression is associated with a discrete pattern of angiogenesis in ductal carcinomas in situ of the breast. *J Pathol* 182, 414–420.

100. Fox, S. B., Westwood, M., Moghaddam, A., et al. (1996) The angiogenic factor platelet-derived endothelial cell growth factor/thymidine phosphorylase is up-regulated in breast cancer epithelium and endothelium. *Br J Cancer* 73, 275–280.

101. Toi, M., Hoshina, S., Taniguchi, T., et al. (1995) Expression of platelet derived endothelial cell growth factor/thymidine phosphorylase in human breast cancer. *Int J Cancer* 64, 79–82.

102. Toi, M., Ueno, T., Matsumoto, H., et al. (1999) Significance of thymidine phosphorylase as a marker of protumor monocytes in breast cancer. *Clin Cancer Res* 5, 1131–1137.

103. Yang, Q., Barbareschi, M., Mori, I., et al. (2002) Prognostic value of thymidine phosphorylase expression in breast carcinoma. *Int J Cancer* 97, 512–517.

104. Kanzaki, A., Takebayashi, Y., Bando, H., et al. (2002) Expression of uridine and thymidine phosphorylase genes in human breast carcinoma. *Int J Cancer* 97, 631–635.

105. Nagaoka, H., Iino, Y., Takei, H., Morishita, Y. (1998) Platelet-derived endothelial cell growth factor/thymidine phosphorylase expression in macrophages correlates with tumor angiogenesis and prognosis in invasive breast cancer. *Int J Oncol* 13, 449–454.

106. Toi, M., Yamamoto, Y., Inada, K., et al. (1995) Vascular endothelial growth factor and platelet-derived endothelial growth factor are frequently co-expressed in highly vascularized breast cancer. *Clin Cancer Res* 1, 961–964.

107. O'Brien, T., Fox, S., Dickinson, A., et al. (1996) Expression of the angiogenic factor thymidine phosphorylase/platelet derived endothelial cell growth factor in primary bladder cancers. *Cancer Res* 56, 4799–4804.

108. O'Brien, T. S., Smith, K., Cranston, D., Fuggle, S., Bicknell, R., Harris, A. L. (1995) Urinary basic fibroblast growth factor in patients with bladder cancer and benign prostatic hypertrophy. *Br J Urol* 76, 311–314.

109. Adams, J., Carder, P. J., Downey, S.,et al. (2000) Vascular endothelial growth factor (VEGF) in breast cancer: comparison of plasma, serum, and tissue VEGF and microvessel density and effects of tamoxifen. *Cancer Res* 60, 2898–2905.

110. Freeman, A., Morris, L. S., Mills, A. D., et al. (1999) Minichromosome maintenance proteins as biological markers of dysplasia and malignancy. *Clin Cancer Res* 5, 2121–2132.

111. Stoeber, K., Swinn, R., Prevost, A. T., et al. (2002) Diagnosis of genito-urinary tract cancer by detection of minichromo-

some maintenance 5 protein in urine sediments. *J Natl Cancer Inst* 94, 1071–1079.

112. Eberhard, A., Kahlert, S., Goede, V., Hemmerlein, B., Plate, K. H., Augustin, H. G. (2000) Heterogeneity of angiogenesis and blood vessel maturation in human tumors: implications for antiangiogenic tumor therapies. *Cancer Res* 60, 1388–1393.

113. Schadendorf, D., Heidel, J., Gawlik, C., Suter, L., Czarnetzki (1995) Association with clinical outcome of expression of VLA-4 in primary cutaneous malignant melanoma as well as P-selectin and E-selectin on intratumoral vessels. *J Natl Cancer Inst* 87, 366–371.

114. Kageshita, T., Hamby, C. V., Hirai, S., Kimura, T., Ono, T., Ferrone, S. (2000) Alpha(v)beta3 expression on blood vessels and melanoma cells in primary lesions: differential association with tumor progression and clinical prognosis. *Cancer Immunol Immunother* 49, 314–318.

115. Kageshita, T., Yoshii, A., Kimura, T., et al. (1993) Clinical relevance of ICAM-1 expression in prmary lesions and serum of patients with malignant melanoma. *Cancer Res* 53, 4927–4932.

116. Banks, R. E., Gearing, A. J., Hemingway, I. K., Norfolk, D. R., Perren, T. J., Selby, P. J. (1993) Circulating intercellular adhesion molecule-1 (ICAM-1), E-selectin and vascular cell adhesion molecule-1 (VCAM-1) in human malignancies. *Br J Cancer* 68, 122–124.

117. Brooks, P. C., Stromblad, S., Klemke, R., Visscher, D., Sarkar, F. H., Cheresh, D. A. (1995) Antiintegrin 3 v blocks human breast cancer growth and angiogenesis in human skin. *J Clin Invest* 96, 1815–1822.

118. Gasparini, G., Brooks, P. C., Biganzoli, E., et al. (1998) Vascular integrin alpha(v) beta3: a new prognostic indicator in breast cancer [in process citation]. *Clin Cancer Res* 4, 2625–2634.

119. Pepper, M. S. (2001) Role of the matrix metalloproteinase and plasminogen activator-plasmin systems in angiogenesis. *Arterioscler Thromb Vasc Biol* 21, 1104–1117.

120. John, A., Tuszynski, G. (2001) The role of matrix metalloproteinases in tumor angiogenesis and tumor metastasis. *Pathol Oncol Res* 7, 14–23.

121. Haas, T. L., Madri, J. A. (1999) Extracellular matrix-driven matrix metalloproteinase production in endothelial cells: implications for angiogenesis. *Trends Cardiovasc Med* 9, 70–77.

122. Lochter, A., Bissell, M. J. (1999) An odyssey from breast to bone: multi-step con-

trol of mammary metastases and osteolysis by matrix metalloproteinases. *APMIS* 107, 128–136.

123. Parfyonova, Y. V., Plekhanova, O. S., Tkachuk, V. A. (2002) Plasminogen activators in vascular remodeling and angiogenesis. *Biochemistry (Mosc)* 67, 119–134.

124. Nielsen, B. S., Sehested, M., Kjeldsen, L., Borregaard, N., Rygaard, J., Dano, K. (1997) Expression of matrix metalloprotease-9 in vascular pericytes in human breast cancer. *Lab Invest* 77, 345–355.

125. Fox, S., Taylor, M., Grondahl-Hansen, J., Kakolyris, S., Gatter, K., Harris, A. (2001) Plasminogen activator inhibitor-1 as a measure of vascular remodelling in breast cancer. *J Pathol* 195, 236–243.

126. Grøndahl-Hansen, J., Christensen, I. J., Rosenquist, C., et al. (1993)High levels of urokinase-type plasminogen activator and its inhibitor PAI-1 in cytosolic extracts of breast carcinomas are associated with poor prognosis. *Cancer Res* 53, 2513–2521.

127. Grøndahl-Hansen, J., Peters, H. A., van Putten, W. L., et al. (1995) Prognostic significance of the receptor for urokinase plasminogen activator in breast cancer. *Clin Cancer Res* 1, 1079–1087.

128. Grøndahl-Hansen, J., Hilsenbeck, S. G., Christensen, I. J., Clark, G. M., Osborne, C. K., Brünner, N. (1997) Prognostic significance of PAI-1 and uPA in cytosolic extracts obtained from node-positive breast cancer patients. *Breast Cancer Res Treat* 43, 153–163.

129. Janicke, F., Pache, L., Schmitt, M., Ulm, K., Thomssen, C., Prechtl, A., Graeff, H. (1994) Both the cytosols and detergent extracts of breast cancer tissues are suited to evaluate the prognostic impact of the urokinase-type plasminogen activator and its inhibitor, plasminogen activator inhibitor type 1. *Cancer Res* 54, 2527–2530.

130. Foekens, J. A., Look, M. P., Peters, H. A., van Putten, W. L., Portengen, H., Klijn, J. G. (1995) Urokinase-type plasminogen activator and its inhibitor PAI-1: predictors of poor response to tamoxifen therapy in recurrent breast cancer. *J Natl Cancer Inst* 87, 751–756.

131. Duffy, M. J. (2002) Urokinase plasminogen activator and its inhibitor, PAI-1, as prognostic markers in breast cancer: from pilot to level 1 evidence studies. *Clin Chem* 48, 1194–1197.

132. Harbeck, N., Schmitt, M., Kates, R. E., et al. (2002) Clinical utility of urokinase-type plasminogen activator and plasminogen activator inhibitor-1 determination in primary breast cancer tissue for individualized therapy concepts. *Clin Breast Cancer* 3, 196–200.

133. Warren, B., Greenblatt, M., Kommineni, V. (1972) Tumor angiogenesis: ultrastructure of endothelial cells in mitosis. *Br J Exp Pathol* 53, 216–224.

134. Dvorak, H. F., Nagy, J. A., Feng, D., Brown, L. F., Dvorak, A. M. (1999) Vascular permeability factor/vascular endothelial growth factor and the significance of microvascular hyperpermeability in angiogenesis. *Curr Top Microbiol Immunol* 237, 97–132.

135. Vaupel, P., Kallinowski, F., Okunieff, P. (1989) Blood flow, oxygen and nutrient supply, metabolic microenvironment of human tumors: a review. *Cancer Res* 49, 6449–6465.

136. Harris, A. L. (2002) Hypoxia—a key regulatory factor in tumour growth. *Nat Rev Cancer* 2, 38–47.

137. Talks, K. L., Turley, H., Gatter, K. C., et al. (2000) The expression and distribution of the hypoxia-inducible factors HIF-1alpha and HIF-2alpha in normal human tissues, cancers, and tumor- associated macrophages. *Am J Pathol* 157, 411–421.

138. Wykoff, C. C., Beasley, N. J., Watson, P. H., et al. (2000) Hypoxia-inducible expression of tumor-associated carbonic anhydrases. *Cancer Res* 60, 7075–7083.

139. Loncaster, J. A., Harris, A. L., Davidson, S. E., et al. (2001) Carbonic anhydrase (CA IX) expression, a potential new intrinsic marker of hypoxia: correlations with tumor oxygen measurements and prognosis in locally advanced carcinoma of the cervix. *Cancer Res* 61, 6394–6399.

140. Bos, R., van der Groep, P., Greijer, A. E., et al. (2003) Levels of hypoxia-inducible factor-1alpha independently predict prognosis in patients with lymph node negative breast carcinoma. *Cancer* 97, 1573–1581.

141. Swinson, D. E., Jones, J. L., Richardson, D., et al. (2003) Carbonic anhydrase IX expression, a novel surrogate marker of tumor hypoxia, is associated with a poor prognosis in non-small-cell lung cancer. *J Clin Oncol* 21, 473–482.

142. Qin, C., Wilson, C., Blancher, C., Taylor, M., Safe, S., Harris, A. L. (2001) Association of ARNT splice variants with estrogen receptor-negative breast cancer, poor induction of vascular endothelial growth factor under hypoxia, poor prognosis. *Clin Cancer Res* 7, 818–823.

143. Schindl, M., Schoppmann, S. F., Samonigg, H., et al. (2002) Overexpression of hypoxia-inducible factor 1alpha is associated with an unfavorable prognosis in lymph node-positive breast cancer. *Clin Cancer Res* 8, 1831–1837.

144. Hasebe, T., Sasaki, S., Imoto, S., Mukai, K., Yokose, T., Ochiai, A. (2002) Prognos-
tic significance of fibrotic focus in invasive ductal carcinoma of the breast: a prospective observational study. *Mod Pathol* 15, 502–516.

145. Hasebe, T., Tsuda, H., Hirohashi, S., et al. (1996) Fibrotic focus in invasive ductal carcinoma: an indicator of high tumor aggressiveness. *Jpn J Cancer Res* 87, 385–394.

146. Jitsuiki, Y., Hasebe, T., Tsuda, H., et al. (1999) Optimizing microvessel counts according to tumor zone in invasive ductal carcinoma of the breast. *Mod Pathol* 12, 492–498.

147. Colpaert, C., Vermeulen, P., Fox, S., AL., H., Dirix, L., Van Marck, E. (2003) The presence of a fibrotic focus in lymph node-negative breast cancer correlates with expression of carbonic anhydrase IX angiogenesis and is a marker of hypoxia and poor prognosis. *Breast Cancer Res Treat*, 81(2): 137–147.

148. de Jong, J. S., van Diest, P. J., Baak, J. P. (1995) Heterogeneity and reproducibility of microvessel counts in breast cancer. *Lab Invest* 73, 922–926.

149. Martin, L., Holcombe, C., Green, B., Leinster, S. J., Winstanley, J. (1997) Is a histological section representative of whole tumour vascularity in breast cancer? *Br J Cancer* 76, 40–43.

150. Kononen, J., Bubendorf, L., Kallioniemi, A., et al. (1998) Tissue microarrays for high-throughput molecular profiling of tumor specimens. *Nat Med* 4, 844–847.

151. LeBaron, M. J., Crismon, H. R., Utama, F. E., et al. (2005) Ultrahigh density microarrays of solid samples. *Nat Methods* 2, 511–513.

152. Generali, D., Buffa, F., Berruti, A., et al. (2006) Phosphorylated ERα, HIf-1α and MAPK Signaling, as predictors of primary endocrine treatment response and resistance in breast cancer patients. *J Clin Oncol* 2008 in press.

153. Liu, X., Minin, V., Huang, Y., Seligson, D. B., Horvath, S. (2004) Statistical methods for analyzing tissue microarray data. *J Biopharm Stat* 14, 671–685.

154. Zhang, D. H., Salto-Tellez, M., Chiu, L. L., Shen, L., Koay, E. S. (2003) Tissue microarray study for classification of breast tumors. *Life Sci* 73, 3189–3199.

155. Protopapa, E., Delides, G. S., Revesz, L. (1993) Vascular density and the response of breast carcinomas to mastectomy and adjuvant chemotherapy. *Eur J Cancer* 29A, 1141–1145.

156. Fox, S., Engels, K., Comley, M., et al. (1997) Relationship of elevated tumour thymidine phosphorylase in node positive breast carcinomas to the effects of adjuvant CMF. *Annal Oncol* 8, 271–275.

Chapter 4

Immunohistochemical Methods for Measuring Tissue Lymphangiogenesis

Steven Clasper and David G. Jackson

Abstract

The field of lymphatic research has benefited enormously from the recent discovery of "marker" proteins that permit not only the identification and quantitation of lymphatic vessels in tissue sections for tumor pathology but also the isolation of primary lymphatic endothelial cells for basic research. This chapter focuses on the use of these markers for the immunohistochemical analysis of lymphangiogenesis in both frozen and paraffin-embedded tissue sections and discusses current protocols and their associated problems.

Key words: Lymphangiogenesis, lymphatic vessel, LYVE-1, metastasis, tumor.

1. Introduction

The measurement of lymphangiogenesis is of great significance in understanding the role of this process during many pathological conditions. Clearly, the most prominent example has been cancer, specifically the metastatic spread of tumors through lymphatic vessels; other examples include lymphedema, wound healing, and inflammation *(1, 2)*. The measurement of lymphangiogenesis within a tissue is usually approached by assessing the density of lymphatic vessels, although in the case of tumor vessels, it should be borne in mind that isolated measurements of this kind can mask artifactual increases that are caused by tissue compression and serendipitous growth next to pre-existing lymphatic networks as well as genuine proliferation *(3)*.

Traditionally, the microscopic identification of lymphatic vessels has relied on skilled analysis of morphology, absence of red

S. Martin and C. Murray (eds.), *Methods in Molecular Biology, Angiogenesis Protocols, Second edition, Vol. 467*
© Humana Press, a part of Springer Science+ Business Media, LLC 2009
DOI: 10.1007/978-1-59745-241-0_4

blood cells within the lumen, and negative staining for blood vascular markers. However, the discovery in the last decade of several lymphatic marker proteins has simplified this task *(3–5)*. Here, we describe the use of two of these markers, LYVE-1 *(6, 7)* and podoplanin *(8)*, to identify lymphatic vessels in frozen or paraffin sections of human tissue. The level of lymphangiogenesis itself is determined by measurement of lymphatic vessel density (LVD) either throughout the tissue specimen or within vessel "hot spots" using the Chalkley point graticule method (see **refs.** *9* and *10)*, combined with estimation based on markers of nuclear division *(3)*.

2. Materials

2.1. General

1. Slide staining tray with lightproof lid.
2. Coplin jars.
3. Phosphate-buffered saline (PBS).
4. Hydrophobic pen (Abcam).
5. Normal goat serum.
6. Foetal calf serum (FCS).
7. Polyclonal rabbit anti-human LYVE-1 (Abcam, R&D Systems, Reliatech).
8. Mouse monoclonal anti-human podoplanin D2-40 (Signet Laboratories).
9. Mouse monoclonal anti-human Ki67 (BD Pharmingen).
10. Clear nail varnish.
11. Chalkley eyepiece graticule.

2.2. Frozen Sections

1. Acetone.
2. Paraformaldehyde: Dissolve 8 g of paraformaldehyde in 90 mL water on a heated stirring block in a fume hood. Add a few drops of $10M$ NaOH to help the powder dissolve. Return the solution to neutral pH using $1M$ HCl and indicator paper. Store at –20°C in small aliquots. To use, melt the solution in a heated water bath, then add an equal volume of 2X PBS.

2.3. Paraffin Sections

1. Microwave-safe dish and slide rack.
2. Citroclear® (HD Supplies).
3. Ethanol.

4. 10 mM sodium citrate, pH 6.0.

2.4. Immunohisto-chemistry

1. Bovine serum albumin (BSA).
2. Envision kits (horseradish peroxidase [HRP] anti-mouse, HRP anti-rabbit, and G|2 double stain) (Dako).
3. Aquamount (BDH).

2.5. Immunofluorescence

1. Goat anti-rabbit immunoglobulin G (IgG) Alexafluor 488-conjugated (Molecular Probes).
2. Goat anti-mouse IgG Alexafluor 568-conjugated (Molecular Probes).
3. Vectashield with Dapi (Vector Laboratories).

3. Methods

3.1. Pre-treatment of Frozen Sections

3.1.1. Fixing

1. Allow slides of sections approximately 10-µm thick to equilibrate to room temperature.
2. Label slides with a pencil, noting specimen and primary antibody to be used.
3. If the podoplanin antibody D2-40 is to be used, then cover section with 4% paraformaldehyde for 10 min. Rinse slides carefully over a sink using a wash bottle of PBS (*see* **Note 1**).
4. If D2-40 is not being used, then place slides in a coplin jar containing 100% acetone for 2 min to fix, then remove and air-dry.
5. Place slides in a jar of PBS for approximately 5 min to allow embedding compound around the sections to dissolve.

3.1.2. Blocking

1. Carefully dry the slides using tissue or a paper towel and draw around each specimen with a hydrophobic pen to retain the small antibody volumes on the section.
2. Place slides in a staining tray and block non-specific antibody-binding sites by applying approximately 200 µL of 5% goat serum in PBS to each section and incubating for 20 min (*see* **Note 2**).
3. Rinse slides carefully over a sink using a wash bottle of PBS (*see* **Note 1**) and place in a jar of PBS for 5 min.
4. Proceed to the appropriate section for either immunohistochemistry or immunofluorescence.

3.2. Pre-treatment of Paraffin Sections

3.2.1. De-waxing

1. Label slides with a pencil, noting specimen and primary antibody to be used.

2. Place slides in a coplin jar containing Citroclear for approximately 5 min. Remove and drain.

3. Place slides in a second coplin jar containing Citroclear for approximately 5 min. Remove and drain (*see* **Note 3**).

4. Place slides in a coplin jar containing 100% ethanol for approximately 5 min. Remove and drain.

5. Place slides in a second coplin jar containing 100% ethanol for approximately 5 min. Remove and drain.

6. Place slides in a coplin jar containing 50% ethanol for approximately 5 min. Remove and drain.

7. Place slides in a coplin jar containing water for approximately 5 min. Remove and drain.

3.2.2. Antigen Retrieval

1. Pre-heat to 100°C in a microwave a covered microwave-safe dish containing enough 10 mM citrate buffer (pH 6.0) to cover a rack of slides.

2. Place the rack of slides into the heated buffer, re-cover, and simmer for 10 min.

3. Place sections in a jar of PBS for approximately 5 min to cool.

3.2.3. Blocking

1. Carefully dry the slides using tissue or a paper towel and draw around each specimen with a hydrophobic pen to retain the small antibody volumes on the section.

2. Place slides in a staining tray and block non-specific antibody-binding sites by applying approximately 200 µL of 5% goat serum in PBS to each section and incubating for 20 min (*see* **Note 2**).

3. Rinse slides carefully over a sink using a wash bottle of PBS (*1*) and place in a jar of PBS for 5 min.

4. Proceed to the appropriate section for either immunohistochemistry or immunofluorescence.

3.3. Immunohisto-chemistry

3.3.1. Primary Antibody

1. Carefully dry slides and place in the staining tray.

2. Apply peroxidase block from the Dako Envision kit (blocks endogenous peroxidase activity that is present in certain tissues) to cover each section. Incubate for approximately 5 min.

3. Rinse slides carefully over a sink using a wash bottle of PBS and place in a jar of PBS for 5 min.

4. Carefully dry slides and place in the staining tray.

5. Apply primary antibody (anti-LYVE-1 or D2-40) at 5 µg/mL in 1% BSA in PBS. Use approximately 200 µL per section. Incubate for 30 min *(4)*.

6. Rinse slides carefully over a sink using a wash bottle of PBS and place in a jar of PBS for 5 min.

3.3.2. Secondary Antibody and Mounting

1. Carefully dry slides and place in the staining tray.

2. Apply enough of the appropriate Dako Envision secondary HRP conjugate to cover each section (anti-rabbit for LYVE-1, anti-mouse for D2-40). Incubate for 30 min.

3. Rinse slides carefully over a sink using a wash bottle of PBS and place in a jar of PBS for 5 min.

4. Mix the Envision peroxidase substrate per the instructions and apply to each section. Incubate for 5–10 min.

5. Wash slides using distilled water from a wash bottle.

6. Place slides in a jar of haemotoxylin solution for approximately 2 min.

7. Wash slides with normal tap water.

8. Carefully dry slides and place in the staining tray.

9. Apply a few drops of Aquamount medium to each section and place a coverslip on top.

10. Invert each slide and press down firmly on a flat pile of paper towels to evenly spread the mounting medium.

11. Seal each slide by painting around the edge of the coverslip with clear nail varnish.

3.4. Immunofluorescence

3.4.1. Primary Antibodies

1. Carefully dry slides and place in the staining tray.

2. Apply primary antibodies (anti-LYVE-1 and D2–40) at 10 µg/mL each in 5% v/v FCS/PBS. Use approximately 200 µL per section. Incubate for 30 min (*see* **Note 4**).

3. Rinse slides carefully over a sink using a wash bottle of PBS and place in a jar of PBS for 5 min.

3.4.2. Secondary Antibodies and Mounting

1. Carefully dry slides and place in the staining tray.

2. Dilute both fluorescently labelled secondary antibodies together 1 in 500 in 5% v/v FCS/PBS and apply approximately 200 µL to each section. Incubate for 30 min, ensuring that the lightproof lid is in place for this step.

3. Rinse slides carefully over a sink using a wash bottle of PBS and place in a jar of PBS for 5 min.

4. Carefully dry slides and place in the staining tray.

5. Apply a few drops of Vectashield to each section and place a coverslip on top.

6. Invert each slide and press down firmly on a flat pile of paper towels to evenly spread the mounting medium.

7. Seal each slide by painting around the edge of the coverslip with nail varnish.

8. Store slides at 4°C in a lightproof box.

3.5. Identification of Lymphatic Vessels

3.5.1. Detection of Lymphatic Vessels

Lymphatic vessels should be clearly visible as stained structures within the tissue. However, there may be a wide variation in size and morphology. Initial lymphatics may appear as small structures (diameter range 10–50 µm), while larger capillaries may appear as elongated structures, sometimes with a collapsed lumen (diameter range 100–200 µm). Tumor-associated lymphatic vessels may appear as small basket-like structures within the tumor mass or as a continuous endothelium surrounding a tumor embolus (11). **Figure 4.1** shows typical lymphatic morphologies in normal and tumor tissue detected by immunohistochemistry with antiserum against LYVE-1. In addition, the contrast in morphology between a lymphatic and a blood vessel can be seen.

3.5.2. Confirmation of Lymphatic Identity

Positive staining of a structure for a single lymphatic marker protein should not be regarded as definitive proof of lymphatic identity. For sections stained with antibodies to LYVE-1 and podoplanin by immunofluorescence, it is a simple matter to check the co-expression of marker proteins. For specimens stained by immunohistochemistry, we recommend the staining of consecutive serial sections with each antibody to confirm identification. Two-colour staining is also possible using immunohistochemistry, for example, by combining peroxidase- and alkaline phosphatase-conjugated antibodies. However, it is our opinion that the procedure is more satisfactory for mutually exclusive staining of distinct cell types or when the two markers localize to different regions of the same cell as described later. Regardless of whether immunohistochemistry or immunofluorescence staining is chosen, the results should still be interpreted with caution (*see* **Note 5**). Discrimination between lymphatic vessels and blood vessels is rarely a problem when using these markers together (*see* **Note 6**).

3.5.3. Vessel Hot Spots

In normal tissue (e.g., dermis), lymphatic vessels may be evenly distributed. However, in tumors these vessels may be concentrated within discrete areas termed *hot spots* induced by agents enriched in the local microenvironment, such as lymphangiogenic growth factors (vascular endothelial growth factor [VEGF] C, VEGF-D, platelet-derived growth factor [PDGF], etc.). Hot spots rather than randomly chosen areas are frequently targeted for tumor vessel counts in the assessment of lymphangiogenesis, although the validity of this practice has been disputed (*see,* e.g., **ref.** *12*

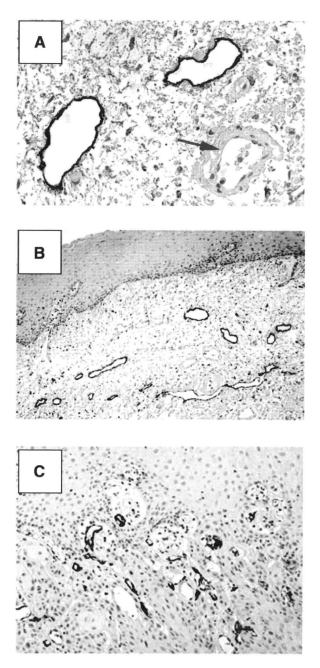

Fig. 4.1. Immunohistochemical staining of lymphatic vessels in soft tisssue. Patent lymphatic vessels of normal human tongue show strong staining of LYVE-1, while the erythrocyte-containing blood vessel (red arrow) shows no staining and a typical thickened vessel wall (×32 objective) (**A**). Abundant LYVE-1-positive lymphatic vessels can be seen beneath the epithelium (×10 objective) (**B**). LYVE-1-positive intratumoral lymphatic vessels within squamous carcinoma of the tongue (×20 objective) (**C**). (Adapted from **ref.** *11*.).

3.6. Measurement of Mean Lymphatic Vessel Density

The method for measurement of mean LVD relies on the ability of the observer to distinguish discrete lymphatic vessels and count them within a known area, thus giving an actual measurement of LVD within the plane of the section.

3.6.1. Calculation of Area of View

Consult the microscope manufacturer's handbook to obtain the field-of-view distance in millimetres for each objective lens to be used. Use this number to calculate the area of view with the formula

$$Area(mm^2) = \pi(\text{Field of view}/2)^2$$

3.6.2. Counting of Lymphatic Vessels

1. Using a low-power objective, a field of view containing lymphatic vessels (e.g., hot spot) should be identified, and a suitable objective lens should be chosen to magnify this area. This size of objective should be constant throughout all of the samples.

2. Each discrete individual stained lymphatic structure (see above), irrelevant of size, is counted as a vessel, and the total within the area of view is recorded.

3. A different region within the same section is then chosen, and the vessel number is again recorded. This is repeated at least three times, and the mean vessel number is calculated.

4. The process is repeated for each section, making sure to keep the objective lens constant.

3.6.3. Calculation of Lymphatic Vessel Density

1. The vessel density is calculated for each section:
Lymphatic vessel density(mm^{-2})=Mean vessel number/Area (mm^2)

2. The application of a suitable statistical method can be used to compare MVDs between tissues and hence levels of lymphangiogenesis (*see* **Note** 7).

3.7. Chalkley Counting

Chalkley counting does not rely on the observer's assessment of individual vessels but instead effectively measures the area covered by vessels *(9)*. A Chalkley eyepiece graticule is required to fit the microscope. This is a rotatable graticule marked with 25 randomly placed spots. Due to the requirement to be able to see the graticule markings against the background of the illuminated slide, this method is only suitable for sections stained by immunohistochemistry, not immunofluorescence.

3.7.1. Determining the Chalkley Count

1. Using a low-power objective, a field of view containing lymphatic vessels (i.e., hot spot) should be identified, and a suitable objective lens should be chosen to magnify this area. This size of objective should be constant throughout all of the samples.

2. The graticule is carefully rotated until the maximum number of dots overlap lymphatic vessels (note that these do not need to be separate vessels), and the number of these dots (maximum score 25) is recorded. See Figure 4.2 for an example of this.

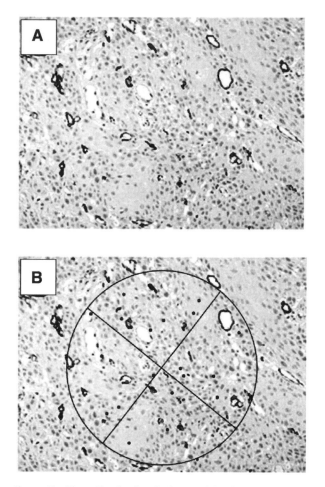

Fig. 4.2. The Chalkley counting method for estimating lymphatic vessel density. Intratumoral lymphatic vessels of tongue squamous carcinoma stained for expression of LYVE-1 (**A**) are overlayed with a representation of a chalkley grid (**B**). Dots that overlap lymphatic vessels are highlighted in red (×20 objective). (Adapted from ref. *11*.).

3. The procedure is repeated a minimum of three times using different regions within the same section.

4. The mean number of dots overlapping lymphatic vessels on a section is the Chalkley count.

5. The application of a suitable statistical method can be used to compare Chalkley counts between tissues (applying cutoffs as appropriate) and hence levels of lymphangiogenesis (*see* **Note** 7).

3.8. Lymphatic Endothelial Cell Proliferation

The basis of this method is that by co-staining with both a lymphatic marker and a proliferation marker it allows identification of actively dividing lymphatic endothelial cells and hence an accurate measurement of ongoing lymphangiogenesis at the time the tissue was taken. As this method requires two-colour staining, it is perhaps most suitable for immunofluorescence; however, two-colour immunohistochemistry may be used as the lymphatic marker and proliferation marker proteins are located on the cell surface and nucleus, respectively. Extreme care must be taken in the identification of lymphatic vessels as only a single lymphatic marker antibody is used.

3.8.1. Frozen Sections

1. Frozen sections must be fixed and may be permeabilised to improve access to the nuclear compartment. Follow the instructions described in **Subheading 3.1.1.** for fixing with paraformaldehyde.

2. Before blocking, rinse the section and incubate for 5 min with 0.5% Triton X-100 in PBS.

3. Rinse with PBS and proceed to blocking.

3.8.2. Immunofluorescence

The protocol for frozen or paraffin sections should be followed with the exchange of the podoplanin D2-40 antibody with a mouse monoclonal raised against the proliferation marker Ki-67.

3.8.3. Immunohistochemistry

The Dako EnVision G|2 double-stain system should be used with anti-LYVE-1 and anti-Ki-67 according to the manufacturer's instructions.

3.8.4. Evaluation of Staining

1. Proliferating lymphatic vessels will contain cells that are stained positively for both LYVE-1 and Ki-67 expression (*see* **Fig. 4.3**). The observer should be aware that tumor tissue and areas of inflammation or wound healing are likely to contain an abundance of proliferating non-lymphatic cells.

2. Mean proliferating LVD may be calculated in a way analogous to the calculation of mean lymphatic density as described in **Subheading 3.6.** when only vessels containing proliferating cells are scored.

3. Immunohistochemistry may be combined with a Chalkley graticule to give a Chalkley count for proliferating vessels.

4. Using either method, a measurement of the ratio of proliferating to non-proliferating vessels gives insight into the degree of lymphangiogenesis in the tissue.

3.9. Measuring Lymphangiogenesis in Mouse Tissues

While the analysis of lymphangiogenesis within human tissue samples is clearly important in a diagnostic and prognostic role, it is appreciated that experimental models are widely used during the study of both adult disease and embryonic development. The techniques described are equally valid using mouse tissues if the correct antibodies are used (*see* **Note 7**). However, the lack of a Dako Envision kit for rat antibodies limits the immunohistochemistry that can be performed using the particular protocols described.

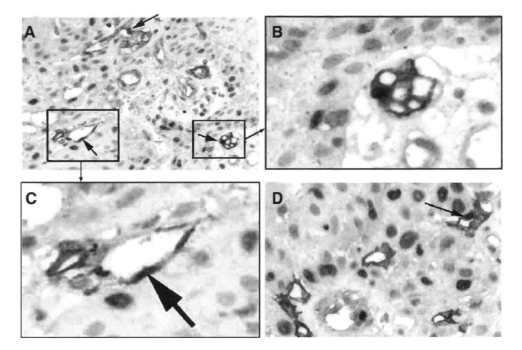

Fig. 4.3. Immunohistochemical measurement of lymphatic vessel proliferation. Newly dividing intratumoral lymphatic vessels are detected by double staining for the Ki-67 proliferation antigen (brown nuclear staining) and LYVE-1 (pink membrane staining). Panels **A-D** show examples of dividing LYVE-1/Ki-67 double-positive small lymph vessels (black arrows) surrounded by large numbers of LYVE-1-negative/Ki-67-positive squamous carcinoma cells. Panel **A**, ×20 objective; panels **B** and **C**, ×100 objective; panel **D**, ×40 objective. (Adapted from ref. *11*).

4. Notes

1. Use a wash bottle with a wide-bore spout. Hold the slide almost vertically and aim the jet at a point above the section, allowing the PBS to flow gently over the section.

2. It is beneficial from this stage onward to ensure that the sections do not dry out. Therefore, the slides should be dried in small batches before adding the next solution.

3. When processing large numbers of slides, it may be easier to place the de-waxing solutions in individual small glass tanks and transfer the slides in a rack from tank to tank. De-waxing solutions may be stored and used for several cycles before replacement with fresh solutions.

4. The use of a negative control antibody is necessary to confirm the validity of the staining. This should either be an isotype-matched antibody or a preimmune serum from the relevant species.

5. LYVE-1, while a widely used marker protein for lymphatic endothelium, is expressed by other cell types, including certain macrophages, liver sinusoidal endothelium, and certain

lung epithelial cells. In light of this, single cells staining positive for LYVE-1 should be identified with care; the use of LYVE-1 as a marker for hepatic lymphatic endothelium is not recommended. It is also worth noting that the expression of LYVE-1 on lymphatic endothelium is down-regulated during inflammation *(13)*. This may clearly lead to an under-estimation of LVD during inflammatory conditions. Similarly, podoplanin is expressed by several different cell types (e.g., epithelia and fibroblasts) in addition to lymphatic endothelia and is present in several tumor types, particularly at the invasive front. Care is therefore required in the interpretation of tumor lymphangiogenesis using this marker. Should problems be encountered with the specificity of LYVE-1 and podoplanin expression in the chosen tissue, antibodies to other marker proteins may be tried. These include the nuclear transcription factor Prox1 and VEGFR3, although again it should be borne in mind that these proteins are also not exclusively expressed by lymphatic endothelium, the former being expressed by hepatocytes and the latter by macrophages and blood vessels associated with tumors and wound healing.

6. Identification of blood vessels in human tissues can be confirmed by positive staining with the antibody PAL-E.

7. It is recommended that all slides are evaluated either single or double blindedly by two independent observers to prevent bias.

8. Polyclonal antisera to mouse LYVE-1 are commercially available (R & D Systems, Reliatech, etc.), as is a hamster monoclonal antibody to podoplanin. Most rabbit antisera to human Prox1 appear to cross-react with the mouse protein. The rat antibody MECA32 works well as a mouse blood endothelial-specific marker.

References

1. Alitalo, K., Tammela, T. Petrova, T. V. (2005) Lymphangiogenesis in development and human disease. *Nature* **438**, 946–953.

2. Stacker, S., Hughes, R. A., Williams, R. A. Achen, M. G. (2006) Current strategies for modulating lymphangiogenesis signalling pathways in human disease. *Curr Med Chem* **13**, 783–792.

3. Van der Auwera, I., Cao, Y., Tille, J. C., et al. (2006) First International consensus on the methodology of lymphangiogenesis quantification in solid human tumours. *Br J Cancer* **95**, 1611–1625.

4. Jackson, D. G. (2001) New molecular markers for the study of tumour lymphangiogenesis. *Anticancer Res* **21**, 4279–4283.

5. Sleeman, J. P., Krishnan, J., Kirkin, V. Baumann, P. (2001) Markers for the lymphatic endothelium: in search of the holy grail ? *Microsc Res Tech* **55**, 61–69.

6. Banerji, S., Ni, J., Wang, S. X., et al. (1999) LYVE-1, a new homologue of the CD44 glycoprotein, is a lymph-specific receptor for hyaluronan. *J Cell Biol* **144**, 789–801.

7. Jackson, D. G. (2004) Biology of the lymphatic marker LYVE-1 and applications in

research into lymphatic trafficking and lymphangiogenesis. *APMIS* **112**, 526–538.

8. Breiteneder-Geleff, S., Soleiman, A., Kowalski, H., et al. (1999) Angiosarcomas express mixed endothelial phenotypes of blood and lymphatic capillaries: podoplanin as a specific marker for lymphatic endothelium. *Am J Pathol* **154**, 385–394.

9. Fox, S. B., Leek, R. D., Weekes, M. P., Whitehouse, R. M., Gatter, K. C. Harris, A. L. (1995) Quantitation and prognostic value of breast cancer angiogenesis: comparison of microvessel density, Chalkley count, and computer image analysis. *J Pathol* **177**, 275–283.

10. Vermeulen, P. B., Gasparini, G., Fox, S. B., et al. (1996) Quantification of angiogenesis in solid human tumours: an international consensus on the methodology and criteria of evaluation. *Eur J Cancer* **32**, 2474–2484.

11. Beasley, N. J., Prevo, R., Banerji, S., et al. (2002) Intratumoral lymphangiogenesis and lymph node metastasis in head and neck cancer. *Cancer Res* **62**, 1315–1320.

12. Shields, J. D., Borsetti, M., Rigby, H., et al. (2004) Lymphatic density and metastatic spread in human malignant melanoma. *Br J Cancer* **90**, 693–700.

13. Johnson, L. A., Prevo, R., Clasper, S., Jackson, D. G. (2007) Inflammation-induced uptake and degradation of the lymphatic endothelial hyaluronan receptor LYVE-1. *J Biol Chem* **282**, 33671–33680.

Section III

Cell Isolation Techniques

Chapter 5

Vascular Endothelial Cells from Human Micro- and Macrovessels: Isolation, Characterisation and Culture

Peter W. Hewett

Abstract

The endothelium residing in different vascular beds displays high-degree phenotypic heterogeneity at morphological, functional, biochemical, and molecular levels. Endothelial cells (ECs) can be easily harvested from large vessels by mechanical removal or collagenase digestion. In particular, the umbilical vein has been used due to its wide availability, and study of the ECs derived from it have undoubtedly greatly advanced our knowledge of vascular biology. However, the majority of the body's endothelium (>95%) forms the microvasculature, and it is these cells providing the interface between the blood and tissues that play a critical role in the development of new blood vessels. This has led to the establishment of techniques for the isolation of microvascular ECs from different tissues to provide more physiologically relevant *in vitro* models of angiogenesis and EC function. The main focus of this chapter is the use of superparamagnetic beads (Dynabeads) coupled to anti- platelet endothelial cell adhesion molecule-1 (PECAM-1) antibodies (PECA beads) to isolate microvessel ECs from human adipose and methods for the characterisation and maintenance of ECs in culture. Adipose tissue is an ideal source of microvessel ECs as it is composed mainly of adipocytes with a rich microvasculature and is easy to disaggregate. Adipose obtained at reduction mammoplasty or adominoplasty is first dissected free of connective tissue and subjected to collagenase type II digestion. The adipocytes are removed by several rounds of centrifugation and separated from the microvessel-rich pellet, which is then further disaggregated with trypsin/EDTA (ethylenediaminctetraacetic acid) solution. Following filtration to remove fragments of connective tissue, the pellet is incubated with PECA beads, and microvessel fragments/ECs are washed and harvested using a magnet. In addition, the adaptation of this basic technique to isolate microvessel ECs from human lung and stomach is also described along with methods for the preparation of large-vessel endothelial cells.

Key words: Dynabeads, endothelial cells, E-selectin, microvascular adipose, PECAM 1/CD31, von Willebrand factor.

S. Martin and C. Murray (eds.), *Methods in Molecular Biology, Angiogenesis Protocols, Second edition, Vol. 467*
© Humana Press, a part of Springer Science+ Business Media, LLC 2009
DOI: 10.1007/978-1-59745-241-0_5

1. Introduction

Human endothelial cells (ECs) derived from large vessels, including the aorta and umbilical *(1)* and saphenous veins, have proven an abundant, convenient, and useful tool for the investigation of many aspects of endothelial biology. However, the endothelium demonstrates a high degree of functional, morphological, biochemical, and molecular diversity between organs and within the different vascular beds of a given organ *(2–6)*. This phenotypic heterogeneity has highlighted the need for reliable techniques for microvessel EC isolation and culture from a variety of tissues to establish more realistic *in vitro* models.

Many techniques have been developed to enrich ECs from tissue homogenates, either directly or after a period in culture *(6)*. Most methods use as their starting point tissue homogenisation and digestion and are often hampered by low EC yield and problems of contaminating cell populations that readily adapt to culture. Some tissues are inherently better suited to microvessel EC isolation, such as brain and adipose, which have high microvascular densities and can be easily disaggregated *(2, 6–9)*. It was Wagner and Matthews *(7)* who first utilised adipose from the rat epididymal fat pad for the isolation of microvascular endothelium. The major advantage of this tissue is the difference in buoyant densities between the adipocytes and stromal component that permits their separation by centrifugation.

The use of superparamagnetic beads (Dynabeads™) coupled to endothelial-specific ligands represented a major advance in the purification of ECs from mixed-cell populations. The original technique described by Jackson and colleagues *(10)* utilised the lectin *Ulex europaeus* agglutinin-1 (*UEA*-1), which binds specifically to α-fucosyl residues of EC glycoproteins. However, *UEA*-1 also binds to some epithelial and mesothelial cells *(1, 11)*. We refined this technique by coupling antibodies raised against platelet EC adhesion molecule-1 (PECAM-1/CD31), a pan-endothelial marker *(3, 4)*, to Dynabeads (PECA beads) and have used these to prepare microvascular ECs from various human tissues *(12, 13)*. Other endothelial markers, including CD34, have similarly been used to isolate microvessel ECs. Many EC markers, including PECAM-1 and CD34, cross-react with subpopulations of haematopoietic cells—highlighting the common developmental origins of these cell types. However, due to their limited viability in culture, haematopoietic cells do not represent a significant problem for EC isolation. Mesothelial cells are a potentially difficult contaminant of EC cultures isolated from tissues such as omentum and lung that are surrounded by mesothelium as they exhibit similar morphology and marker expression *(6, 11)*. However, the absence of PECAM-1 and E-selectin expression in human mesothelial cells readily distinguishes them from ECs *(11)*. Furthermore, the use of PECA beads should eliminate mesothelial cell contamination of endothelial isolates.

In this chapter, we describe in detail the use of PECA beads to isolate microvessel ECs from human adipose tissue and methods for the routine culture of these cells. In addition, the adaptation of this purification technique for the culture of microvessel ECs from human lung and stomach *(12, 13)* is provided alongside routine methods for the preparation of large-vessel ECs from the umbilical vein and aorta. Some of the key characteristics of ECs are considered that can be used to confirm endothelial identity.

2. Materials

2.1. Microvessel EC Isolation and Culture

All solutions should be warmed to 37°C prior to use.

1. 10% bovine serum albumen (BSA) solution: Dissolve 10 g of BSA in 100 mL of calcium-magnesium-free Dulbecco's phosphate-buffered saline (PBS/A), 0.22-μm filter sterilise, and store at 4°C.

2. Antibiotic/antimycotic solution: Dilute 100X antibiotic/ antimycotic solution (Sigma, Poole, Dorset, UK) in PBS/A to give a final concentration of 100 U/mL penicillin, 0.1 mg/mL streptomycin, and 0.25 μg/mL amphotericin B. Aliquot and store the 100X stock solution at –20°C and dilute prior to use.

3. Collagenase solution: Dissolve type II collagenase (Sigma) at 2000 U/mL in Hank's balanced salt solution (HBSS) containing 0.5% BSA, 0.22-μm filter sterilise, aliquot, and store at –20°C.

4. Trypsin/EDTA (ethylenediaminetetraacetic acid) solution: Dilute 10X stock solution of porcine trypsin (2.5%) in PBS/A and add 1 mM (0.372 g/L) EDTA; 0.22-μm filter sterilise, aliquot, and store at –20°C.

5. Gelatin solution: Dilute stock (2%) porcine gelatin solution (Sigma) in PBS/A to give a 0.2% solution and store at 4°C. To coat tissue culture dishes, add 0.2% gelatin solution and incubate for 1 h at 37°C or overnight at 4°C. Remove the gelatin solution from the flasks immediately prior to plating the cells.

6. Growth medium: Many different growth media have been described for the culture of microvessel ECs (*see* **Note 1**). Supplement M199 (with Earle's salts) with 14 mL/L of 1M N-[2-hydroxyethyl] piperazine-N′-[2-hydroxy-propane] sulphonic acid (HEPES) solution, 20 mL/L of 7.5% sodium

hydrogen carbonate solution, and 20 mL/L 200 mM L-glutamine solution and mix 1:1 with Ham F12 nutrient mix (Sigma). To 680 mL of M199/Ham F12 solution, add 20 mL of penicillin (100 U/mL)/streptomycin (100 µg/mL) solution (Sigma), 1500 U/L of heparin, 300 mL of iron-supplemented calf serum (CS) (Hyclone, Logan, UT, USA) (*see* **Note 2**), 1 µg/mL hydrocortisone, 5 ng/mL basic fibroblast growth factor (bFGF), and 20 ng/mL epithelial growth factor (EGF) (PeproTech EC, London) (*see* **Note 3**). Store at 4°C for no more than 1 mo.

7. Cryopreservation medium: Growth medium containing 10% (v/v) dimethylsulphoxide (DMSO, Sigma).

8. Dispase solution: Dissolve 2 U/mL dispase in Medium M199 containing 20% CS, 0.22-µm filter sterilise, aliquot, and store at –20°C. *Note:* This is only required for the isolation of human lung microvessel ECs.

9. Preparation of PECA beads: Mix 0.1–0.2 mg of mouse anti-PECAM-1 monoclonal antibody (e.g., clone 9G11, R&D Systems, Abingdon, Oxon, UK) in sterile PBS/A containing 0.1% BSA (PBS/A + 0.1% BSA) per 10 mg of Dynabeads-M450 (Dynal UK, Wirral, UK) pre-coated with pan anti-mouse immunoglobulin G$_2$ (IgG$_2$) (*see* **Note 4**). Incubate on a rotary stirrer for 16 h at 4°C. Remove free antibody by washing four times for 10 min and then overnight in PBS/A + 0.1% BSA. PECA beads maintain their activity for more than 6 months if sterile and stored at 4°C. However, it is necessary to wash the beads with PBS/A + 0.1% BSA to remove any free antibody prior to use.

2.3. Equipment for EC Isolation

1. A class II laminar flow cabinet is essential for all procedures involving the handling of tissue and cultured cells to maintain sterility and protect the operator.

2. Scalpels, scissors, and forceps are required for the isolation procedures and should be sterilised by autoclaving at 121°C for 30 min.

3. 100-µm nylon filters: Cover the top of a polypropylene funnel about 10 cm with 100-µm nylon mesh (Lockertex, Warrington, Cheshire, UK) and sterilise by autoclaving.

4. Magnet: A suitable magnet is required for the magnetic cell selection system employed (e.g., the Dynal Magnetic Particle Concentrator-15), which accepts 15-mL tubes (*see* **Note 5**).

2.3.1. Disposable Plastics

1. 25 and 75 cm² tissue culture flasks.

2. Large plastic dishes (e.g., Bioassay dishes, Nunc, Naperville, IL, USA).

3. 30-mL universal tubes.

4. 50-mL centrifuge tubes.

5. Lab-Tek multiwell glass chamber slides (Nunc, Naperville, IL, USA).

6. 20-mL Leur-Lok syringes.

7. Luer-Lok three-way stopcocks (Nipro).

2.4. Antibodies for EC Characterisation

There are many commercial antibodies available against endothelial markers.

1. Monoclonal antibodies against human PECAM-1, E-selectin (e.g., clones 9G11 and 5D11, respectively, R8D Systems) and vWF (e.g. clone F8/86; Dako, High Wycombe, Bucks, UK).

2. Fluorescein isothiocyanate (FITC)-conjugated anti-mouse secondary antibodies (Sigma, Poole, Dorset, UK).

3. Nuclear counterstain: Dissolve Hoescht 33342 (Sigma) in PBS/A 10 µg/mL to give a 10 µg/mL solution, aliquot, and store at –20°C.

3. Methods

3.1. Isolation and Culture of Human Adipose Microvessel ECs

A suitable large sterile container is required for the collection of adipose tissue obtained during breast or abdominal reductive surgery (*see* **Note 6**). The fat can be processed immediately or stored for up to 48 h at 4°C.

3.1.1. Collection of Tissue

3.1.2. Isolation of Adipose Microvessel ECs

1. Working under sterile conditions in a class II cabinet, place the tissue on a large sterile dish (e.g., Bioassay dish, Nunc) and wash with 2% antibiotic/antimycotic solution. Avoiding areas of dense (white) connective tissue (which are often prevalent in breast tissue) and visible blood vessels, scrape the fat free from the connective tissue with two scalpel blades.

2. Chop the fat finely and aliquot 10–20 g into sterile 50-mL centrifuge tubes. Add 10 mL of PBS/A and 5 10 mL of the type II collagenase solution. Shake the tubes vigorously to further break up the fat and incubate with end-over-end mixing on a rotary stirrer at 37°C for approximately 1 h. Following digestion, the fat should have broken down, and no spicules should be evident.

3. Centrifuge the digests at 500g for 5 min, discard the fatty (top) layer and retain the cell pellet with some of the lower (aqueous) layer. Add PBS/A and re-centrifuge 500 g for 5 min.

4. Re-suspend the cell pellet in 10% BSA solution and centrifuge (200g, 10 min). Discard the supernatant and repeat the

centrifugation in 10% BSA solution. Wash the pellet with 50 mL of PBS/A. Viewed under the light microscope, the tissue digest should contain obvious microvessel fragments in addition to single cells and debris.

5. Re-suspend the pellet obtained in 5 mL of trypsin/EDTA solution and incubate for 10–15 min with occasional agitation at 37°C. Add 20 mL HBSS containing 5% CS (HBSS + 5% CS) and mix thoroughly to neutralise the trypsin. We have found it advantageous to break up the microvessel fragments and cell clumps further with trypsin/EDTA as this reduces the number of contaminating cells co-isolated with the ECs during PECA bead purification.

6. Filter the suspension through 100-μm nylon mesh to remove fragments of sticky connective tissue. Centrifuge the filtrate at 700g for 5 min and re-suspend the resulting pellet in about 1–2 mL of ice-cold HBSS + 5% CS.

7. Add approximately 50 μL of PECA beads and incubate for about 20 min at 4°C with occasional agitation (*see* **Note 7**). Add HBSS + 5% CS to a final volume of about 12 mL, mix thoroughly, and select the microvessel fragments/ECs using a suitable magnet (*see* **Note 5**) for 3 min. Repeat the cell selection process three to five more times by washing the magnetically separated material in approximately 12 mL of HBSS + 5% CS and reselecting the microvessel fragments with the magnet.

8. Suspend the magnetically separated cells in growth medium (*see* **Note 8**) and seed at high density onto 0.2% gelatin-coated 25-cm² tissue culture flasks and incubate at 37°C in a humidified atmosphere of 5% CO_2.

3.1.3. Adipose Microvessel ECs in Culture

Following the PECA bead selection procedure, small microvessel fragments and single cells coated with Dynabeads should be evident under light microscopy (*see* **Note 9**). After 24 h, the cells adhere to the flasks and start to grow out from the microvessel fragments present to form distinct colonies. Human mammary microvessel ECs (HuMMECs) isolated using this technique grow to confluence forming cobblestone contact-inhibited monolayers within 10–14 d depending on the initial seeding density. We have successfully cultured these cells to passage 8 without observable change in their morphology but routinely use them in experiments between passages 3 and 6.

3.1.4. Subculture of Adipose Microvessel ECs

Maintain the microvessel ECs at 37°C, 5% CO_2, changing the medium every 3–4 d. When confluent, ECs can be passaged using trypsin/EDTA solution onto 0.2% gelatin-coated dishes at a split ratio of 1:4 as follows:

1. Discard the old medium and wash the cell monolayer twice with 5–10 mL of PBS/A. Add a few millilitres of trypsin/

EDTA solution, wash it over the cell monolayer, remove the surplus to leave the cells just covered, and incubate at 37°C for 1–2 min. Monitor the cells regularly under the microscope until they round up and detach (*see* **Note 10**). Striking the flask sharply helps to dislodge the cells and break up cell aggregates.

2. Add sufficient growth medium to achieve a split ratio of about 1:4 and plate the cells onto gelatin-coated flasks.

3.1.5. Microcarrier Beads

Microcarrier beads can be used to continuously culture ECs without the need to use trypsin. Following hydration and sterilisation according to the manufacturer's instructions, add gelatin-coated Cytodex 3™ microcarrier beads (Sigma) to the medium and allow the ECs to crawl onto and attach to the microcarriers. Agitate the flasks occasionally to facilitate seeding carriers over a period of 2–3 d. Remove the beads and place into a fresh gelatin-coated flasks containing growth medium and allow the cells to attach to the flask, agitating occasionally to ensure good distribution of the cells. Once sufficient cells have attached, the beads may be removed and placed into a fresh flask, and the process is repeated.

3.1.6. Cryopreservation of ECs

Following trypsinization (see **Subheading 3.1.4.**) Suspend ECs at about $2 \times 10^6/mL$ in the cryopreservation medium and dispense into suitable cryovials. Cool the vials to –80°C at 1°C/min and store under liquid nitrogen.

3.1.7. Maintaining the Purity of Microvessel EC Cultures

It may be necessary to reselect the ECs with PECA beads or perform minor manual "weeding" to maintain the purity of cultures. Reselection with PECA beads can be performed as described in **Subheading 3.1.2., step** 7 following removal of the cells from flasks using trypsin/EDTA solution (*see* **Subheading 3.1.4.**). Provided that there are clear morphological differences between contaminating cell populations and the ECs (*see* **Note 11**), it is relatively straightforward, although time consuming, to physically remove them. Manual weeding should be performed with the stage of a phase contrast microscope within a class II cabinet to ensure sterile conditions. A needle or Pasteur pipet is used to carefully remove contaminating cells from around EC colonies. The medium is discarded, and the adherent cells are washed with several changes of sterile PBS/A to remove the dislodged contaminating cells.

3.2. Characterisation of ECs

3.2.1. EC Morphology in Culture

Cobblestone morphology is very typical of ECs derived from many tissues, and these cells are usually readily distinguished from the typical fibroblastoid contaminating cell populations. However, a more elongated morphology has been reported for human microvessel ECs derived from some tissues, and similar elongated phenotypes forming "swirling" monolayers are often observed following stimulation of ECs with growth factors. ECs will form "capillary-like" tube networks within a few hours of plating on matrices such as growth factor-reduced Matrigel™. This phenomenon also occurs in many types of microvessel ECs if cultures are left for several days at confluence (**Fig. 5.1**). However, the formation of "capillary-like" structures is not an exclusive property of ECs in culture.

3.2.2. Key EC Markers

There are many criteria on which EC identification may be based, and these have been reviewed extensively (*see* **refs.***2–4* and 6). Many endothelial markers/properties are not unique to ECs, and several may be required to confirm endothelial identity. ECs isolated from different vascular beds may also display phenotypic heterogeneity, and lack of a particular marker may not preclude the endothelial origin of isolates *(2)*. It is often useful to demonstrate the absence of markers characteristic of potential contaminating cell populations such as smooth muscle α-actin-positive stress fibres and the intermediate filament protein desmin, which are expressed by smooth muscle cells and pericytes *(14)*.

A number of good endothelial markers have been identified, including intercellular adhesion molecule 2 (ICAM-2/CD102), endothelial selective adhesion molecule (ESAM), and vascular endothelial cadherin (VE cadherin) *(4)*. Here we focus on von Willebrand Factor (vWF), PECAM-1 *(15)*, and E-selectin (endothelial-leukocyte adhesion molecule-1/CD62E) *(16)* that we believe to be useful for the rapid identification of ECs.

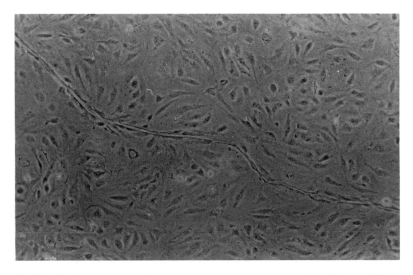

Fig. 5.1. Photomicrograph of human mammary microvessel endothelial cell (EC) (at passage 2) demonstrating typical cobblestone morphology and tube formation on the surface of the post-confluent EC monolayer.

vWF is only expressed at significant levels in ECs, platelets, megakaryocytes, and syncytiotrophoblasts. In ECs, it is stored in the rod-shaped Weibel-Palade bodies, which produce characteristic punctate perinuclear staining. These organelles are present in large-vessel ECs but have been reported to be scarce or absent in the capillary endothelium of various species *(1–4, 6)*. However, typical granular perinuclear staining for vWF has been reported in cultured human kidney, dermis *(10)*, synovium, lung *(12)*, stomach *(13)*, decidua, heart, adipose *(9, 12)*, and brain *(8)* microvessel ECs.

PECAM-1 is constitutively expressed on the surface of ECs ($>10^6$ molecules/cell) and to a lesser extent in platelets, granulocytes, and a subpopulation of CD8+ lymphocytes *(11, 15)*. PECAM-1 staining of ECs *in vitro* is characterised by typical intense membrane fluorescence at points of cell–cell contact (*see* **Fig. 5.2**).

Strong expression of E-selectin following stimulation with pro-inflammatory cytokines appears to be a unique characteristic of ECs *(12)*. It is not expressed constitutively by the majority of ECs, but stimulation with tumor necrosis factor-α (TNF-α) or interleukin-1β (IL-1β) leads to intense E-selectin staining of the EC membrane, reaching a maximum after 4–8 h (*see* **Fig. 5.3**).

3.2.3. Immunocytofluorescent Characterisation of ECs

Outlined next is a simple protocol for the immunofluorescent detection of EC markers.

1. Preparation of ECs on glass slides: Multiwell glass chamber slides are extremely useful for this purpose as multiple tests can be performed on the same slide, conserving both reagents

Fig. 5.2. Immunofluorescent staining of platelet endothelial cell adhesion molecule-1 (PECAM-1/CD31) in human mammary adipose microvessel endothelial cell.

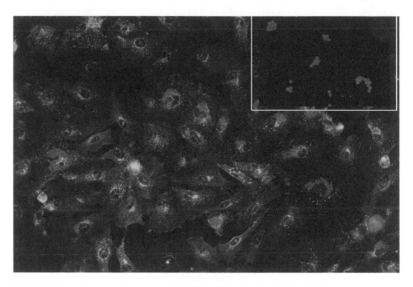

Fig. 5.3. Intense immunofluorescent staining of E-selectin detected in human mammary microvessel endothelial cell following 6 h exposure to tumor necrosis factor-α (10 ng/ mL), which is absent in unstimulated control cells (*inset*).

and cells. ECs are cultured on chamber slides that have been pre-treated for 1 h with 5 μg/cm² bovine fibronectin (Sigma) in PBS/A or 0.2% gelatin. When sufficient cells are present, discard the medium and wash twice with PBS/A prior to fixation. Different fixatives can be employed depending on the activity of the antibody used. Acetone fixation is suitable for most antibodies: Place the slides (*see* **Note 12**) in cold acetone (–20°C, 10 min), air-dry, and store frozen at –80°C. Alternatively, fix cells in 3% formaldehyde solution for 30 min at room temperature. Formaldehyde does not permeabilise the plasma membrane, and further treatment with 0.1% Nonidet P-40 or Triton X-100 is required to detect cytoplasmic/nuclear antigens.

2. Immunocytofluorescent staining: Warm slides to room temperature and wash with PBS/A (twice for 5 min). Block slides for 20 min with 10% normal serum from the species in which the secondary antibody was raised to prevent non-specific binding of the secondary antibody.

3. Incubate slides with a predetermined or the manufacturer's recommended concentration of primary antibody in PBS/A for 60 min at room temperature.

4. Wash slides with PBS/A (three times for 5 min) and incubate with the appropriate FITC-labelled secondary antibody at 1:50 dilution in PBS/A for 30 min to 1 h at room temperature and protect from direct light.

5. Counterstain cells with Hoescht 33342 (10 µg/mL) in PBS/A for 10 min to facilitate assessment of EC purity.

6. Wash slides in PBS/A (three times for 5 min); mount in 50% (v/v) glycerol in PBS/A. Stained slides can be stored for several months in the dark at 4°C.

To avoid false positives generated by non-specific binding of secondary antibodies, it is essential to include negative controls of cells treated as described but with an isotype-matched control antibody or PBS/A substituted for the primary antibody. It is also useful to include other cell types, such as fibroblasts and smooth muscle cells as negative controls and previously characterised ECs as positive controls.

Cells are incubated with 1–10 ng/mL TNF- or IL-1 in growth medium for 4–6 h prior to fixation to induce E-selectin expression. Unstimulated cells should also be included as controls.

3.2.4. Other Properties of Human Adipose Microvessel ECs

Human adipose microvessel ECs cells possess typical EC characteristics, including scavenger receptors for acetylated low-density lipoprotein, expression of the transforming growth factor-β co-receptor endoglin (CD105), and high levels of angiotensin-converting enzyme activity. All the microvessel EC types that we have cultured also express the vascular endothelial cell growth factor (VEGF) receptors Flt-1 and KDR/Flk-1 (14) and proliferate and express tissue factor in response to VEGF. Similarly, the angiopoietin receptors (Tie-1 and Tie-2/Tek) are also expressed on these cells.

3.3. Isolation of Microvessel ECs from Other Vascular Beds

We have adapted the basic method for the selection of adipose microvessel ECs to isolate ECs from other tissues. Here we outline briefly modifications that have been made for the isolation of human lung (12) and stomach ECs (18).

3.3.1. Microvessel ECs from Human Lung

Although it has a high microvascular density, lung is composed of many diverse cell types that readily adapt to culture and is often more difficult to obtain than adipose tissue. We have successfully isolated ECs from normal lung from transplant donors and diseased tissue from transplant recipients (12). To ensure that microvascular ECs are harvested, a thin strip of tissue at the periphery of the lung is used. As the amount of tissue available is usually limited and subsequent yield of cells is low after direct Dynabead selection, it is better to allow the cells to proliferate in culture for a few days and then perform the magnetic purification with PECA beads before they became overgrown with contaminating cells.

1. Cut small peripheral sections of lung (3–5 cm long, about 1 cm from the periphery) and wash in antibiotic/antimycotic solution.

2. Dissect the underlying tissue from the pleura and chop it very finely using a tissue chopper.

3. Wash the "mince" above sterile 20-μm nylon mesh (prepared as described for 100-μm nylon mesh filters; *see* **Subheading 2.3.**) to filter out red blood cells and fine debris.

4. Incubate the retained material overnight in dispase solution on a rotary stirrer at 37°C.

5. Pellet the digest, resuspend in about 5 mL of trypsin/EDTA solution, and incubate at 37°C for 15 min.

6. Add growth medium and remove fragments of undigested tissue by filtration through 100-μm nylon mesh.

7. Pellet and resuspend the cells in growth medium and plate onto gelatin-coated dishes.

8. Monitor the cultures daily; trypsinize and select the ECs using PECA beads before they become overgrown by contaminating cells (*see* **Subheading 3.1.7.**).

3.3.2. Human Stomach Microvessel ECs

Microvessel ECs can be cultured from stomach mucosa obtained from biopsies or organ donors *(13)*.

1. Expose the stomach mucosa and wash with antibiotic/antimycotic solution.

2. Dissect the mucosa from the underlying muscle, chop into 2- or 3-mm pieces, and incubate in 1 m*M* EDTA in HBSS at 37°C in a shaking water bath for 30 min.

3. Transfer the pieces of mucosa to collagenase type II solution for 60 min and then trypsin/EDTA solution for 15 min in a shaking water bath at 37°C.

4. Using a blunt dissecting tool, scrape the mucosa and submucosa from white fibrous tissue.

5. Suspend the mucosal tissue in HBSS + 20% CS and wash through 100-μm nylon mesh.

6. Centrifuge the filtrate (700*g*, 5 min) and resuspend the pellet in about 12 mL of HBSS + 5% CS. Proceed with PECA bead selection (*see* **Subheading 3.1.2., step 7**).

3.4. Human Large-Vessel EC Isolation

3.4.1. Preparation of Human Umbilical Vein Endothelial cells (HUVEC)

The umbilical vein provides an abundant source of large-vessel ECs, which have been used extensively due to the availability of this post-natally redundant tissue and ease with which pure human umbilical vein endothelial cell (HUVEC) preparations can be obtained. The procedure given here has been adapted from the method originally reported by Jaffe and colleagues *(1)*.

1. Collect umbilical cords in a suitable covered container and store at 4°C (*see* **Note 13**).

2. Working over a large dish or tray to contain any spills, wash off any blood from the surface of the cord with antibiotic/antimycotic solution and inspect the length very carefully for the presence of clamp marks and damaged areas, which must

be avoided to prevent smooth muscle cell contamination. Cut a length (>20 cm) of cord free of clamp marks using cord scissors.

3. Cannulate the umbilical vein using a three-way disposable Luer-Lok stopcock and secure using ligature. The umbilical vein with its larger and thinner walls is relatively easy to distinguish from the two narrow umbilical arteries surrounded by smooth muscle.

4. Attach a syringe to the stopcock and wash the cord gently with about 30 mL of PBS/A until all the residual blood is removed.

5. Perfuse the cord with about 5–10 mL of 0.1% (v/v) type I collagenase (e.g., CLS-Collagenase, Worthington) solution in HBSS (prepared as described for type II collagenase; *see* **Subheading 2.1.**), carefully displacing any residual air, and clamp off the free end.

6. Place the cord on a large plastic tray (e.g., Nunc bioassay dish) cover and place in a humidified incubator at 37°C for about 15 min.

7. Gently massage the cord and collect the collagenase solution from the vein into a 30-mL universal tube. Wash thoroughly with about 20 mL of HBSS + 5% FCS (foetal calf serum).

8. Centrifuge at 200*g* for 10 min and resuspend the pellet in growth medium (*see* **Note 1**); plate cells onto gelatin-coated 25-cm² flasks. Under the phase contrast microscope, sheets of ECs in addition to many single cells, red blood cells, and debris should be apparent.

9. Allow cells to attach to the flask for at least 4 h or overnight and then wash the cells twice with PBS/A to remove non-adherent cells and debris. Add fresh growth medium and incubate the flask at 37°C in a humidified 5% CO_2 incubator, changing the medium every 2–3 d until the HUVECs reach confluence. Passage the cells at a 1:3 ratio as described in **Subheading 3.1.4.** HUVECs de-differentiate quite rapidly in culture and should not be used beyond passage 4.

3.4.2. Human Aortic ECs

A relatively simple method for the isolation of human aortic ECs (HAEC) is outlined as described previously *(18)*. It is also possible simply to physically remove the ECs by gentle scraping of the luminal surface of the aorta rather using collagenase digestion (*see* **Note 14**). This method can be directly applied to isolate ECs from the aortas of other large mammals.

1. Sections of human thoracic aorta obtained at post-mortem are washed thoroughly in antibiotic/antimycotic solution.

2. Cut the vessel open longitudinally to expose the surface of the luminal endothelium and lay flat on a sterile Petri dish.

Add collagenase solution (*see* **Note 14**) over the exposed luminal surface, cover with lid, and incubate at 37°C for about 15 min in a humidified incubator.

3. Wash the surface of the aorta with about 5 mL of HBSS + 5% CS, drain, and collect into a 30-mL universal tube. Centrifuge at 200g for 10 min, resuspend the cell pellet in about 5 mL of growth medium, plate onto gelatin-coated T25 flasks, and incubate at 37°C in a humidified 5% CO_2 incubator.

4. After 12–24 h, discard the medium, wash twice with PBS/A (10 mL), add fresh medium, and subculture the cells as described, passaging with trypsin at confluence (*see* **Subheading 3.1.4.**).

4. Notes

1. Many different growth media have been described for the culture of ECs. Some microvessel ECs have very specific requirements, such as the presence of human serum, while large-vessel ECs tend to be far less fastidious in their growth requirements. This M199/Ham F12 nutrient mix-based recipe is relatively inexpensive and works well for a range of ECs, but researchers may wish to optimise their medium further. MCDB 131 *(19)* containing the supplements described (**Subheading 2.1., item 6**) represents an excellent alternative and can be used with lower concentrations of serum. There are now also several commercial optimised EC media that are based on MCDB 131 available (e.g., EGM^{Tm}-1/2 and EGM^{Tm}-MV-2, Clonetics, San Diego, CA, USA); although more expensive, they are very convenient and give outstanding results.

2. Iron-supplemented CS provides an economical alternative to FCS and well supports the proliferation and survival of the ECs we have worked with (*see* **Note 1**). However, it is necessary to batch test all bovine sera to ensure the optimal growth of ECs, and some microvessel ECs isolated from some vascular beds, such as the endometrium, require the presence of human serum *(20)*.

3. We routinely use recombinant bFGF and EGF, but VEGF (5–10 ng/ml) can also be added and provides excellent EC support for the majority of cultured ECs but is more expensive. These growth factors may also be substituted with EC growth supplement (ECGS) derived from bovine brain, which is rich in acidic and basic FGF.

4. Precoated Dynabeads (and CELLelection™ beads; *see* **Note 8**) carrying various secondary antibodies (e.g., pan anti-mouse) are available from Dynal and are very convenient. However, anti-immunoglobulin-coated beads can be prepared as follows: Incubate the secondary antibody (150 µg/mL) in 0.17M sodium tetraborate buffer (pH 9.5; 0.22-µm sterile filtered) with tosyl-activated Dynabeads-M450 for 24 h on a rotary stirrer at room temperature. Wash the beads four times for 10 min and then overnight in PBS/A + 0.1% BSA on a rotary stirrer at 4°C before coating them with the primary antibody as described (*see* **Subheading 2.2.**). We have found PECA beads to be more reliable for purification of ECs *(12)* than using tosyl-activated Dynabeads directly coated with *UEA*-1 *(9)*.

5. The MPC-1 magnet that we use is no longer manufactured, but suitable alternatives from Dynal include the MPC-15, which accepts 15-mL tubes, and MPC-50, which takes 50-mL centrifuge tubes.

6. Omental adipose tissue obtained through general abdominal surgery can also be used. However, care should be taken to remove the fat from the omental membranes, which are covered with a layer of mesothelium prior to dissection. Using PECA beads, we have not found mesothelial cell contamination to be a problem.

7. The cell PECA bead suspension is incubated at 4°C during the purification steps to minimise non-specific phagocytosis of Dynabeads.

8. We do not routinely remove Dynabeads following cell selection. However, it may be necessary to remove the Dynabeads if, for example, the cells are required for flow cytometry. This problem can be overcome by using CELLelection Dynabeads (Dynal) coated with the anti-PECAM-1 antibody to select the ECs. In this system, antibodies are conjugated to the Dynabeads via a DNA linker that can be cleaved with deoxyribonuclease 1 (DNase-1) to release the beads from the cells following selection.

9. Dynabeads are internalised within about 24 h of selection and are diluted to negligible numbers/cell by the first passage through cell division. Consistent with the original observations of Jackson and colleagues *(10)* using *UEA*-I-coated Dynabeads, we have not observed any adverse effects on the adherence, proliferation, or morphology of EC following PECA bead selection *(15)*.

10. To maintain cell viability, it is important to rapidly remove the ECs from the flasks as they are very sensitive to trypsin exposure. The use of a trypsin inhibitor to neutralise the tryptic activity immediately following detachment from the

flask may prolong viability of EC cultures. Mung bean trypsin inhibitor (Sigma) can be used for this purpose, although we have not quantified its effect on endothelial viability.

11. The major contaminating cell population observed in unselected adipose EC cultures demonstrates a distinct fibroblastic morphology.

12. The plastic wells must be removed from the multiwell slides as acetone rapidly dissolves the plastic. The gasket should be left in place to keep reagents separate during the staining procedure.

13. In our hands, storage of the cords up to 48 h after delivery does not appear to adversely affect the yield or viability of isolated cells.

14. The luminal surface of the aorta can be simply scraped very gently to remove the aortic ECs. However, we have observed greater contamination of cultures with smooth muscle cells using this approach compared with collagenase digestion. Following isolation of ECs, smooth muscle cells can be easily explanted from these aortic segments.

608   **References**

1. Jaffe, E. A., Nachman, R. L., Becker, C. G., Minidi, C.R. (1973) Culture of human endothelial cells derived from umbilical veins: identification by morphological and immunological criteria. *J Clin Invest* 52, 2745–2756.

2. Kumar, S., West, D. C., Ager, M. (1987) Heterogeneity in endothelial cells from large vessels and microvessels. *Differentiation* 36, 57–70.

3. Kuzu, I., Bicknell, R., Harris, A. M., Jones, M., Gatter, K. G., Mason, D. Y. (1992) Heterogeneity of vascular endothelial cells with relevance to diagnosis of vascular tumors. *J Clin Pathol* 45, 143–148.

4. Garlanda, C., Dejana, E. (1997) Heterogeneity of endothelial cells: specific markers. *Arterioscler Thromb Vasc Biol* 17, 1193–1202.

5. Trepel, M., Arap, W., Pasqualini, R. (2002) *In vivo* phage display and vascular heterogeneity: implications for targeted medicine. *Curr Opin Chem Biol* 6, 399–404.

6. Hewett, P. W., Murray, J. C. (1993) Human microvessel endothelial cells: isolation, culture and characterization. *In Vitro Cell Dev Biol* 29A, 823–830.

7. Wagner, R. C., Matthews, M. A. (1975) The isolation and culture of capillary endothelium from epididymal fat. *Microvasc Res* 10, 286–297.

8. Dorovini-Zis, K., Prameya, R., Bowman, P. D. (1991) Culture and characterization of microvessel endothelial cells derived from human brain. *Lab Invest* 64, 425–436.

9. Hewett, P. W., Murray, J. C., Price, E. A., Watts, M. E., Woodcock, M. (1993) Isolation and characterization of microvessel endothelial cells from human mammary adipose tissue. *In Vitro Cell Dev Biol* 29A, 325–331.

10. Jackson, C. J., Garbett, P. K., Nissen, B., Schrieber, L. (1990) Binding of human endothelium to *Ulex europaeus*-1 coated Dynabeads: application to the isolation of microvascular endothelium. *J Cell Sci* 96, 257–262.

11. Hewett, P. W., Murray, J. C. (1994) Human omental mesothelial cells: a simple method for isolation and discrimination from endothelial cells. *In Vitro Cell Dev Biol* 30A, 145–147.

.,12. Hewett, P. W.s Murray, J. C. (1993) Immunomagnetic purification of human microvessel endothelial cells using Dynabeads coated with monoclonal antibodies to PECAM-1. *Eur J Cell Biol* 62, 451–454.

13. Hull, M. A., Hewett, P. W., Brough, J. L., Hawkey, C. J. (1996) Isolation and culture of human gastric endothelial cells. *Gastroenterology* 111, 1230–1240.

14. Diaz-Flores, L., Gutiérrez, R., Varela, H., Rancel, N., Valladares, F. (1991) Microvascular pericytes: A review of their morphological and functional characteristics. *Histol Histopathol* 6, 269–286.

15. Newman, P. J., Berndt, M. C., Gorski, J., et al. (1990) PECAM-1 (CD31) cloning and relation to adhesion molecules of the immunoglobulin gene superfamily. *Science* 247, 1219–1222.

16. Bevilacqua, M. P., Pober, J. S., Mendrich, D. L., Cotran, R. S., Grimbone, M. A. (1987) Identification of an inducible endothelial-leukocyte adhesion molecule. *Proc Natl Acad Sci U S A* 84, 9238–9242.

17. Ferrara, N., Gerber, H. P., LeCoulter, J. (2003) The biology of VEGF and its receptors. *Nat Med* 6, 669–676.

18. Antonov, A. S., Nikolaeva, M. A., Klueva, T. S., et al. (1986) Primary culture of endothelial cells from atherosclerotic human aorta. *Atherosclerosis* 59, 1–19.

19. Knedler, A., Ham, R. G. (1987) Optimized medium for clonal growth of human microvascular endothelial cells with minimal serum. *In Vitro Cell Dev Biol* 23, 481–491.

20. Grimwood, J., Bicknell, R., Rees, M. C. P. (1995) The isolation characterisation and culture of human decidual endothelium. *Hum Reprod* 10, 101–108.

Chapter 6

Lymphatic Endothelial Cells: Establishment of Primaries and Characterization of Established Lines

Riccardo E. Nisato, Raphaele Buser, and Michael S. Pepper

Abstract

This chapter describes detailed methods for the isolation of primary human lymphatic endothelial cells from neonatal foreskin. We also provide protocols and information for their characterization and propagation. Isolation of primary human lymphatic endothelial cells requires a two-step process: mechanical and enzymatic digestion of human foreskins and cell sorting by fluorescence-activated cell sorting of $CD31^+$/podoplanin$^+$/CD45$^-$ cells. Characterization of these cells requires an assessment of the expression of several markers specific for lymphatic endothelium. This is determined by fluorescence-activated cell sorting, immunocytochemistry, and polymerase chain reaction. All procedures are based on simple laboratory techniques and, with the exception of a cell sorter and the skills to use it, do not require specialized equipment.

Key words: Endothelial cells, lymphatic, LYVE-1, podoplanin, primary.

1. Introduction

Until relatively recently, lymphatic vessels were identified directly or indirectly by lymphangiography, vital dye uptake (Evans blue, trypan blue, patent blue, and fluorescent-labeled tracers); the absence of membrane components such as laminins, collagen IV and XVIII; the lack of PAL-E staining of CD31-positive endothelial cells (in frozen tissues) and 5′-nucleotide activity (1). One of the major advances in the lymphatic field has been the discovery of several markers specific for lymphatic endothelium (2). These markers have allowed for a precise histological localization of the lymphatic vasculature and a clear discrimination between

S. Martin and C. Murray (eds.), *Methods in Molecular Biology, Angiogenesis Protocols, Second edition, Vol. 467*
© Humana Press, a part of Springer Science+ Business Media, LLC 2009
DOI: 10.1007/978-1-59745-241-0_6

blood and lymphatic vessels (by immunohistochemical analysis), especially in pathological conditions. They have also have allowed for the identification and isolation of pure populations of blood endothelial cells (BECs) and lymphatic endothelial cells (LECs) from human or animal tissues *(3)*. This technical advance is in fact responsible for most of the recent discoveries made on the molecular mechanisms mediating lymphangiogenesis. Several markers have been shown to be reliable in distinguishing lymphatic from blood vessel endothelium, although none are strictly endothelial specific, with precise patterns of expression for several markers remaining to be confirmed *(2)*.

We provide a detailed procedure for the isolation and culture of primary human LECs as well as simple methods to characterize them. Isolation of LECs requires enzymatic and mechanical dissociation of the human tissue (foreskins) and fluorescent-activated cell sorting (FACS) for assessment of the expression of membrane antigens specific for lymphatic endothelium. Indeed, isolation of human primary LECs is achieved by sorting cells that are CD31[+]/podoplanin[+]/CD45[-]. The characterization of the LECs is also discussed based on the expression of several LEC-specific markers *(3–6)*.

Since almost the entire procedure is based on the expression of LEC markers, it is worthwhile to summarize the pattern of expression of the most studied molecules that are believed to discriminate between LECs and BECs (summarized in **Table 6.1**) *(2)*.

In the discussion that follows, we have limited ourselves to the expression of markers on endothelial cells. Expression by other cells is summarized in **Table 6.1**

1.1. CD34 and Vascular Endothelial Growth Factor Receptor 3

CD34 expression was initially believed to be confined to the blood vasculature and not to be expressed by lymphatic vessels *(4)*. However, recent reports have demonstrated that CD34 is also expressed in normal or pathologic lymphatic endothelium *(7, 8)*. Furthermore, Kriehuber et al. were able to isolate a subpopulation of LECs that co-expressed CD34 and podoplanin from adult human skin *(3)*, indicating that the repertoire of CD34 expression on endothelial cell populations is broader than originally anticipated.

Originally, the expression of vascular endothelial growth factor receptor 3 (VEGFR-3) was believed to be restricted to lymphatic endothelium in adult tissues. Its expression is predominantly in lymphatic endothelium in normal adult tissues, but it is now clear that VEGFR3 is also expressed in some fenestrated blood vessels, in tumor-induced blood vessels, and during tissue repair *(9)*.

Even if used previously to isolate BECs and LECs, we would no longer encourage the use of these antigens for LEC isolation.

1.2. CD31

CD31, also called platelet-endothelial cell adhesion molecule-1 (PECAM-1), is expressed by leukocytes and endothelial cells. CD31

Table 6.1
Molecular markers in the blood and lymphatic vasculature

Markers	Blood Endothelium	Lymphatic Endothelium	Expression in Non-endothelial Cells
CD44	+	–	Lymphocytes, not expressed by HUVECs *in vivo*
CD34	+	–*	Hematopoietic progenitor cells
VEGFR3	–*	+	Macrophages, dendritic cells
vWF	+	Low/–	Megacariocytes
CD31	+	Low	Leucocytes, platelets
Neuropilin-2	–	+	Neuronal cells
LYVE-1	–	+	Activated macrophages
Podoplanin	–	+	Podocytes, osteoblasts, epidermal keratinocytes
Prox-1	–	+	Hepatocytes, cells in lens, retina, heart, pancreas
Desmoplakin	–	+	Heart, epithelial cells
Integrin-α9	–	+	Epithelial cells of the crypt in human foetal intestine
Neuropilin-1	+	–	Neuronal cells
Mannose-R	–	+	Neuronal cells

* see chapter 1.1 CD34 and VEGFR-3

is highly expressed on all endothelial cells of existing and newly formed blood and lymphatic vessels, making it a commonly used pan-endothelial cell marker. It has been widely used to isolate endothelial cells from human biopsies and can be used to isolate pure BEC and LEC populations when combined with a lymphatic marker such as podoplanin or LYVE-1 *(3, 5)*.

1.3. LYVE-1   LYVE-1 is almost exclusively expressed by lymphatic endothelium in settings where CD44 is low or not expressed. It is also expressed by sinusoidal endothelium of the liver and the spleen but is not expressed by tumor-induced blood vessels. LYVE-1 is one of the most important and commonly used markers for the detection and isolation (using a magnetic bead strategy) of LECs *(10, 11)*. However, its level of expression is variable and heterogeneous within the same population of cells, thus providing a non-homogeneous FACS signal that discourages its use for cell sorting.

1.4. Podoplanin

Podoplanin is expressed by endothelial and non-endothelial cells. In the case of endothelium, its expression is restricted to lymphatics, and it is considered to be one of the most reliable markers for the detection and isolation of lymphatic endothelium *(12)*.

1.5 Prox-1

In the adult, Prox-1 is the most reliable marker for distinguishing blood from lymphatic endothelium. Given that Prox-1 is a transcription factor (and thus located inside the cell), it is not suitable for LEC isolation. However, it is currently the most reliable selective LEC marker *(2, 13, 14)*.

Isolation and characterization of human LECs requires an assessment of the expression of several LEC-specific markers. Given the short life span of primary LEC cultures, it is important to follow at regular intervals the morphology of the cells and their capacity to maintain the expression of these LEC-specific markers, as well as their functional properties, that is, the capacity to respond to lymphangiogenic stimuli. With regard to the morphological and functional properties of isolated LECs, this will not be dealt with in this chapter. However, we refer the reader to previously published reports *(2–6, 15)*.

2. Materials, Media, Buffers, and Antibodies

2.1. Reagents

1. Sterile forceps, scalpel with a number 10 blade.
2. Hank's balanced salt solution (HBSS) (Gibco no. 24020-091).
3. Solution containing 0.25% trypsin (Gibco no. 15050-014).
4. Dulbecco's modified Eagle's medium (DMEM) (Gibco no. 41965-049).
5. Foetal calf serum (FCS) (Gibco no. 10270-106).
6. Penicillin, streptomycin, amphotericin mix (PSA) (Gibco no. 15290-018; dilute 125X).
7. 70-μm cell strainer (BD-Falcon no. 35–2350).
8. Rat tail tendon-derived collagen type I (Becton Dickinson) (reference 354236).
9. Ethylenediaminetetraacetic acid (EDTA) solution: We prepare phosphate-buffered saline (PBS) without Ca^{2+} and Mg^{2+} in bi-distilled water containing 136.7 mM of NaCl, 2.68 mM of KCl, 8.08 mM of $Na_2HPO_4.2H_2O$, 1.47 mM of KH_2PO_4, and we add 0.537 mM of Titriplex-EDTA and 1.5 mL/L of phenol red 1.5 cc/L.
10. Trypsin/EDTA solution: We prepare a 0.25% trypsin solution (Becton Dickinson) in PBS without Ca^{2+} and Mg^{2+} and

dilute this 10 times in a solution containing 50% PBS without Ca^{2+} and Mg^{2+} and 50% EDTA solution (as in **item 9**).

11. Endothelial cell growth medium: EBM (Cambrex, CC-3121) supplemented with

    20% FCS, 1X glutamine, 1X PSA.

    1 µg/mL hydrocortisone acetate (Sigma H-0396). Prepare a stock solution containing 5 mg/mL hydrocortisone acetate by diluting hydrocortisone acetate with dimethyl sulfoxide (DMSO). The stock solution is stable for several months when stored at 4°C.25 µg/mL N-6,2'-O-dibutyryl-adenosine 3',5'-cyclic monophosphate (dcAMP) (Sigma D-0627). Prepare a stock solution of 2.5 mg/mL of dcAMP by diluting dcAMP in PBS; final concentration is 25 µg/mL. 100 U/mL of penicillin and 100 µg/mL of streptomycin. The stock solution is stable for several months when stored at –80°C.

12. 4'-6-Diamidino-2-phenylindole (DAPI) (Sigma).

13. FacsVantage^{SE} (Becton Dickinson) or equivalent.

14. Non-enzymatic cell dissociation solution (Sigma).

15. Expertise to use FACS.

16. Trizol (Life Technologies).

17. Tandom hexanucleotides (Boehringer Mannheim).

18. Superscript II reverse transcriptase (Life Technologies).

19. *Taq*DNA polymerase (Invitrogen).

2.2. Antibodies

1. Rabbit anti-human podoplanin (serum 201853; see **ref. 6**). We have not used commercial antibodies but recommend using a commercial rat anti-human podoplanin, which has been successfully used by other laboratories (AngioBio, CA, USA, cat. no. 11-009, clone NZ-1).

2. Anti-human podoplanin (Prof. D. Kerjaschki, Institute of Pathology, University of Vienna, Austria) (AngioBio).

3. Anti-human LYVE-1 (Prof. D.R. Jackson, MRC Human Immunology Unit, Institute of Molecular Medicine, Oxford, UK) (AngioBio).

4. Mouse monoclonal biotinylated anti-human CD31 (Ancell, 180-030).

5. Mouse immunoglobulin G$_1$ (IgG$_1$)/biotin-negative control (Ancell, cat. 278-030).

6. RPE-Cy5-conjugated anti-CD45 (Serotec, MCA1719C).

7. Rabbit anti-human Prox-1 (Reliatech or equivalent AngioBio product).

8. Fluorescein isothiocyanate (FITC)-conjugated goat anti-rabbit F(ab')$_2$ (Immunotech).

9. PE-conjugated streptavidin (BD Pharmingen no. 554061).

10. Alexa488-conjugated anti-rabbit IgG (Molecular Probes).

11. Isotype-matched antibodies or preimmune sera can be purchased from many companies.

3. Methods

3.1. Isolation of Cells from Neonatal Foreskins

3.1.1. Day 1

1. Collect neonatal foreskins in 50 mL of sterile HBSS containing 5X PSA, on ice (*see* **Note 1**).

2. Handle each foreskin separately and perform all the following steps on ice to preserve tissue integrity.

3. Place each foreskin into a 10-mm tissue culture dish containing ice-cold HBSS containing 5X PSA and cut each foreskin into small squares of approximately 3 × 3 mm with a no. 10 scalpel blade.

4. Collect the squares in 30 mL of HBSS containing 5X PSA and shake vigorously to remove erythrocytes. Remove HBSS containing 5X PSA and repeat three times.

5. Place the squares (epithelial side facing up) into a 10-mm tissue culture dish and incubate in approximately 5 mL of HBSS containing 5X PSA at 4°C for 1 h. All squares must be totally immersed in HBSS.

6. Remove HBSS and incubate in approximately 5 mL 0.25% trypsin containing 5X PSA at 4°C overnight.

3.1.2. Day 2

1. Transfer tissues into a new 10-mm dish containing DMEM, 10% FBS, and PSA (5X), on ice; keep the epidermis facing up.

2. Using forceps, gently remove the epidermal sheet (*see* **Note 2**).

3. Scrape the dermis repeatedly with a scalpel until the skin is completely disrupted (*see* **Note 3**).

4. Collect medium containing cells with a Pasteur pipet and filter through a sterile 70-µm cell strainer into a 50-mL Falcon tube containing DMEM, 10% FBS, and 5X PSA.

5. Re-filter the homogenate through 70-µm cell strainer and gently homogenise the material remaining on the strainer using a 1-mL syringe plunger into the same 50-mL Falcon tube (*see* **Note 4**).

6. Centrifuge at 1200 rpm, 4°C, for 10 min.

7. Resuspend pellet in DMEM, 10% FBS, and 5X PSA and filter cell suspension through a new sterile 70-µm cell strainer.

8. Centrifuge cells and resuspend pellet in 3 mL of endothelial cell growth medium.

9. Plate cells obtained from one foreskin into a 35-mm tissue culture dish coated with collagen type I and culture the cells at 37°C in 5% CO_2 (*see* **Note 5**).

10. After 4 to 5 h, remove medium, gently wash two or three times with warm PBS to remove non-adherent cells, and then add 3 mL of endothelial cell growth medium (*see* **Note 6**).

11. Change media everyday for 4 d and then every 2–3 d.

12. Allow the cells to grow until 70% to 90% confluence (usually 11–15 d) (*see* **Note 7**).

13. Trypsinize the cells and plate them into one 60-mm tissue culture dish (passage 1).

14. When 70% to 90% confluence is reached (after approximately 2 to 3 d), trypsinize the cells and plate them into one 100-mm tissue culture dish.

15. When 70% to 90% confluence is reached (after approximately 2 to 3 d), trypsinize the cells and plate them into two 100-mm tissue culture dishes.

16. When 70% to 90% confluence is reached (after approximately 2 to 3 d), trypsinize the cells and plate them into four 100-mm tissue culture dishes (*see* **Note 8**).

3.1.3. Isolation of LECs from Bulk Culture by Fluorescence-Activated Cytometric Cell Sorting

1. When the four 100-mm tissue culture dishes reach 70% to 90% confluence, harvest the cells and resuspend them at 10^6/mL in cold PBS containing 0.2% BSA and 5X PSA (*see* **Note 9**).

2. Distribute 10^5 cells into eight calibration tubes and the rest of the cells in another tube (for isolation).

3. Centrifuge at 200g at 4°C for 5 min (*see* **Note 10**).

4. Resuspend the cells in cold PBS containing 0.2% BSA and 5X PSA supplemented with (*see* **Note 11**). In the calibration tubes, resuspend in 100 μL; in the isolation tube, resuspend at 10^6 cells/mL. Add the following to each tube:

   Test tube 1: Nothing.

   Test tube 2: FITC-labelled goat anti-rabbit F(ab')$_2$.

   Test tube 3: PE-labelled streptavidin.

   Test tube 4: Rabbit anti-human podoplanin followed by FITC-labelled goat anti-rabbit F(ab')$_2$.

   Test tube 5: Anti-human CD31 followed by PE-labelled streptavidin.

   Test tube 6: Anti-human podoplanin and anti-human biotinylated CD31 followed by FITC-labelled goat anti-rabbit and PE-labelled streptavidin.

   Test tube 7: Anti-human podoplanin and Cy5-RPE-labelled anti-human CD45.

Test tube 8: Anti-human biotinylated CD31 and Cy5-RPE labelled Cy5-RPE labelled anti-human CD45 followed by PE-labelled streptavidin.

Isolation tube: In parallel with the test tubes, incubate 10^6 cells/ mL with anti-human podoplanin, anti-human biotinylated CD31, and Cy5-RPE-labelled anti-human CD45 followed by FITC-labelled goat anti-rabbit and PE-labelled streptavidin. Recommended antibody concentrations:

Rabbit serum (201853) anti-human podoplanin IgG at a dilution of 1:100 (end user will have to optimize antibody dilution if using another antibody).

Mouse monoclonal biotinylated anti-human-CD31 (4 µg/mL).

Anti-CD45-RPE-Cy5 (4 µg/mL).

Cell suspensions are next exposed to both secondary antibodies together: Goat anti-rabbit F(ab')$_2$FITC (5 µg/mL) and strep-PE (1.25 µg/mL) on ice for 30 min.
See also the Subheading 3.1.1.

5. Use test tubes to calibrate the FACS cell sorter, then sort cells contained in the isolation tube (*see* **Note 12**).

6. Human LECs are sorted into podoplanin$^+$/CD31$^+$/CD45$^-$ (LEC) subset using a FacVantage.

7. Centrifuge sorted cells at 200g, 4°C, for 5 min.

8. Resuspend the cells in endothelial growth factor medium and plate them onto collagen-coated tissue culture dishes at a density of approximately 20,000 to 40,000 cells/cm^2 (*see* **Note 13**).

9. Expand lymphatic the endothelial cell population using a split ratio of 1:3 (*see* **Note 14**).

3.2. Characterization of Established Lines

3.2.1. Characterization of LEC-Specific Antigen Expression by Flow-Activated Cell Sorting

1. Allow the cells to grow on 100-mm tissue culture dishes until they reach confluence.

2. Wash with 15 mL cold PBS (*without* Ca^{2+} and Mg^{2+}) two or three times.

3. Add 10 mL of warm cell dissociation solution and keep the cells at 37°C for 5 to 10 min (*see* **Note 15**).

4. Gently detach the cells by pipetting to obtain a cell suspension (*see* **Note 16**).

5. Centrifuge cells at 200g at 4°C for 5 min and resuspend cells at 10^5 cells/mL in ice-cold PBS containing 0.2% BSA (blocking buffer).

6. Leave cells in blocking buffer at least 30 min on ice.

7. Prepare 10^5 cells/tube in 100 µL PBS containing 0.2% BSA for each of the following conditions (Ab indicates antibody):

	Control Cells	Control 2nd Ab	Irrelevant Ab	Reactive Ab
Incubation 1	0.2% BSA PBS	0.2% BSA PBS	Irrelevant Ab	Primary reactive Ab
Incubation 2	0.2% BSA PBS	2nd Ab	2nd Ab	2nd Ab

8. Centrifuge the cells at $200g$ at 4°C for 5 min.

9. Resuspend the cells in 100 µL PBS containing 0.2% BSA supplemented with the indicated antibodies:

Irrelevant antibody Concentration or dilution	Primary antibody Concentration or dilution	Secondary antibody Concentration or dilution
IgGµ/Biotin-negative control 4 µg/mL	Streptavidin-CD31 4 µg/mL	Biotinylated-PE 1.25 µg/mL
Preimmune serum 1/100	Anti-podoplanin (201853) 1/100	Anti-rabbit F(ab)² 5 µg/mL
Preimmune serum 1/100	Anti-LYVE-1 (D.J.) 1/100	Anti-rabbit F(ab?')² 5 µg/mL

10. Each incubation is performed at 4°C for 30 min, and in between each incubation, cells are washed three times with an excess of PBS containing 0.2% BSA.

11. At the end of the procedure, analyze the FACS profiles from at least 10,000 events (*see* **Note 17**).

3.2.2. Characterization of LEC-Specific Antigen Expression by Immunocytochemistry

1. Grow LECs on glass coverslips until they reach confluence.

2. Wash with PBS three times.

3. Fix cells with 4% cold methanol for 2 min (*see* **Note 18**).

4. Wash cells three times with PBS.

5. Incubate cells in blocking buffer (PBS containing 1% BSA) for 1 h at room temperature.

6. Incubate cells with 5 µg/mL anti-human Prox-1 diluted in blocking buffer, overnight, at 4°C in a humid chamber (*see* **Note 19**).

7. Wash cells three times with PBS.

8. Incubate cells with 2 µg/mL of Alexa488-conjugated anti-rabbit IgG for 1 h at room temperature.

9. Wash cells three times with PBS.

10. Incubate cells with a solution of PBS containing 0.5 µg/mL of DAPI.

11. Wash cells three times with PBS.

12. "Mount" coverslips and analyze under a fluorescent microscope (*see* **Note 20**).

3.2.3. Characterization of LEC-Specific Antigen Expression by Reverse Transcriptase Polymerase Chain Reaction

1. Allow the cells to grow to confluence in 100-mm tissue culture dishes.

2. Wash two or three times with 15 mL cold PBS.

3. Extract RNA using Trizol reagent as instructed by manufacturer.

4. Perform reverse transcription with Superscript II reverse transcriptase following manufacturer's instructions.

5. Perform polymerase chain reactions (PCRs) using *Taq*DNA polymerase using manufacturer's instructions and the primers and conditions listed in **Table 6.2**.

Table 6.2
Human primer sequences, PCR product lengths, optimal annealing temperatures, and amplification cycle number

Molecule	Sense (5′-3′)	Antisense (5′-3′)	Template (bases)	Annealing temperature (°C)	Number of cycles
Integrin-α9	ACTAT-GAAGCCGAC-CACATCCTAC	ATAAACCTT-GCCGAT-GCCTTTGTC	386	60	40
Integrin-β1	CAATGAAG-GGCGTGTTGG-TAGAC	TACAGACAC-CACACTCG-CAGATG	383	55	30
VEGF-C	TCCGGACTC-GACCTCTCG-GAC	CCCCACATC-TATACACAC-CTCC	324	60	40
LYVE-1	GGCAAGGAC-CAAGTT-GACACAG	TGGAGCAG-GAGGAGTAG-TAGTAGG	376	55	35
Podoplanin	AGCCAGAAGAT-GACACTGA-GACTAC	ACCACAAC-GATGATTC-CACCAATG	357	55	35
Prox-1	CCACCT-GAGCCACCAC-CCTTG	GCTTGACGT-GCGTACT-TCTCCATC	333	60	35
CD34	AAGCCTAGCCT-GTCACCT-GGAAATG	GGCAAGGAG-CAGGGAG-CATACC	300	60	35
CD44	TGGGTTCATA-GAAGGGCAT-GTGGTG	ATTCTGTCT-GTGCTGTCG-GTGATCC	420	60	35

6. Analyse PCR products on a 2% agarose gel (*see* **Note 21**).
7. See **Note 22**.

4. Notes

1. The surgical procedure required for circumcision is not a sterile process; use large amounts of PSA to kill bacterial and fungal contaminations. Low temperatures maintain tissue integrity.

2. This is not crucial since keratinocytes will not proliferate under the culture conditions used to grow endothelial cells and will be discarded during the isolation procedure. However, this procedure gives rise to a purer population of endothelial cells (recommended).

3. This procedure is long and fastidious; be patient.

4. Only a few collagen fibres should remain.

5. Given the amount of debris and the heterogeneity of the cell mixture, it is difficult to estimate the number of cells that have to be plated per surface. We recommend seeding the cells onto 35-mm tissue culture dishes and suggest that the end user optimize with experience. For collagen coating of dishes: Dilute rat tail tendon collagen type I (Becton Dickinson) with a solution of 0.1% acetic acid (dilute pure acetic acid with distilled water) to obtain a final concentration of 75 µg/mL. Coat cell culture surfaces with the solution containing 75 µg/mL collagen 1 h at room temperature or overnight at 4°C. Then, wash cell culture surfaces with PBS to eliminate acetic acid prior to seeding the cells.

6. The medium contains a lot of fibres from the extracellular matrix, cell debris, and cells. Changing the medium allows the cells to grow in a cleaner environment.

7. Do not allow the cells to reach confluence; non-endothelial cells, such as fibroblasts, will overgrow the endothelial cell population.

8. These steps are necessary to expand the population of cells isolated from the foreskins and are essential if one is to obtain an adequate number of cells. Indeed, the procedure to isolate a virtually pure population of cells by FACS requires a large number of cells.

9. A large amount of PSA is required since cell sorting is performed under semi-sterile conditions, and large amounts of PSA will avoid contamination. Avoid excessive trypsin usage, which may result in the loss of antigenicity.

10. Always keep cells on ice and make sure to keep the centrifuge at 4°C. Cells often internalize membrane proteins that bound to an antibody with resulting loss of the signal; keeping the cells at low temperature avoids antigen internalization.

11. Incubate each primary antibody separately for 30 min at 4°C. Secondary antibodies can be incubated at the same time if using those indicated here since cross-reactivity has not been observed. If using other secondary antibodies, users will have to test for cross-reactivity. Between each antibody incubation, wash cells two or three times in an excess of PBS containing 0.2% BSA. Centrifuge cells at 200g.

12. Test tubes have to be analyzed to calibrate the apparatus prior to sorting the cells, and cell sorting requires a person with expertise in this technique. Calibration of the cell sorter is not an easy process; we therefore recommend that 2×10^5 cells are added per tube for the first few cell-sorting assays.

13. Cell sorting induces a major stress to the cells, and cell death can be observed after seeding onto cell culture surfaces. Thus, we recommend seeding the cells at high density.

14. It is important to split the cells when they reach 70% to 90% confluence; otherwise, they may stop proliferating due their contact inhibition properties. It is important to avoid over trypsinization; this results in cell damage and eventually cell death.

15. Trypsin can also be used to detach the cells, but antigenicity can be lost.

16. Aggressive pipetting results in cell damage and ultimately cell death.

17. LYVE-1 is an antigen with expression that varies considerably and that only provides a small shift in fluorescence by FACS. Reminder: Always keep cells in ice, use ice-cold solutions, and keep centrifuge at 4°C.

18. Add cold methanol extremely slowly and carefully since methanol disrupts the cell monolayer very easily.

19. In parallel, incubate cells with an irrelevant antibody used at the same concentration. The signal provided by the irrelevant antibody indicates the specificity of the signal.

20. Since the Prox-1 and DAPI staining co-localise in the nucleus, this allows one to count the number of positive nuclei compared to the total number of cells per field.

21. Do not forget controls: Positive control, -RT conditions, water, and house-keeping gene to ensure RNA integrity and reverse transcription efficiency. Check that the temperature of the PCR apparatus is well calibrated. These results might be strengthened by using BECs as a control. The signals provided by the PCR have to be analysed carefully because they are not totally related to the purity of the LEC population. To date, VEGF-C expression seems to be present in BECs but not in LECs. Thus, if the PCR shows VEGF-C expression in the LECs, this would suggest a contamination of the cell population by BECs (*see* **Note 6**).

22. The isolated LECs usually keep expressing LEC-specific antigens up to passages 10 to 13. It is strongly recommended that the capacity of LECs to express LEC-specific markers is assessed every other passage, especially during the first few isolation attempts.

References

1. Sleeman, J. P., Krishnan, J., Kirkin, V., Baumann, P. (2001) Markers for the lymphatic endothelium: in search of the holy grail? *Microsc Res Tech* 55, 61–69.

2. Oliver, G., Alitalo, K. (2005) The lymphatic vasculature: recent progress and paradigms. *Annu Rev Cell Dev Biol* 21, 457–483.

3. Kriehuber, E., Breiteneder-Geleff, S., Groeger, M.,et al (2001) Isolation and characterization of dermal lymphatic and blood endothelial cells reveal stable and functionally specialized cell lineages. *J Exp Med* 194, 797–808.

4. Hirakawa, S., Hong, Y.K., Harvey, N., et al(2003) Identification of vascular lineage-specific genes by transcriptional profiling of isolated blood vascular and lymphatic endothelial cells. *Am J Pathol* 162, 575–586.

5. Podgrabinska, S., Braun, P., Velasco, P., Kloos, B., Pepper, M. S., Skobe, M. (2002) Molecular characterization of lymphatic endothelial cells. *Proc Natl Acad Sci U S A* 99, 16069–16074.

6. Nisato, R. E., Harrison, J.A., Buser, R., et al(2004) Generation and characterization of telomerase-transfected human lymphatic endothelial cells with an extended life span. *Am J Pathol* 165, 11–24.

7. Sauter, B., Foedinger, D., Sterniczky, B., Wolff, K., Rappersberger, K. (1998) Immunoelectron microscopic characterization of human dermal lymphatic microvascular endothelial cells. Differential expression of CD31, CD34, and type IV collagen with lymphatic endothelial cells versus blood capillary endothelial cells in normal human skin, lymphangioma, and hemangioma *in situ*. *J Histochem Cytochem* 46, 165–176.

8. Natkunam, Y., Rouse, R. V., Zhu, S., Fisher, C., van De Rijn, M. (2000) Immunoblot analysis of CD34 expression in histologically diverse neoplasms. *Am J Pathol* 156, 21–27.

9. Partanen, T. A., Arola, J., Saaristo, A., et al (2000) VEGF-C and VEGF-D expression in neuroendocrine cells and their receptor, VEGFR-3, in fenestrated blood vessels in human tissues. *FASEB J* 14, 2087–2096.

10. Jackson, D. G. (2004) Biology of the lymphatic marker LYVE-1 and applications in research into lymphatic trafficking and lymphangiogenesis. *APMIS* 112, 526–538.

11. Banerji, S. J., Wang, S. X., Clasper, S., (1999) LYVE-1, a new homologue of the CD44 glycoprotein, is a lymph-specific receptor for hyaluronan. *J Cell Biol* 144, 789–801.

12. Breiteneder-Geleff, S., Soleiman, A., Kowalski, H., (1999) Angiosarcomas express mixed endothelial phenotypes of blood and lymphatic capillaries: podoplanin as a specific marker for lymphatic endothelium. *Am J Pathol* 154, 385–394.

13. Petrova, T. V., Makinen, T., Makela, T. P., (2002) Lymphatic endothelial re-programming of vascular endothelial cells by the

Prox-1 homeobox transcription factor. *EMBO J.* 21, 4593–4599.

14. Oliver, G., Sosa-Pineda, B., Geisendorf, S., Spana, E. P., Doe, C. Q., Gruss, P. (1993) Prox 1, a prospero-related homeobox gene expressed during mouse development. *Mech Dev* 44, 3–16.

15. Leak, L. V., Jones, M. (1993) Lymphatic endothelium isolation, characterization, and long-term culture. *Anat Rec* 236, 641–652.

Chapter 7

Isolation, Culture and Characterisation of Vascular Smooth Muscle Cells

Adrian T. Churchman and Richard C.M. Siow

Abstract

Smooth muscle cells (SMCs) are key players in the pathogenesis of atherosclerosis and restenosis; however, they are also important in formation and development of *de novo* blood vessels during vasculogenesis and angiogenesis. Vascular SMCs can be formed by proliferation of existing SMCs, maturation of pericytes, or putative smooth muscle progenitor cells, thereby contributing to development of atherosclerotic plaques and angiogenic processes. Modulation of SMC phenotype is now recognised as a key event in the development of vascular diseases. This chapter describes the isolation and culture of vascular SMCs and pericytes from human and animal blood vessels for in vitro studies.

Key words: Angiogenesis, atherosclerosis, explant cultures, pericytes, smooth muscle cells.

1. Introduction

Smooth muscle cells (SMCs) are key players in the pathogenesis of atherosclerosis and restenosis after angioplasty *(1)*; however, they are also important in formation and development of *de novo* blood vessels (vasculogenesis) through differentiation of mesenchymal cells under the influence of mediators secreted by the endothelial cells comprising newly formed vessels *(2)*. In angiogenesis, vascular SMCs can be formed by proliferation of existing SMCs or maturation of pericytes *(2, 3)*. Recent experimental findings suggest a potential role of putative smooth muscle progenitor cells in the circulation or within adult tissues and the perivascular adventitia in the development of atherosclerotic plaques and biology of angiogenesis *(4)*. Modulation of vascular smooth muscle phenotype, SMC

S. Martin and C. Murray (eds.), *Methods in Molecular Biology, Angiogenesis Protocols, Second edition, Vol. 467*
© Humana Press, a part of Springer Science + Business Media, LLC 2009
DOI: 10.1007/978-1-59745-241-0_7

migration, and hypertrophy are now recognised as key events in the development of arterial lesions in vascular diseases (1). This has led to an increase in experimental research on SMC function in response to growth factors, extracellular matrix, modified lipoproteins, and other pro-atherogenic and pro-angiogenic mediators under controlled *in vitro* conditions to address the cellular mechanisms involved. Most of the methodologies used for vascular SMC isolation and culture have been developed to accomplish such studies (5).

Vascular SMCs *in vivo* are capable of existing in a range of different phenotypes, which fall within a continuous spectrum between "contractile" and "synthetic" states (5, 6). The healthy adult vasculature predominantly consists of the contractile-state SMCs; their main function is in the maintenance of vascular tone. They exhibit a "muscle-like" appearance, with up to 75% of their cytoplasm containing contractile filaments. However, in culture, these cells are able to revert to a synthetic phenotype that is normally found in embryonic and young developing blood vessels (7). These proliferative cells synthesise extracellular matrix components such as elastin and collagen and consequently contain large amounts of rough endoplasmic reticulum and Golgi apparatus but few myofilaments in their cytoplasm (6–8). The modulation of vascular SMCs from a contractile to a synthetic phenotype is an important event in atherogenesis and restenosis, resulting in myointimal thickening and arterial occlusion. This may arise from damage to the endothelium and exposure of SMCs to circulating blood components, such as oxidized lipoproteins, and pro-inflammatory stimuli such as reactive oxygen species and growth factors released from endothelial cells, neutrophils, macrophages, and platelets (1).

The methodology used to isolate vascular SMCs can determine the initial phenotype of cells obtained for culture (5, 7). The two main techniques commonly employed in SMC isolation from arterial and venous tissues are enzymatic dissociation, which readily yields a small number of SMCs initially in the contractile phenotype, and explantation, which yields larger amounts of SMCs after 1–3 wk, in the synthetic and proliferative phenotypes. The lower yield of SMCs following enzymatic dissociation is suitable for studies on single dispersed cells, while tissue explantation provides the potential for obtaining longer-term cultures of confluent SMCs. This chapter describes these two alternative techniques for vascular SMC isolation from human arterial tissues, subculture passaging, and characterisation for cell culture studies. In addition, a brief description of the isolation of pericytes from human placental microvessels is also provided since these muscle-like perivascular cells are recognised to play a key role in scaffolding, maturation, and contraction of microvessels (3).

2. Materials

2.1. SMC Cultures

1. The most commonly used growth medium in SMC culture is Dulbecco's modified Eagle's medium (DMEM) containing either 1000 or 4500 mg/L glucose (Sigma); however, Medium 199 (Sigma) is also suitable (*see* **Note 1**). The following additions to the basal medium are necessary prior to use (final concentrations): 2 mM L-glutamine, 40 mM bicarbonate, 0.002% phenol red, 100 U/mL^{-1} penicillin, 100 µg/mL streptomycin, and 10 % (v/v) foetal calf serum (FCS; Sigma). Sterile stocks of these components are usually prepared and stored as frozen aliquots as described in this chapter. The complete medium can be stored at 4°C for up to 1 mo and is prewarmed to 37°C prior to use in routine cell culture. Culture medium without the FCS component is used during the isolation procedures.

2. Hank's balanced salt solution (HBSS; Sigma) is used as a tissue specimen collection medium. The following additions, from sterile stock solutions, are necessary prior to use (final concentrations): 100 µg/mL gentamycin, 0.025M HEPES, 20 mM bicarbonate, and 0.001% phenol red. The HBSS can be stored in aliquots at 4°C for up to 2 wk.

3. L-Glutamine (200 mM) stock solution: Dissolve 5.84 g L-glutamine (Sigma) in 200 mL tissue culture-grade deionised water and sterilise by passing through a 0.22-µm filter. Aliquots of 5 mL are stored at −20°C, and 4 mL are used in 400 mL of medium.

4. Bicarbonate (4.4%, 0.52M)-phenol red (0.03%) solution: Dissolve 44 g NaHCO$_3$ and 30 mg phenol red in 1000 mL tissue culture-grade deionised water and sterilise by autoclaving for 10 min at 115°C. Aliquots of 15 mL are stored at 4°C for up to 6 mo, and two aliquots are used in 400 mL of medium.

5. Penicillin and streptomycin stock solution (80X concentrate): Dissolve 480 mg penicillin (G sodium salt; Sigma) and 1.5 g streptomycin sulphate (Sigma) in 200 mL tissue culture-grade deionised water and sterilise by passing through a 0.5-µm pre-filter and a 0.22-µm filter. Aliquots of 5 mL are stored at −20°C, and one aliquot is used in 400 mL of medium.

6. Gentamycin solution (80X concentrate): Dissolve 750 mg gentamycin sulfate (Sigma) in 100 mL tissue culture-grade deionised water and sterilise by passing through a 0.22-µm filter. Aliquots of 5 mL are stored at −20°C, and one aliquot is used in 400 mL of HBSS.

7. HEPES solution (1M): Dissolve 47.6 g of HEPES (Sigma) in 200 mL tissue culture-grade deionised water and sterilise by passing through a 0.22-µm filter. Aliquots of 5 mL can be stored at –20°C, and two aliquots are used in 400 mL of HBSS.

8. Trypsin solution (2.5%): Trypsin from porcine pancreas (Sigma) is dissolved (2.5 g/100 mL) in phosphate-buffered saline A (PBS-A) and sterilised by passing through a 0.22-µm filter. Aliquots of 10 mL are stored at –20°C.

9. Ethylenediaminetetraacetic acid (EDTA) solution (1%): EDTA disodium salt is dissolved (500 mg/50 mL^{-1}) in tissue culture-grade deionised water and sterilised through a 0.22-µm filter. Aliquots of 5 mL are stored at 4°C.

10. Trypsin (0.1%)-EDTA (0.02 %, 0.5 mM) solution is prepared by adding 10 mL trypsin (2.5%) and 5 mL EDTA (1%) to 250 mL sterile PBS-A. This solution is prewarmed to 37°C before use to detach cells from culture flasks and stored at 4°C for up to 2 mo.

2.2. Enzymatic Dissociation of SMCs

1. Collagenase, type II (Sigma): Dissolve collagenase in serum-free medium (3 mg/mL) on ice. Particulate material is removed by filtering the solution through a 0.5-µm pre-filter and then sterilised by passing through a 0.22-µm filter. This enzyme solution can be stored long term as 5- to 10-mL aliquots at –20°C until use.

2. Elastase type IV from porcine pancreas (Sigma): Immediately before use, elastase is dissolved in serum-free medium (1 mg/mL), and the pH of solution is adjusted to 6.8 with 1M HCl. This solution is then sterilised by passing through a 0.22-µm filter and kept on ice.

2.3. Immunofluorescence Microscopy

1. Antibody to specific SMC antigen (e.g., monoclonal mouse α-smooth muscle actin; Sigma).

2. Normal rabbit serum (Sigma).

3. Fluorescein isothiocyanate conjugated rabbit anti-mouse secondary antibody (Dako).

4. Methanol (100%).

2.3. Immunofluorescence Microscopy

All procedures should be carried out in a class II laminar flow safety cabinet using aseptic technique. Dissection equipment should be thoroughly washed and kept sterilised by immersion in 70% ethanol or by autoclaving at 121°C for 20 min.

1. 25-cm² and 75-cm² tissue culture flasks and 90 mm Petri dishes.

2. Sterile Pasteur 5- and 10-mL pipets.

3. Sterile 30-mL universal containers and 10-mL centrifuge tubes.

4. Scalpel handles and blades, small scissors, watchmaker's forceps, and hypodermic needles.

5. Cork board for dissection covered with aluminium foil, both sterilised by thorough spraying with 70% ethanol.

6. Sterile conical flasks of various sizes.

7. Lab-Tek chamber slides (Nunc).

3. Methods

3.1. Collection of Tissue Samples

In this laboratory, cells are routinely isolated from human umbilical arteries, a readily available source of human vascular SMCs. As soon as possible after delivery, the whole umbilical cord, obtained with prior ethical approval and consent of mothers, is placed in the HBSS collection medium and stored at 4°C. Cords collected and stored in this way can be used for SMC isolation up to 48 h after delivery. SMCs can also be isolated from arteries following harvesting of endothelial cells from the corresponding umbilical vein *(9)*. Immediately prior to proceeding with SMC isolation, 5-cm lengths of the umbilical artery should be carefully dissected out from the cord, ensuring minimal surrounding connective tissue remains, and stored in new collection medium. Other common sources of arterial tissue are from human, mouse, rat, or porcine aortae, which should be carefully dissected out from the body, cleaned of extraneous tissues, and stored in the collection medium at 4°C as soon as possible (*see* **Note 2**). Isolation of SMCs/pericytes from microvessels can be performed using human placental or bovine retinal tissue *(10)*.

3.2. Isolation of Smooth Muscle Cells by Enzymatic Dispersion

1. As much surrounding connective tissue as possible should be dissected away from around the artery and the tissue washed with new HBSS collection medium. The artery is then placed on the sterile dissection board and covered with HBSS to keep it moist.

2. The artery is fixed to the dissection board at one end using a hypodermic needle and then cut open longitudinally, using small scissors, with the luminal surface upward.

3. The endothelium is removed along the whole length of the artery by scraping the cell layer off with a sterile scalpel blade and the tissue re-moistened with HBSS.

4. The thickness and nature of the arterial wall will vary depending on the source of tissue; however, in general the arterial intima and media are peeled into 1- to 2-mm wide transverse strips using watchmaker's forceps and a scalpel. Muscle strips are transferred to a 90-mm Petri dish containing HBSS. This procedure is repeated for the whole surface of the vessel.

5. Most of the HBSS medium is then aspirated off, and the muscle strips are cut into 1- to 2-mm cubes using scissors or a scalpel blade. The cubes are then washed in new HBSS and transferred into a sterile conical flask of known weight; the mass of tissue is measured to determine volume of enzyme solution needed for digestion.

6. Collagenase solution in serum-free culture medium is next added to the tissue to give a ratio of tissue (g) to enzyme solution (mL) of 1:5 (w/v). The flask is then covered with sterile aluminium foil and shaken in a water bath at 37°C for 30 min.

7. Elastase solution is then prepared and directly added to the solution containing the tissue and collagenase. The flask is returned to the shaking water bath at 37°C, and every 30 min during the following 2–5 h the suspension is mixed by pipetting in the sterile safety cabinet.

8. At each 30-min interval, a 10-µL sample of suspension is transferred to a haemocytometer to check for the appearance of single cells. This is repeated until the tissue is digested and there are no large cell aggregates visible. Digestion should not proceed for longer than 5 h to avoid loss of cell viability (*see* **Note 3**).

9. The cell suspension is finally divided into 10-mL centrifuge tubes and centrifuged at 50–100g for 5 min. The supernatants are carefully aspirated, and 2–5 mL of pre-warmed complete culture medium are added to resuspend the cells using a sterile pipet. Cells are then seeded into 25-cm² culture flasks at a density of about 8×10^5 cells/mL and placed into a 37°C, 5% CO_2 incubator with the flask cap loose.

10. Viable SMCs should adhere to the flask wall within 24 h, and all the medium is then replaced with fresh pre-warmed complete culture medium. Half of the culture medium is replaced every 2–3 d until a confluent SMC monolayer is obtained.

3.3. Isolation of Smooth Muscle Cells by Explant Culture

Explant cultures are suitable if limited vascular tissue is available, for example, from human aortas or carotid arteries. The sample is first treated exactly as described in **Subheading 3.2., steps 1–3**, and then the following procedure is adopted:

1. The artery is cut into 2-mm cubes with a scalpel blade and placed on the surface of a 25-cm^2 culture flask using a sterile Pasteur pipet, ensuring that the luminal surface is in contact with the flask wall. The artery should be kept continually moist with the HBSS, and a small drop of serum-containing medium should be placed on each cube when placed in the flask.

2. The cubes are distributed evenly on the surface with a minimum of 12–16 cubes per 25-cm^2 flask. The flask is then placed upright, and 5 mL serum containing medium are added directly to the bottom of the flask before transferring into a 37°C, 5% CO_2 incubator with its cap loose. To facilitate adherence of the explanted tissue to the culture plastic substrate, the flask is kept upright for 2–4 h in the incubator before the flask is carefully placed horizontal such that the medium completely covers the attached muscle cubes.

3. The explants should be left undisturbed for 4 d, inspecting daily for infections (*see* **Note 4**). Every 4 d, any unattached explant cubes should be removed and half the medium replaced with fresh pre-warmed serum-containing medium. Cells will initially migrate out from the explants within 1 to 2 wk.

4. After 3 to 4 wk, there should be sufficient density of SMCs around the explants for removal of the tissue. Using a sterile Pasteur pipet, the cubes are gently dislodged from the plastic flask surface and aspirated off with the culture medium. The culture medium is replaced, and cells are then left for a further 2–4 d to proliferate and form a confluent SMC monolayer (*see* **Note 5**).

3.4. Isolation of Pericytes from Microvessels

Due to the relative availability of human placentas, isolation from this tissue is described. Use of placental tissue yields larger quantities of pericytes due to the high amount of villi present. Alternatively, bovine retinas are a suitable tissue source for pericyte isolation.

1. Dissect a central section of the placenta and wash thoroughly in serum-free medium. Ensure that the section chosen is distant from any large blood vessels and the outer membrane.

2. Manually dissect and cut the tissue into small 5-mm^2 pieces and incubate in serum-free media containing 3 mg/mL collagenase for 3 h at 37°C in a shaking water bath.

3. Separate the microvessels by passing the suspension through a 70-µm mesh filter (Falcon) and wash through two times with serum-free media.

4. Remove the microvessels from the mesh filter and place into a 25-cm^2 culture flask containing medium supplemented with 20% serum.

5. After 24 h, the medium is removed, and attached cells are washed once with warmed sterile PBS to remove floating debris. Fresh growth medium is added, and after 5–6 d, pericytes and endothelial cells proliferate out from the microvessel fragments.

6. As the culture medium does not contain endothelial cell growth supplements, any endothelial cells present initially no longer survive after two rounds of trypsinisation, leaving a pure culture of pericytes, which can be characterised by their morphology and antigen expression as detected by immunofluorescence.

3.5. Subculture of Smooth Muscle Cells

1. Once a confluent monolayer has been attained in a 25-cm² flask by either isolation method, the SMCs can be subcultured (passaged) into further 25-cm² flasks or a 75-cm² flask. The culture medium is removed, and cells are washed twice with pre-warmed sterile PBS-A to remove traces of serum.

2. Pre-warmed trypsin/EDTA solution (0.5 mL for 25-cm² flask or 1 mL for a 75-cm² flask) is added to cover the cells, and the flask is incubated at 37°C for 2–4 min. The flask is then examined under the microscope to ensure cells have fully detached. This can also be facilitated by vigorous tapping of the side of the flask three to six times to break up cell aggregates (*see* **Note 6**).

3. Serum-containing medium (5 mL) is added to stop the action of the trypsin, which can reduce SMC viability through prolonged exposure. The cell suspension is then drawn up and down a sterile Pasteur pipet four to six times to further break up any cell clumps.

4. The cells are then transferred into new culture flasks at a split ratio of 1:3, and sufficient serum-containing medium is added to the new flasks (5 mL in a 25-cm² and 10 mL in a 75-cm² flask). The flasks are returned to the 37°C, 5% CO_2 incubator, and the culture medium is changed as described in **Subheading 3.2., step 10**.

The SMCs can be passaged between 10 and 20 times, depending on species, before their proliferation rate significantly decreases. Phenotypic changes of enzyme-dispersed SMCs to the "proliferative" state occurs following passaging, the extent of which depends on the vessel type and species from which the SMCs are derived, the culture medium, and their seeding density *(5–7, 11)*.

3.6. Characterisation of Smooth Muscle Cells

Confluent monolayers of vascular SMCs exhibit a characteristic "hill-and-valley" morphology in culture *(5, 7)*. Isolated primary cells can be identified by their positive antigen reaction with antibodies against specific marker proteins for differentiated SMCs such as α-smooth muscle actin *(12)*. Other specific proteins have since been identified and can be used as markers of differentiated SMCs through use of specific antibodies in both cell cultures and tissue sections *(7, 12–14)* as outlined in **Table 7.1**. Smooth muscle myosin heavy chain (SMMHC) is the most specific marker; however, expression is not always maintained in long-term SMC cultures. The absence of endothelial cells in cultures can be confirmed by negative staining for von Willebrand factor or the lack of uptake of acetylated low-density lipoproteins, both endothelial cell-specific markers *(9, 15)*. The following procedure, described here in brief, can be used to identify SMCs by their positive staining with a fluorescein isothiocyanate (FITC)-labelled primary antibody against smooth muscle α-actin.

1. Smooth muscle cells are subcultured into Lab-Tek slide wells and characterised after 48 h.

2. The culture medium is removed from the wells, and cells are gently washed three times with serum-free culture medium before fixing with ice-cold methanol (100%) for 45 s and then further washed three times with ice-cold PBS-A.

3. Cells are then incubated with a mouse monoclonal anti-smooth muscle α-actin antibody at 1:50 dilution with PBS-A for 60 min at room temperature. As a negative control, some cells are incubated with PBS-A only at this stage.

4. The primary antibody or PBS-A is then removed, and cells are washed three times with PBS-A and incubated for 5 min with normal rabbit serum at 1:20 dilution.

5. After a single wash with PBS, cells are then further incubated for 30 min at room temperature with FITC-conjugated rabbit anti-mouse immunoglobulin G (IgG; Santa Cruz) diluted 1:50 in PBS-A.

Table 7.1
Markers for characterisation of vascular smooth muscle cells (SMCs) (*see 7, 11–14*)

Specific SMC Protein Markers	Protein Function
Smooth muscle myosin heavy chain	SMC contractile protein
Calponin	Contractile regulator
SM22α	Cytoskeletal protein
Desmin	Intermediate filament
H-Caldesmon	Actin-binding protein
Metavinculin	Actin-binding protein
Smoothelin	Cytoskeletal protein
α-Smooth muscle actin	Contractile protein

t0.1
t0.2
t0.3
t0.4
t0.5
t0.6
t0.7
t0.8
t0.9
t0.10
t0.11

6. Finally, cells are washed three times with PBS-A and viewed under a microscope equipped for epifluorescence with appropriate filters for FITC.

Visualisation of positive staining with this technique should reveal cells with a three-dimensional network of long, straight, and uninterrupted α-actin filaments running parallel to the longer axis of the cells and an underlying row of parallel filaments along the smaller cell axis, with no cytoplasmic staining between filaments.

3.7. Cryopreservation

Vascular SMCs can be cryopreserved with a recovery greater than 50%. Explant cultures of SMCs do not appear to be adversely affected by freezing; however, enzyme-dispersed SMCs may have a reduced proliferation rate and passaging efficiency on thawing. The following protocol is suggested; however, other techniques of cryopreservation are also available.

1. Confluent SMC cultures should be detached from one 75-cm² flask as described in **Subheading 3.5.**

2. Following centrifugation of the cell suspension for 5 min at 1000 rpm, the supernatant is aspirated, and the cell pellet is resuspended well in serum-containing culture medium with an additional 10% (v/v) dimethyl sulfoxide (DMSO) and transferred to a suitable cryovial.

3. To facilitate gradual freezing, the cryovial is then stored at 4°C for 1 h, transferred to –20°C for 1 h min and –70°C for 1 h before immersing into liquid nitrogen for long-term storage. Alternatively, cryovials can be placed in a dedicated "freezing chamber" containing isopropanol that lowers their temperature by 1°C per minute when placed directly in a –70°C freezer.

4. To defrost cells, the cryovial should be rapidly warmed to room temperature by placing at 37°C and the cell suspension transferred to a 25-cm² culture flask. Then add 20 mL of pre-warmed serum-containing culture medium to the flask; change the medium after 24 h.

5. Alternatively, on defrosting, the cells can be centrifuged at 50–100g for 5 min in serum-containing medium. The supernatant is aspirated to remove the DMSO, the cell pellet is resuspended well in pre-warmed serum-containing medium and then transferred into a 25-cm² flask. Cells should be passaged once prior to use in experiments.

3.8. Summary

The three techniques for SMC isolation described yield cells with very different proliferative properties in culture. If larger quantities of SMCs in culture are required, the explant isolation technique is recommended, although initially slower to yield cells. Enzymatic dispersion may provide more cells in the contractile phenotype, but the initial yield may be low, and subcultures less readily proliferate through as many passages. Future studies are likely to address the isolation and characterisation of stem cells that can differentiate into smooth muscle progenitor cells. It remains to be elucidated whether these cells participate in processes leading to angiogenesis and the pathogenesis of vascular diseases.

4. Notes

1. Other "smooth muscle cell optimised" culture media are commercially available, such as Smooth Muscle Basal and Growth Media (TCS Cellworks), which are based on MCDB131 medium. These media may help to promote more rapid outgrowth of cells from tissue explants, but it should be noted that they may contain components that could alter SMC function, such as insulin, basic fibroblast growth factor (bFGF), and hydrocortisone.

2. For maximal SMC yield and viability, cells should be isolated as soon as possible after harvesting the vessels. This also reduces the risk of infections since tissues are often handled and excised under non-sterile conditions.

3. If the SMC yield and viability is low following enzyme dispersion, soybean trypsin inhibitor, at a final concentration of 0.1 mg/mL^{-1}, can be added to the enzyme solution to inhibit the action of non-specific proteases, which may contaminate commercial elastase.

4. Should infections frequently occur following explantation, additional antibiotics and fungicides can be supplemented to the culture medium for the initial 24 h following isolation and then the medium replaced. Gentamycin (25 µg/mL) and amphotericin B (2 µg/mL^{-1}) are commonly used, and 5-mL aliquots of these can be stored at −20°C as 2.5-mg/mL^{-1} and 0.2 mg/mL^{-1} stocks, respectively.

5. When the explanted tissue is removed, SMCs can also be detached by trypsinisation and redistributed evenly in the same flask as described in **Subheading 3.4.** This is advisable if cells have grown in a very dense pattern around the explants and will facilitate obtaining the confluent monolayer of cells.

6. The trypsin/EDTA solution should be prewarmed to 37°C only immediately prior to use and not left in a heated water bath for extended periods to prevent loss of activity. Cells

should not need incubation with trypsin/EDTA at 37°C for longer than 5–7 min to detach from the flask, and this may indicate that a fresh solution should be prepared.

Acknowledgements

The vascular SMC research in this laboratory is supported by the British Heart Foundation and Heart Research UK. We are grateful to all the midwives in the Maternity Unit and Birth Centre at St. Thomas' Hospital, Guy's and St. Thomas' NHS Foundation Trust for supplying human umbilical cords obtained with prior ethical approval and consent of mothers.

References

1. Lusis, A. J. (2000) Atherosclerosis. *Nature* 407, 233–241.

2. Carmeliet, P. (2003) Angiogenesis in health and disease. *Nat Med* 9, 653–660.

3. Bergers, G., Song, S. (2005) The role of pericytes in blood-vessel formation and maintenance. *Neuro-oncology* 7, 452–464.

4. Liu, C., Nath, K. A., Katusic, Z. S., Caplice, N. M. (2004) Smooth muscle progenitor cells in vascular disease. *Trends Cardiovasc Med* 14, 288–293.

5. Campbell, J. H., Campbell, G. R. (1993) Culture techniques and their applications to studies of vascular smooth muscle. *Clin Sci* 85, 501–513.

6. Thyberg, J., Hedin, U., Sjolund, M., Palmberg, L., Bottger, B. A. (1990) Regulation of differentiated properties and proliferation of arterial smooth muscle cells. *Arteriosclerosis* 10, 966–990.

7. Campbell, J. H., Campbell, G. R. (1987) Phenotypic modulation of smooth muscle cells in culture, in Campbell, J. H., Campbell, G. R.eds., *Vascular Smooth Muscle Cells in Culture*, Vol. 1, pp. 39–55. CRC Press, Boca Raton, FL.

8. Owens, G. K. (1995) Regulation of differentiation of vascular smooth muscle cells. *Physiol Rev* 75, 487–517.

9. Morgan, D. L. (1996) Isolation and culture of human umbilical vein endothelial cells, in Jones, G. E ed.., *Methods in Molecular Medi-cine: Human Cell Culture Protocols*, pp. 101–109.Humana Press, Totowa, NJ.

10. Schlingemann, R. O., Rietveld, F. J. R., W. deWaal, R. M, Ferrone, S., Ruiter, D. J. (1990) Expression of the high molecular weight melanoma-associated antigen by pericytes during angiogenesis in tumours and in healing wounds. *Am J Pathol* 136, 1393–1405.

11. Thyberg, J., Nilsson, J., Palmberg, L., Sjolund, M. (1985) Adult human arterial smooth muscle cells in primary culture. Modulation from contractile to synthetic phenotype. *Cell Tissue Res* 239, 69–74.

12. Skalli, O., Ropraz, P., Trzeciak, A., Benzonana, G., Gillessen, D., Gabbiani, G. (1986) A monoclonal antibody against -smooth muscle actin. *J Cell Biol* 103, 2787–2796.

13. Proudfoot, D., Shanahan, C. M. (2001) Vascular smooth muscle, in Koller, M. R., Palsson, B. O Masters, J. R. W.eds. *Human Cell Culture*, Vol. 5, pp. 43–64. Kluwer Academic Press, Dordrecht, The Netherlands.

14. Hao, H., Gabbiani, G., Bochaton-Piallat, M. (2003) Arterial smooth muscle cell heterogeneity: implications for atherosclerosis and restenosis development. *Arterioscler Thromb Vasc Biol* 23, 1510–1520.

15. Voyta, J. C., Via, D. P., Butterfield, C. E., Zetter, B. R. (1984) Identification and isolation of endothelial cells based on their increased uptake of acetylated–low density lipoprotein. *J. Cell Biol* 99, 2034–2040.

Chapter 8

Circulating Endothelial Progenitor Cells

Michael C. Schmid and Judith A. Varner

Abstract

Recent studies have found that bone marrow-derived cells give rise to endothelial cells during states of tissue repair and disease. We have found that one key integrin, integrin-α4β1, promotes the homing of circulating endothelial progenitor cells (EPCs) to sites of ongoing tissue repair. This integrin facilitates the adhesion of EPCs to the vascular endothelium in inflamed tissue or within tumors. We demonstrate how to identify, isolate, purify, and characterize EPCs. We also demonstrate *in vivo* analysis of the roles of bone marrow-derived cells in tumor growth and angiogenesis by demonstrating adoptive transfer, bone marrow transplantation, tumor models, and immunohistochemistry for markers of blood and endothelial vessels. Finally, we show how to characterize cell adhesion mechanisms regulating bone marrow-derived progenitor cell trafficking.

Key words: Angiogenesis, CD34, endothelial progenitor cell, Lin⁻, lymphangiogenesis, peripheral blood mononuclear cell, Sca1, stem cell.

1. Introduction

Bone marrow-derived precursor cells have recently been shown to regulate neovascularization of tumors and ischemic tissues *(1)*. Progenitor cells home to ischemic or hypoxic tissues (angiogenic niches), where they incorporate into sprouting blood vessels at low, but measurable, frequencies. The nature of this cell type remains unclear, however. Isner and colleagues first isolated mononuclear cells (MNCs) from human peripheral blood that were enriched for expression of the hematopoietic stem cell marker CD34 *(2)*. On culture in endothelial growth media, these cells expressed endothelial lineage markers, such as CD31, Tie2, and vascular endothelial growth factor receptor 2 (VEGFR2),

S. Martin and C. Murray (eds.), *Methods in Molecular Biology, Angiogenesis Protocols, Second edition, Vol. 467*
© Humana Press, a part of Springer Science+ Business Media, LLC 2009
DOI: 10.1007/978-1-59745-241-0_8

and incorporated into blood vessels in ischemic tissues. These cells were therefore described as bone marrow-derived endothelial progenitor cells (EPCs) or "hemangioblasts." Since then, Rafii and colleagues identified a VEGFR2 and AC133-positive subpopulation of CD34-positive circulating cells that form endothelial cell-like colonies *in vitro* (3, 4).

Recently, other studies have shown that mononuclear progenitor cells expressing monocyte markers such as CD11b or CD14 also give rise to endothelial cell-like colonies *in vitro* and possibly *in vivo* (5–10). Hildbrand and colleagues described a $CD34^+CD11b^+$ cell population from cord blood that gives rise to endothelial cell colonies *in vitro* (9). Interestingly, $CD34^-/CD14^+$ human peripheral blood monocytes have also been shown to express endothelial cell markers such as von Willebrand factor and vascular endothelial cadherin (VE cadherin) after several weeks in culture in the presence of endothelial growth media *(10)*. Interestingly, a novel Gr^+CD11b^+ bone marrow-derived cell population was shown to contribute to neovascularization *(11)*. These myeloid cells comprised 5% of the total cell population within colorectal carcinoma and Lewis lung carcinoma tumors. In fact, tumor angiogenesis was increased in the presence of Gr^+CD11b^+ cells, which also exhibited high levels of MMP9. Deletion of MMP9 abolished the angiogenesis-promoting activity. Tumor-derived Gr^+CD11b^+ cells expressed the endothelial cell markers VE cadherin and VEGFR2 and incorporated into tumor endothelium *(11)*. These findings suggest that endothelial-like cells arise from bone marrow-derived hematopoietic progenitor cell populations ($CD34^+$ $AC133^+$ cells) or myeloid progenitor cell-like populations ($CD14^+$ or $CD11b^+$ cells). It is possible that endothelial cells and myeloid cells share a common precursor.

A number of studies therefore suggest a common link between monocytic cells and microvascular endothelial cells and further support the idea that tumor vasculogenesis occurs through the recruitment of bone marrow-derived endothelial precursor cells, which may include subsets of myeloid lineage cells. Continued characterization of the bone marrow-derived cells that contribute to neovascularization and analysis of the mechanisms by which these cells modulate neovascularization will clarify the cell types that promote vasculogenesis. Together, these studies suggest that myeloid lineage bone marrow-derived cells give rise to vascular endothelium, and that several lineage markers, including Gr, CD11c, and CD11b, identify putative EPCs.

Some studies suggest that macrophages, rather than EPCs, are the key bone marrow-derived cell type that contributes to adult neovascularization. Studies by de Palma and colleagues identified a $Tie2^+$ subpopulation of $CD11b^+$ tumor-associated myeloid cells that promotes angiogenesis in tumors *(12)*. Although these cells stimulate angiogenesis, they do not actively incorporate into blood vessels. Further studies from the same group showed that Tie2-expressing monocytic cells are specifically recruited to spontaneously arising mouse pancreatic tumors and to orthotopic xenografts of human glioma *(12)*. From these studies, it is clear that myeloid cells play key roles in tumor angiogenesis; importantly, these cells may include EPCs, macrophages, and other myeloid lineage cells.

Our lab recently characterized the specific mechanisms by which myeloid and endothelial precursor cells adhere to tumor endothelium and extravasate into tumor tissue. Bone marrow-derived cells express a number of functional integrins ($\alpha2\beta1$, $\alpha4\beta1$, $\alpha5\beta1$, $\alpha v\beta3$, $\alpha v\beta5$, $\alpha M\beta2$ [CD11b], and $\alpha X\beta2$ [CD11c]) that may play roles in these cells within neovascular microenvironments *(13, 14)*. Our studies revealed that integrin-$\alpha4\beta1$, a receptor for vascular cell adhesion molecule (VCAM) and fibronectin, selectively promotes the homing of both EPCs *(15)* and monocytes *(14)* to neovascular tissue, and that this integrin is essential for the participation of these cells in angiogenesis and tumor growth. Human $CD34^+$ and murine Lin^-Sca1^+ progenitor cells as well as bone marrow-derived myeloid cells ($CD14^+$ $CD11b^+$) adhered to endothelial cells *in vitro* and tumor endothelium *in vivo* via integrin-$\alpha4\beta1$. Treatment of mice bearing Lewis lung carcinoma tumors with antagonists of integrin-$\alpha4\beta1$ significantly suppressed the number of monocytes and EPCs within tumors and reduced blood vessel density. Our studies indicated that suppression of monocyte/macrophage homing to tumors by the application of an integrin-$\alpha4\beta1$ antagonist could be a useful supplementary approach to suppress tumor angiogenesis and growth.

In this chapter, we describe how to isolate peripheral blood or bone marrow-derived MNCs and from them, $CD34^+$ human progenitor cells and Lin^- murine progenitor cells; how to culture endothelial cells and progenitor cells; how to characterize cell surface markers, such as integrins; how to evaluate cell adhesion mechanisms *in vitro*; and how to study cell trafficking and contributions to tumor growth *in vivo*.

2. Materials

2.1. Progenitor Cell Isolation

1. 50mL conical tubes (human cell isolation).
2. 15-mL conical tubes (rodent cell isolation).
3. Phosphate-buffered saline, Ca^{2+} and Mg^{2+} free, with 0.5 mM EDTA (ethylenediaminetetraacetic acid) (PBS-E).

4. Histopaque-1077 (Sigma-Aldrich, St. Louis, MO) for human cell isolation.

5. Histopaque 1083 (Sigma-Aldrich) for mice, rat, or rabbit cell isolation.

6. 10-cc syringe with 18-gage bevel-tip needles.

7. Ammonium chloride red blood cell lysis solution: 155 mM NH$_4$Cl, 10 mM NaHCO$_3$, and 0.1 mM EDTA.

8. Blood draw: 100 mL from humans (or two buffy coats from blood bank), 5–40 mL from rabbit, 1–2 mL from mice (or mouse bone marrow), 5 mL from rat.

2.2. Progenitor Cell Purification

1. Fc-receptor blocking agent.

2. Fluorescein isothiocyanate (FITC)-conjugated anti-CD34 (clone AC136, Miltenyi Biotech, Auburn, CA).

3. PE-conjugated anti-CD133 (AC133/1, Miltenyi Biotech).

4. Anti-c-kit (clone 2B8) from eBioscience (San Diego, CA).

5. Anti-Sca-1 (clone E13-161.7) from eBioscience.

6. Antibiotin-FITC (Miltenyi Biotech).

7. Lineage cell depletion kit from Miltenyi Biotech.

8. Anti-CD34 conjugated supramagnetic colloidal beads from Miltenyi Biotech.

9. MACs buffer: PBS pH 7.2 supplemented with 0.5 % bovine serum albumin (BSA) and 2 mM EDTA.

10. Magnetic cell separator (MidiMACS, Miltenyi Biotech).

11. LS$^+$/VS$^+$ columns (Miltenyi Biotech).

2.3. FACs Analysis

1. PE-conjugated mouse antihuman α4β1 (clone 9F10, BD Biosciences, San Jose, CA) .

2. PE-conjugated anti-α5β1 (clone IIA1, BD Biosciences).

3. PE-conjugated anti-β2 (clone CLB-LFA 1/1, eBiosciences).

4. Mouse antihuman αvβ3 (FITC-conjugated LM609, Chemicon International, Temecula, CA).

5. Mouse antihuman αvβ5 (PE-conjugated P1F6) (Chemicon International).

6. PE-conjugated rat antimouse α5β1 (5H10-27) (BD Biosciences).

7. PE-conjugated rat antimouse αv (RMV-7) (BD Biosciences).

8. PE-conjugated rat antimouse β2 (M18/2) from (BD Biosciences).

9. PE-conjugated antimouse α4β1 (R1–2) (BD Biosciences).

10. Anti-VCAM (clone P8B1) (Chemicon International).

11. Antihuman β1 (clone P4C10) was from Chemicon International.

12. Rat antihuman β7 (clone FIB504) (BD Biosciences).

13. Alexa 488-conjugated secondary antibodies from Invitrogen (Carlsbad, CA).

14. PBS containing 1% BSA (radioimmunoassay grade).

15. FcR blocking reagent.

2.4. Endothelial Progenitor Cell Culture

1. Plasma fibronectin (BD Biosciences).

2. EPC growth medium: EBM-2 containing 5% fetal calf serum (FCS), human vascular endothelial growth factor 1 (VEGF1), human fibroblast growth factor 2 (FGF2), human epidermal growth factor, insulin-like growth factor, gentamicin, and ascorbic acid but no prednisone (Cambrex, East Rutherford, NJ).

3. 6-well plastic culture plates.

4. PBS, Ca^{2+} and Mg^{2+} free, with 0.5 mM EDTA (PBS-E).

5. Glass chamber slides.

6. Gelatin (Sigma-Aldrich).

7. DiI acetylated-LDL (Biomedical Technologies, Stoughton, MA).

8. FITC-UEA-1 lectin (Vector Laboratories, Burlingame, CA) for human cells.

9. FITC-*Bandeira simplicifolia* lectin (Vector Laboratories) for rat, mice, rabbit.

10. 1% paraformaldehyde.

11. Vectashield (Vector Laboratories).

2.5. Endothelial Cell Culture

1. 7.5% sodium bicarconate (e.g., 9330, Irvine Scientific, Irvine, CA).

2. 1M HEPES buffer (e.g., 9319, Irvine Scientific).

3. Fetal bovine serum.

4. M199 powdered medium (10 × 1 L, Gibco brand 31100-035, Invitrogen).

5. To prepare M199 basal medium: Mix the contents of one M199 powder package in 950 mL sterile, endotoxin-free water. Add 16.7 mL 7.5% sodium bicarbonate and 10 mL 1M HEPES buffer. Mix, adjust to pH 7.4 if necessary, and sterile filter using 0.45-μm sterile filter units.

6. Endothelial cell growth supplement (ECGS) (Upstate Biotechnology 02-102) stock: Add 5 mL basal medium to one 15-mg vial of ECGS.

7. Heparin (500,000 units; Sigma-Aldrich): For 100X heparin, pour powdered heparin into a 50-mL conical tube. Rinse the heparin bottle with 5 mL basal medium and add 27.7 mL basal medium. Mix and store stock at –20°C.

8. M199 complete endothelial cell growth medium: Using a 500-mL 0.45-µm sterile filter unit, combine and filter 390 mL basal M199 medium, 100 mL fetal bovine serum, 5 mL 100X heparin, and 5 mL ECGS.

9. Gentamicin (10 mg/mL).

10. Alternatively, use endothelial growth medium (EGM-2: EBM-2 containing 2% FCS, human VEGF1, human FGF2, human epidermal growth factor, insulin-like growth factor, gentamicin, ascorbic acid, prednisone) (Cambrex).

11. Trypsin/EDTA (CC-5012) and trypsin-neutralization solution (CC-5002, Cambrex).

2.6. Adhesion Assays

1. Plasma fibronectin (BD Biosciences).

2. Vitronectin (BD Biosciences).

3. Human dermal microvascular endothelial cells (Cambrex).

4. Human umbilical vein endothelial cells (HUVECs; Cambrex).

5. Costar 48-well, nontissue culture-treated plates (Costar cat. no. 3547).

6. Adhesion buffer: Hank's balanced salt solution, 10 mM HEPES, pH 7.4, 2 mM MgCl$_2$, 2 mM CaCl$_2$, 0.2 mM MnCl$_2$, 1% BSA.

7. 3% heat-denatured BSA: Heat a solution of 3 g/100 mL BSA on a stirring heating plate until it begins to turn cloudy; remove immediately and allow to cool at room temperature. Sterile filter through a 0.45-µm filter unit.

8. 3.7% paraformaldehyde in PBS: Heat 3.7 g paraformaldehyde in 50 mL H$_2$O to 60°C in fume hood. Add 1 N NaOH in drops to dissolve thoroughly; let cool, then add 50 mL 2X PBS; bring to pH 7.4.

9. Antihuman α5β1 (JBS5, Chemicon International).

10. Antimurine α5β1 (5H10-27, BD Bioscience).

11. Antihuman αvβ5 (P1F6, Chemicon International).

12. Antimurine αv (RMV-7, BD Bioscience).

13. Antihuman αvβ3 (LM609, Chemicon International).

14. Antihuman α4β1 (HP2/1, Chemicon).

15. Antimurine α4β1 (PS/2, BD Biosciences).

16. Antihuman β1 integrin (P4C10, Chemicon International).

17. Antihuman β2 integrin (BD Biosciences).

18. Antimurine β2 (M18/2, BD Biosciences).

19. Recombinant soluble VCAM (rsVCAM, R&D Systems, Minneapolis, MN).

20. 5-and-6-4-chloromethylbenzoylamino-tetramethyl-rhodamine (CMTMR; Invitrogen). Prepare stock solution of 10 mM CMTMR in dimethyl sulfoxide (DMSO).

2.7. Bone Marrow Transplant

1. FVB/N-Tie2LacZ, FVB/N-Tie2GFP, or C57Bl/6 ActEGFP donor mice (Jackson Lab, Bar Harbor, ME); FVB/N and C57Bl6 recipient mice.

2. Sulfamethoxazole and trimethoprim antibiotics.

3. Prepare MAC buffer: PBS without $CaCl_2$ or $MgCl_2$ containing 2 mM EDTA and 0.5% BSA (ice cold).

4. Two 5-mL syringes filled with 25-gage needle for each mouse.

5. Prewarmed Histopaque 1083 and ammonium chloride red blood cell lysing solution (155 mM NH_4Cl, 10 mM $NaHCO_3$, and 0.1 mM EDTA).

2.8. Tumor and Matrigel Studies

1. Lewis lung carcinoma and other tumor cell lines.

2. Growth factor-reduced Matrigel (BD Biosciences).

3. Recombinant human bFGF or VEGF.

4. Rat antimouse α4β1 (low-endotoxin PS/2, BD Biosciences).

5. Rat antimouse β2 integrin (low-endotoxin M1/70, BD Biosciences).

6. Rat anti-murine CD31 (MEC13.3, BD Bioscience).

7. Low-endotoxin anti-αvβ3 (LM609) or antihuman α4β1 (HP2/1).

8. Purified CD34$^+$ cells.

9. EGFP$^+$Lin$^-$ or EGFP$^+$Lin$^+$ cells.

10. FITC-*Bandiera simplicifolia* lectin.

11. Low-endotoxin immunoglobulin G_{2b} (IgG2b) (BD Biosciences).

12. Low-endotoxin anti-α4β1 (PS/2, BD Biosciences).

13. Low-endotoxin anti-VCAM (MK-1, BD Biosciences).

14. rsVCAM (R&D Systems).

2.9. Immunohisto-chemistry

1. Radioimmunoassay-grade BSA (Sigma-Aldrich).

2. Alexa 488 (green)-conjugated secondary antibody (Invitrogen).

3. Alexa 568 (red)-conjugated secondary antibody (Invitrogen).

4. Various antibodies, such as antifibronectin (clone TV-1) from Chemicon International; rat antimouse VCAM (M/K-2, Chemicon International); antipan species VCAM (H-276, product number sc-8304, Santa Cruz Biotechnology); anti-F4/80 (AbD Serotec, Raleigh, NC).

5. Rat antimouse CD31 (MEC 13.3, BD Biosciences).

6. Mouse antihuman Ki67 (Chemicon International, Temecula, CA).

7. Tissue Tek OCT (Electron Microscopy Sciences, Hatfield, PA).

8. Mounting medium (e.g., Vectashield, Vector Laboratories).

9. Phosphate-buffered saline.

10. Glass coverslips.

11. Glass slides.

3. Methods

3.1. Progenitor Cell Isolation

1. Put 15 mL room-temperature Histopaque 1077 into 50-mL tubes (or 5 mL Histopaque1083 in 15-mL tubes for rodent cells) for density gradient centrifugation.

2. Dilute freshly isolated blood in 2–4 volumes PBS-E (or dilute buffy coat from blood bank at same proportions in PBS-E).

3. Carefully layer 35 mL diluted blood suspension on top of Histopaque layer (*see* **Note 1**).

4. Centrifuge immediately at 400*g* for 25 min at room temperature without braking at centrifuge (*see* **Note 2**).

5. After centrifugation, carefully aspirate about half of the upper layer and discard. The opaque interface contains the MNCs.

6. Prepare clean centrifuge tubes (conical 50 or 15 mL) containing 10 mL PBS-E for the collection of the cells.

7. Carefully remove the mononuclear layer with a 10-cc syringe and 18-gage needle. Transfer cells to tubes containing PBS-E, pooling MNCs.

8. Centrifuge at 300*g* for 10 min at 20°C to pellet and to remove Histopaque.

9. Aspirate supernatant, resuspend cells in a small amount of PBS-E (5–10 mL), and pool cells from each two or three tubes into one 50-mL tube. Bring volume up to 50 mL with

PBS-E. Mix gently with pipet, and centrifuge again at 200g for 10 min at 20°C to remove platelets.

10. Aspirate supernatant and resuspend each pellet, pooling all cells in total volume of 5 mL PBS-E.

11. Add 15–20 mL of ammonium chloride red blood cell lysis solution and incubate on ice for 5–10 min. In case of large amounts of red blood cells, add 20 mL ammonium chloride and keep in ice for 10 min. In case of animal blood, use a ratio of 4:1 ammonium chloride to PBS-E in a 15-mL tube.

12. Add PBS-E to 50 mL (human cells) or 15 mL tubes (murine cells) and centrifuge at 1200 rpm for 5 min at 5°C to remove residual ammonium chloride.

13. Aspirate supernatant, resuspend in 5 mL PBS-E (or in 1 mL for murine cells) and determine number of cells/milliliter using a hemocytometer (*see* **Note 3**).

3.2. Progenitor Cell Purification

3.2.1. Human Progenitor Cell Purification

1. Add 100 µL FcR blocking reagent per 10^8 isolated human MNCs to prevent unspecific or Fc-receptor-mediated binding of magnetic beads to nontarget cells.

2. Label cells by adding 100 µL CD34 MicroBeads per 10^8 total cells, mix well, and incubate for 30 min in the refrigerator at 6–12°C.

3. To assess the purity of the isolated human progenitor cells, add 10 µL FITC-conjugated anti-CD34 antibody recognizing a separate epitope from the QBEND/10 (e.g., CD34-PE, Clone: AC136) and 10 µL PE-conjugated anti-CD133 and incubate for a further 10 min in the refrigerator at 6–12°C.

4. Wash cells carefully in wash buffer and resuspend in 1–10 mL of buffer.

5. Pass cells through 30-µm nylon mesh or preseparation filter (Miltenyi Biotech) to remove clumps. Collect flowthrough cells (*see* **Note 4**).

6. Place the LS$^+$/VS$^+$ column on the magnetic separator; rinse with buffer. Apply cells to the LS$^+$/VS$^+$ column on the separator; wash the column three times with 3 mL MACs buffer. CD34$^+$ cells bound to the magnetic beads will be retained on the column. The flowthrough will contain CD34$^-$ cells.

7. Remove the LS$^+$/VS$^+$ column from the separator, place column on a tube, and pipet 5 mL buffer on top of column to elute bound CD34$^+$ cells.

8. Firmly flush out retained cells with pressure using the plunger supplied with the column.

9. Repeat magnetic separation twice by applying the eluted cells to a new prefilled positive selection column, wash, and elute retained cells in 2.5 mL buffer.

10. Perform FACs analysis on a sample of the purified human progenitor cells to assess purity (*see* **Note 5**).

3.2.2. Murine Progenitor Cell Isolation

1. Add 100 µL FcR blocking reagent per 10^8 isolated murine MNCs to prevent unspecific or Fc-receptor-mediated binding of magnetic beads to nontarget cells.

2. Centrifuge cells at 300g. Wash one time. Resuspend the pellet in 400 µL of MACs buffer. For depletion of Lin$^+$ cells, cells will be magnetically labeled with a cocktail of biotinylated antibodies against a panel of so-called lineage antigens (CD5, CD45R [B220], CD11b, Gr-1 (Ly-6G/C), 7-4, and Ter-119). Label cells by adding 100 µL Lin antibodies per 10^8 total cells in 400 µL, mix well, and incubate for 30 min in the refrigerator at 6–12°C.

3. Wash cells carefully and resuspend in 1–10 mL of buffer. Add antibiotin MicroBeads and incubate 30 min in the refrigerator.

4. Pass cells through 30-µm nylon mesh or preseparation filter (Miltenyi Biotech) to remove clumps. Collect flowthrough cells (*see* **Note 4**).

5. Apply cells to the LS$^+$/VS$^+$ column mounted on the magnetic separator; wash the column three times with 3 mL buffer, collecting the flowthrough. These are the Lin$^-$ progenitor cells.

6. Repeat magnetic separation twice by applying the eluted cells to a new prefilled negative selection column, collecting flowthrough and washes each time.

7. To assess the purity of the murine cells, incubate aliquots of 50,000–100,000 purified murine cells with FcR block, then add 10 µL antibiotin-FITC (Miltenyi Biotech), anti-c-kit-FITC or anti-Sca-1-FITC antibodies. Incubate 20 min on ice, wash twice with PBS, and analyze by flow cytometry. Progenitor cells will be Lin$^-$ and Sca1$^+$ (*see* **Note 5**).

3.3. FACs Analysis

1. To evaluate the expression of integrins on EPCs, aliquot 5 × 10^5 cells into polypropylene snap-cap tubes, one tube per antibody to be tested. Prepare also one tube for a positive control and one tube for a negative control. Centrifuge cells at 200g and remove supernatant. Resuspend in PBS plus BSA.

2. Incubate cells in FcR Blocking Reagent.

3. Add 100 µL of antibody per tube (e.g., a 1/30 dilution of a hybridoma supernatant or a 1/100 dilution of a rabbit polyclonal, prepared in tissue culture medium containing 1% BSA). Incubate 20–60 min on ice.

4. Wash cells twice by centrifugation with medium containing 1% BSA.

5. For antibodies that are not directly labeled, add 100 µL of secondary antibody (FITC conjugated or rhodamine conjugated) per tube of cells (e.g., 0.25 mg/mL of rabbit antimouse, goat antirat, or goat antirabbit IgG in tissue culture medium). Incubate for 20–60 min on ice, in the dark).

6. Wash cells twice by centrifugation with cold PBS containing 1% BSA. Fix cells with 200 µL of cold 0.5% paraformaldehyde in PBS. Store in the dark at 4°C until analysis.

3.4. Progenitor Cell Culture

1. Under sterile conditions, coat 6-well tissue culture plates with 10 µg/mL fibronectin.

2. Plate 10×10^6 mononuclear or CD34+ cells per 6-well plastic plate in 2 mL culture medium. Do not remove media until d 4 (*see* **Note 6**).

3. Wash gently with PBS, Ca^{2+} and Mg^{2+} free.

4. Add PBS containing 1 mM EDTA (PBS-E) and incubate 10–15 min at 37°C.

5. Collect the detached cells by repeated flushing of plate with PBS-E (*see* **Note 7**).

6. Centrifuge cells at 200g for 5 min at 5°C.

7. Resuspend cells in EPC growth medium and replate on fresh fibronectin-coated plastic dishes at 1×10^6 cells/6-well dish.

8. On d 7, change media.

9. To assess the number of endothelial-like cells in these cultures, plate 2×10^6 MNCs/well of a glass chamber slide coated with fibronectin plus 0.5% gelatin or vitronectin plus 0.5% gelatin.

10. On d 4–14, add 5 µL of an undiluted solution of DiI-Ac-LDL to one well (500 µL medium) and incubate for 1 h at 37°C.

11. Remove media containing DiI-Ac-LDL from the culture

12. Wash cells carefully with PBS.

13. Fix with 1% paraformaldehyde at 4°C (500 µL each).

14. Wash once in PBS.

15. Dilute Ulex lectin to 30 µL/300 µL in PBS and add 300 µL to each well for 1 h at room temperature in the dark. For rodents, use BS1-lectin at 5 µL/500 µL PBS.

16. Gently wash the slide with PBS two or three times.

17. Apply 1 drop of Vectashield mounting medium on the well of the slide and cover with coverslip.

18. Perform fluorescence microscopy, counting double-stained cells under a fluorescence microscope.

3.5. Endothelial Cell Culture

1. Thaw one vial of endothelial cells and place into warmed medium immediately at a density of 5×10^3 cells/cm^2. Use 1 mL medium/cm^2 until cells are more than 40% confluent, then use 2 mL/cm^2.

2. Change medium every other day.

3. Split cells 1:4 when they are 60–80% confluent using 0.025% trypsin, 0.01% EDTA (this is 10X lower than usual concentrations) (*see* **Note 8**).

4. Freeze several vials after first passage. Cells are useful for up to six passages before they begin to senesce.

3.6. Adhesion Assays

3.6.1. Adhesion of EPC to Endothelial Cells

1. To prepare HUVEC monolayers: seed 200,000 HUVECs in EGM- per well of a 48-well plate 24 h before adhesion assay.

2. To prepare EPC or BMDC population, pellet purified cells by centrifugation and resuspend in warmed adhesion medium containing 25 µ*M* CMTMR. Incubate at 37°C for 60 min. Wash the cells twice by centrifugation and resuspend in prewarmed adhesion medium. Rinse and resuspend cells at 800,000 cells/mL in adhesion buffer.

3. Remove supernatant from HUVEC monolayers, using only a Pipetman to aspirate (*see* **Note 9**).

4. Add 200,000 EPCs or other bone marrow-derived cells in 0.25 mL per well of HUVEC monolayer.

5. Incubate in 37°C incubator 30 min to 2 h.

6. Aspirate supernatant with Pipetman (*see* **Note 10**).

7. Wash three times with warm adhesion buffer: Add warm adhesion buffer to wells and remove gently with Pipetman.

8. Remove all medium with Pipetman (*see* **Note 11**).

9. Wash once with PBS.

10. Fix with 3.7% paraformaldehyde.

11. Store at 4°C until viewing on microscope.

3.6.2. Inhibition of Adhesion of EPCs to Endothelial Cells

1. Prepare HUVEC monolayers: Seed 200,000 HUVECs per well of a 48-well plate 24 h before adhesion assay.

2. To prepare EPC or BMDC population, pellet purified cells by centrifugation and resuspend in warmed adhesion medium containing 25 µ*M* CMTMR. Incubate at 37°C for 60 min. Wash the cells twice by centrifugation and resuspend in prewarmed adhesion medium. Rinse and resuspend cells at 800,000 cells/mL in adhesion buffer.

3. Prepare 1-mL aliquots of murine cells. Add 25 µg rsVCAM, 25 µg antimurine α4β1, 25 µg antimurine α5β1, 25 µg antimurine αv, or 25 µg antimurine β2 integrin antibody cells. Alternatively, prepare similar concentrations of antihuman integrin antibodies when using human CD34+ cells.

4. Remove supernatant from HUVEC monolayers using only Pipetman to aspirate (*see* **Note 11**).

5. Place 0.25 mL progenitor cells (200,000) per well (six wells per antibody). Incubate in 37°C incubator. Aspirate supernatant of three wells per antibody with Pipetman after 15 min; wash once with warm adhesion buffer. Add warm adhesion buffer to wells and replace in incubator.

6. At 30 min after start of adhesion, aspirate supernatant of remaining three wells per antibody with Pipetman; wash once with warm adhesion buffer.

7. Remove all medium with Pipetman, wash once with PBS; fix with 3.7% paraformaldehyde. Store at 4°C until viewing on microscope.

3.6.3. Adhesion of EPCs to ECM

1. Coat 48-well plates with 5 µg/mL fibronectin, vitronectin, rsVCAM, or BSA. Incubate overnight at 4°C.

2. Discard coating solution and rinse with protein-free PBS (divalent cation free). Block nonspecific binding sites with heat-treated 3% BSA in PBS. Block wells with 200 µL per well at room temperature 1 h at 37°C incubator.

3. Wash once with PBS.

4. Resuspend EPCs in adhesion buffer, 800,000 cells per milliliter. Label cells as 3.6.2.3.

5. Prepare 5-mL aliquots of cells. Add 25 µg/mL antimurine α4β1, 25 µg/mL antimurine α5β1, 25 µg/mL antimurine αv, 25 µg/mL antimurine β2 integrin antibodies to aliquots of cells.

6. Aliquot triplicates per ECM plate per each antibody condition. Add three 250-µL aliquots of anti-α4β1, anti-α5β1, anti-αv, anti-β2, rsVCAM, and no antibody per ECM protein. Incubate 15–30 min.

7. Aspirate supernatant of remaining three wells per antibody with Pipetman; wash once with warm adhesion buffer.

8. Remove all medium with Pipetman, wash once with PBS; fix with 3.7% paraformaldehyde. Store at 4°C until viewing on microscope.

3.7. Bone Marrow Transplant

1. Treat recipient mice with a suspension of sulfamethoxazole and trimethoprim (SMZ) for 1 wk prior to irradiation and 2 wk after bone marrow transplantation.

2. Irradiate female mice of FVB/N or C57Bl6 genotype (4–6 wk) with a single dose of 1000 rad (10 Gy). Bone marrow should be transplanted within the same day.

3. Euthanize mice. Sterilize with ethanol.

4. Asceptically remove femurs, using separate sterile surgical tools for cutting soft tissues and bones. Remove muscle tissue from femurs. Cut off ends of each bone and flush with MAC buffer (2 mL/femur and 1 mL/tibia) into plastic centrifuge tubes on ice. Triturate to break up clumps and form single cell suspension.

5. Centrifuge at 400g for 10 min at room temperature.

6. Resuspend in 10 mL ammonium chloride buffer and incubate for 4–5 min at room temperature.

7. Transfer to 50-mL centrifuge tube and stop reaction by adding 30 mL PBS without $MgCl_2$ or $CaCl_2$.

8. Centrifuge at 400g for 10 min at room temperature.

9. Resuspend in 4 mL MAC buffer and transfer to 15-mL Falcon tube. Add 4 mL of RT Histopaque 1083 to the bottom of the tube. Centrifuge at 400g for 25 min at room temperature without brakes.

10. Aspirate surface layer, then transfer cloudy layer containing the purified bone marrow cells into a new 15-mL centrifuge tube. Wash twice with PBS by resuspending in PBS and centrifuging at 400g for 10 min at 4°C.

11. Add ice-cold PBS without $MgCl_2$ or $CaCl_2$.

12. Resuspend at 2.5×10^7 cells/mL in PBS (store on ice until ready for injection into recipient mice). Warm bone marrow cells to room temperature before injecting into recipient mice.

13. Inject room temperature bone marrow cells into irradiated recipient mice by tail vein injection (5×10^6 cells in 200 μL/animal).

14. Monitor animals for 4 wk after bone marrow transplantation; animals will be fully recovered and ready for *in vivo* angiogenesis studies.

3.8. Tumor Homing Studies

1. Label human CD34$^+$ cells with CMTMR: Pellet purified cells by centrifugation and resuspend in warmed culture medium containing 25 μM CMTMR. Incubate at 37°C for 60 min. Wash the cells twice by centrifugation and resuspend in prewarmed culture medium. Rinse and resuspend cells at 800,000 cells/mL in culture medium.

2. To test the roles of various integrins in progenitor cell homing to tumor neovasculature, incubate 0.5×10^6 CMTMR-labeled

CD34$^+$ human progenitor cells, CMTMR-labeled cultured EPCs, in EBM basal culture medium, medium with 50 µg/mL of low-endotoxin anti-αvβ3 (LM609), or antihuman α4β1 (either HP2/1 from Biogen or 9F10 from BD Bioscience) on ice for 30 min.

3. Inject cells by tail vein injection in a final volume of 100 µL into nude mice bearing 0.5-cm LLC or other tumors. After 1 h, sacrifice animals. Excise tumors.

4. Alternatively, inject 1×10^6 purified EGFP$^+$Lin$^-$ or Lin$^+$ cells into C57Bl6 mice bearing 0.5-cm LLC tumors.

5. Treat animals injected with EGFP$^+$Lin$^-$ cells for 5 d by intravenous administration of 100 µL saline, 200 µg low-endotoxin anti-αMβ2 (Pharmingen) or 200 µg low-endotoxin anti-α4β1 (PS/2, Biogen) every other day.

6. After 5 d, sacrifice animals and excise tumors plus surrounding connective tissue. Embed in OCT to cryopreserve. Section into 10-µm thick cryosections.

7. Determine microvascular density by immunostaining to detect CD31 and CMTMR or EGFP. Count CMTMR$^+$ CD31$^+$ or EGFP$^+$ CD31$^+$ vessels per ×200 microscopic field in 10 randomly selected fields per tumor. Calculate average microvessel density in all tumors per treatment group.

3.9. Immunohisto-chemistry

1. Remove frozen section on glass microscope slides from −80°C freezer.

2. Dry the glass portion of the slide surrounding the sections with a Kimwipe and draw a circle around the section with the hydrophobic pen (PAP-Pen).

3. Fix slides for 30 s in cold acetone (place in slide-washing container filled with acetone for 30 s, then remove).

4. Wash twice for 5 min each in PBS. After this step, never let the section dry out.

5. Block the sections for 2 h with 8% normal goat serum or 2.5% BSA in a humidified chamber created in a Tupperware container. To create a humidified chamber, place a damp paper towel in the bottom of a Tupperware or other container and then place a pipet tip container lid on top of the paper towel. Place the slide flat on the lid and place a drop (100 µL) of block buffer on each encircled section. Replace the Tupperware lid and incubate at room temperature for 2 h.

6. Rinse the slides twice in PBS by dipping five times into container filled with PBS.

7. Dry the glass portions of the slide and remove excess liquid from the section by touching the corner of a Kimwipe to the edge of the section.

8. Apply 100 µL of the primary antibody dilution, typically at 1–10 µg/mL in PBS-containing block buffer, to the encircled section on the left of the slide. Apply block buffer to the section on the right of the slide (negative control). Incubate for 2 h at room temperature.

9. Wash the slides by dipping in PBS-filled containers five times for 10 dips, then ending with one 5-min PBS wash. Dry the slide and wick the excess liquid from the section as described 3.9.7.

10. Incubate in secondary antibody dilution (1/250–1/1000) in 2.5% BSA in PBS by applying 100 µL to the encircled section. Incubate 1 h at room temperature. If the primary antibody is directly labeled, skip this step.

11. Wash the slide by dipping five times for 10 dips in PBS, then ending with one 5-min wash.

12. Dry the slide and wick the excess liquid from the section. Apply 20 µL of Vectashield (or other mounting medium) and place a coverslip gently on the section.

13. Place a drop of nail polish on the corners of the coverslip to fix it in place. When dry, store slide in refrigerator in a flat slide holder or in a slide box.

4. Notes

1. This is best accomplished using a 25-mL pipet for human cells.

2. Centrifugation at low temperatures may result in cell clumping and lower recovery.

3. Using this method, 3×10^8 EGFP$^+$ bone marrow-derived MNCs were purified by gradient centrifugation from the femurs and tibias of six ACTB-EGFP mice.

4. Wet filter with buffer before use.

5. Using this method, 3×10^6 CD34$^+$ human cells were purified from 7×10^9 peripheral blood MNCs at 98% purity; 4.8×10^6 murine Lin$^-$ cells were isolated from bone marrow-derived MNCs by two rounds of negative selection for a panel of lineage markers (CD5, B220, CD11b, Gr-1, 7-4, and Ter119; Miltenyi Biotech). The selected population was 90% positive for lineage markers and 80% positive for c-kit expression. Lin$^+$ cells were 100% positive for lineage markers. The 0.4×10^6 Lin-Sca 1$^+$ cells (70% Sca1$^+$) were isolated from Lin$^-$ cells by three rounds of immune selection for the stem cell marker Sca1$^+$ (Miltenyi Biotech).

6. "EPCs" become spindle shaped; remove floating cells. It is important to note that several investigators have found that these "EPCs" express monocyte markers in addition to endothelial markers; thus, the exact nature of these *in vitro* differentiated cells remains to be established.

7. Add trypsin at 500 µL per well and incubate 2 min at 37°C only if necessary to detach the remaining cells. Avoid trypsin when FACs analysis will be performed.

8. Do not allow cells to become confluent as they are sensitive to contact inhibition.

9. Do not use vacuum aspiration, which may disturb the monolayer.

10. Do not use vacuum aspiration as it will disturb the monolayer.

11. Do not use vacuum aspiration.

References

1. Schmid, M. C., Varner, J. A. (2006) Myeloid cell trafficking and tumor angiogenesis. *Cancer Lett* Epub Oct 16.

2. Asahara, T., Murohara, T., Sullivan, A., et al. (1997) Isolation of putative progenitor endothelial cells for angiogenesis. *Science* 275, 964–967.

3. Peichev, M., Naiyer, A. J., Pereira, D., et al. (2000) Expression of VEGFR-2 and AC133 by circulating human CD34(+) cells identifies a population of functional endothelial precursors. *Blood* 95, 952–958.

4. Gill, M., Dias, S., Hattori, K., et al. (2001) Vascular trauma induces rapid but transient mobilization of VEGFR2(+)AC133(+) endothelial precursor cells. *Circ Res* 88, 167–174.

5. Rehman, J., Li, J., Orschell, C. M., March, K. L. (2003) Peripheral blood "endothelial progenitor cells" are derived from monocyte/macrophages and secrete angiogenic growth factors. *Circulation* 107, 1164–1169.

6. Gulati, R., Jevremovic, D., Peterson, T. E., et al. (2003) Diverse origin and function of cells with endothelial phenotype obtained from adult human blood. *Circ Res* 93, 1023–1025.

7. Chavakis, E., Aicher, A., Heeschen, C., et al. (2005) Role of beta2-integrins for homing and neovascularization capacity of endothelial progenitor cells. *J Exp Med* 201, 63–72.

8. Urbich, C., Heeschen, C., Aicher, A., et al. (2005). Cathepsin L is required for endothelial progenitor cell-induced neovascularization. *Nat Med* 11, 206–213.

9. Hildbrand, P., Cirulli, V., Prinsen, R. C., et al. (2004). The role of angiopoietins in the development of endothelial cells from cord blood CD34+ progenitors. *Blood* 104, 2010–2019.

10. Schmeisser, A., Garlichs, C. D., Zhang, H., et al. (2001) Monocytes coexpress endothelial and macrophagocytic lineage markers and form cord-like structures in Matrigel under angiogenic conditions. *Cardiovasc Res* 49, 671–680.

11. Yang, L., DeBusk, L. M., Fukuda, K., et al. (2004) Expansion of myeloid immune suppressor Gr+CD11b+ cells in tumor-bearing host directly promotes tumor angiogenesis. *Cancer Cell* 6, 409–421.

12. De Palma, M., Venneri, M. A., Roca, C., Naldini, L. (2003) Targeting exogenous genes to tumor angiogenesis by transplantation of genetically modified hematopoietic stem cells. *Nat Med* 9, 789–795.

13. Garmy-Susini, B., Varner, J. (2005) Circulating endothelial progenitor cells. *Br J Cancer* 93, 855–858.

14. Jin, H., Su, J., Garmy-Susini, B., Kleeman, J., Varner, J. (2006) Integrin alpha4beta1 promotes monocyte trafficking and angiogenesis in tumors. *Cancer Res* 66, 2146–2152.

15. Jin, H., Aiyer, A., Su, J., Borgstrom, P., Stupack, D., Friedlander, M., Varner, J. (2006) A homing mechanism for bone marrow-derived progenitor cell recruitment to the neovasculature. *J Clin Invest* 116, 652–662.

Section IV

In Vitro Techniques

Chapter 9

In Vitro Assays for Endothelial Cell Functions Related to Angiogenesis: Proliferation, Motility, Tubular Differentiation, and Proteolysis

Suzanne A. Eccles, William Court, Lisa Patterson, and Sharon Sanderson

Abstract

This chapter covers the breakdown of the process of angiogenesis into simple assays to measure discrete endothelial cell functions. The techniques described are suitable for studying stimulators or inhibitors of angiogenesis and determining which aspect of the process is modulated. The procedures outlined are robust and straightforward but cannot cover the complexity of the angiogenic process as a whole, incorporating as it does myriad positive and negative signals, three-dimensional interactions with host tissues and many accessory cells, including fibroblasts, macrophages, pericytes, and platelets. The extent to which *in vitro* assays predict responses *in vivo* (e.g., wound healing, tumor angiogenesis, or surrogate techniques such as Matrigel plugs, sponge implants, corneal assays, etc.) remains to be determined.

Key words: Chemotaxis, endothelial metalloproteinases, haptotaxis, matrix, migration, tube formation cells, zymography.

1. Introduction

Since Folkman's seminal work in the early 1970s *(1)* and the later identification of key angiogenic (and lymphangiogenic) mediators, there has been an explosion of interest in all aspects of neoangiogenesis *(2–10)*. In particular, it is recognised that successful inhibition of tumor angiogenesis (on which sustained malignant growth and spread is critically dependent) could have significant therapeutic benefits *(11–25)*. Such strategies should have wide application in most solid tumor types and may be less susceptible to the development of resistance *(26, 27)*. Other conditions that

S. Martin and C. Murray (eds.), *Methods in Molecular Biology, Angiogenesis Protocols, Second edition, Vol. 467*
© Humana Press, a part of Springer Science+ Business Media, LLC 2009
DOI: 10.1007/978-1-59745-241-0_9

may also be treatable by antiangiogenic agents include diabetic retinopathy, arthritis, and psoriasis *(28–34)*. On the other hand, the ability to induce angiogenesis in a controlled fashion may assist wound healing following injury or surgery *(35)* or ameliorate certain cardiovascular pathologies *(36, 37)*

To develop effective angiogenesis modulators, it is essential to gain a deeper understanding of the complex gene expression changes and molecular mechanisms underlying the process. We also need to develop informative *in vitro* (and *in vivo*) assays for hypothesis testing and screening of potential therapeutic agents *(38–44)*. Endothelial cells (ECs) in the adult are normally quiescent unless activated by angiogenic factors such as vascular endothelial growth factors (VEGFs), basic fibroblast growth factor (bFGF), hepatocyte growth factor (HGF), cytokines including interleukin 8 (IL-8), and so on *(3, 8, 9, 12, 45–50)*. EC activation results in enhanced proliferation, migration and "invasion" of surrounding tissues. This is mediated by upregulation of specific integrins (such as αvβ3 and αvβ5) *(51–54)* and secretion of matrix metalloproteinases (MMPs) *(55–57)* and urokinase plasminogen activator (uPA) *(58, 59)*. The capillary buds then organise into tubules and differentiate into new vessels, finally acquiring pericyte coverage and growth factor independence as they mature *(60)*.

Angiogenesis can potentially be inhibited at many levels, such as suppression of production (or sequestration) of angiogenic factors, interference with their binding to EC receptors such as VEGFR2, inhibition of integrin binding or targeting downstream signalling pathways *(51, 61–66)*.

Depending on the point of intervention in the angiogenic process, different assays may be required *(38, 67, 68)*, but (following biochemical assays to establish target inhibition) the simplest and most informative of them aim to assess the ability of EC to perform one or more of the following key functions: proliferation in response to an angiogenic stimulus (e.g., VEGF, bFGF); chemotaxis (directional motility) in response to an attractant gradient; haptotaxis (integrin-mediated non-directional motility) on matrix proteins; proteolytic enzyme production (e.g., gelatinase MMPs by gelatin zymography); tubularisation (differentiation) on matrix proteins.

Assays to measure each of these functions are described in turn.

2. Materials

2.1. Cell Culture and Maintenance

1. Normal human umbilical vein endothelial cells (HUVECs) (pooled) (ZHC-2101) (TCS Cellworks, Buckingham (http://www.tcscellworkscatalogue.co.uk/) (*see* **Notes 1** and **2**).

2. Large-vessel endothelial cell growth medium package, which contains TCS large-vessel endothelial cell basal medium and large-vessel endothelial cell growth supplement (ZHM-2953).

3. Phosphate buffered saline (PBS) containing 1 mm ethylenedi aminetetraacetic acid (EDTA).

4. Trypsin/EDTA solution, 0.025%/1 m*M* in PBS. Alternatively, TrypLE™ Express trypsin for cell detachment (Invitrogen, Paisley, UK) (*see* **Note 3**).

5. Trypan blue 0.4% (w/v) in PBS.

6. Tissue culture flasks.

7. Disposable pipets.

8. Haemocytometer.

9. Humidified incubator set at 37°C with an atmosphere of 5% CO_2 in air.

2.2. Cell Proliferation Assay

1. 96-well tissue culture plates.

2. Multichannel pipet.

3. Alkaline phosphatase substrate containing buffered *p*-nitrophenylphosphate (pNPP) (Sigma Aldrich UK). Protect from light and keep at 4°C.

4. Solution of 1*M* sodium hydroxide (NaOH).

5. Solution of trypsin/EDTA or TrypLE Express trypsin.

6. Plate reader or spectrophotometer capable of reading absorption at 405 nm.

2.3. Chemotaxis Assays (HUVECs)

1. HUVEC pooled donors (TCS CellWorks Buckingham UK) (cat. no. ZHC-2101).

2. Large-vessel endothelial cell growth medium (cat. no. ZHM-2953).

3. Solution of trypsin/EDTA or TrypLE Express (Invitrogen).

4. Olympus Ix70 inverted fluorescent microscope fitted with both an ×20 and ×40 objective and cooled charged-coupled device (CCD) digital camera.

5. U-MWB filter set cube to detect the fluorophore (Olympus UK).

6. Image acquisition software: Image-Pro Plus (Media Cybernetics UK).

7. BD FluoroBlok™ 24-well plates (BD Biosciences) 3-μm pore (cat. no. 351156) (*see* Fig. 9.1).

8. Fibronectin matrix protein-coated 24-well 3-μm FluoroBlok plates (BD Biosciences, cat. no. 354543). Store at 4°C.

Part of fluoroblok plate   Single insert

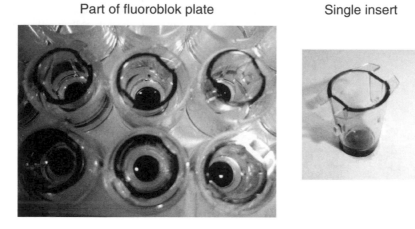

Fig. 9.1. Photograph of a FluoroBlok plate (BD Biosciences) showing one of the removable filter inserts. The blue (poly-ethylene terephthalate-coated filter) is specifically designed to absorb visible light within the 490- to 700-nm range, which prevents transmission of fluorescence from the cells on the top of the filter.

9. Recombinant human epidermal growth factor (EGF) for use as chemoattractant (Sigma Aldrich cat. no. E-9644). Make a stock of 1 mg/mL EGF by dissolving lyophilised EGF in 0.2-μm filtered 10 m*M* acetic acid containing 0.1% bovine serum albumin (BSA). Store small aliquots at –20°C.

10. CellTracker™ Green dye 5-chloromethylfluorescein diac-etate (CMFDA, Molecular Probes, Invitrogen).

2.4. Invasion Assay   BD BioCoat™ FluoroBlok™ Matrigel-coated 24-well plates (BD Biosciences cat. no. 354141).

2.5. Haptotaxis Assay
1. Endothelial cells (e.g., HUVECs pooled donors; TCS Cell-Works, Buckingham, UK).

2. Large-vessel endothelial cell growth medium (ZHM-2953).

3. Solution of trypsin/EDTA or TrypLE Express (Invitrogen).

4. Sterile 0.1- to 10-μL pipet tips for Gilson P20.

5. Inverted microscope fitted with ×20 objective and digital camera.

6. Mitomycin C (Sigma Aldrich, M0503) prepared immedi-ately before use.

7. 24-well tissue culture plates.

8. Sterile Gilson pipet tip for scraping a wound in the cell monolayer.

2.6. Tubular Differentiation	1. Matrigel™ (BD Biosciences) stored at –20°C. Place the gel on ice in a 4°C refrigerator to thaw overnight before use.
	2. 24-well tissue culture plates.
	3. Gilson pipet tips for P1000.
	4. CellTracker Green dye CMFDA (Molecular Probes).
2.7. Quantitation of Migration and Invasion (see Figure 9.2)	1. Olympus IX70 inverted fluorescence microscope fitted with an environmental chamber (Solent Scientific UK).
	2. Retiga EXi cooled CCD camera (Q Imaging), ×20 and ×40 objectives.
	3. Image-Pro Plus 5.1 (Media Cybernetics Europe).
2.8. Quantitation of Tubularisation	1. Olympus IX70 inverted fluorescence microscope fitted with an environmental chamber (Solent Scientific Limited UK).
	2. Retiga EXi cooled CCD camera (Q Imaging), ×20 and ×40 objectives.
	3. Image-Pro Plus 5.1 (Media Cybernetics Europe).
2.9. Gelatin Zymography	1. 10% Novex® Tris-glycine sodium dodecyl sulfate (SDS) gels containing either casein or collagen (Invitrogen).
	2. Tris-glycine SDS running buffer (10X). Prepare 1X Tris-glycine SDS running buffer by adding 100 mL of

Fig. 9.2. Diagram of the image analysis setup used to quantify cell migration, invasion, haptotaxis, and tubular differentiation. The example shown is the tubular differentiation. (*See Color Plates*)

10X Novex Tris-glycine SDS running buffer to 900 mL deionised water.

3. Novex Tris-glycine SDS sample buffer (2X) stored at –20°C.

4. Protein molecular weight markers stored at –20°C.

5. Matrix metalloproteinase MMP2/MMP9 standards (Chemicon Europe).

6. Novex Zymogram renaturing buffer (10X). Prepare 1X renaturing buffer by adding 100 mL of 10X Novex renaturing buffer to 900 mL of deionised water.

7. Novex Zymogram developing buffer (10X). Prepare 1X developing buffer by adding 100 mL of 10X Novex developing buffer to 900 mL of deionised water.

8. Colloidal blue staining kit (Invitrogen).

9. Methanol.

10. Distilled water.

11. XCell II Mini cell electrophoresis tank (Novex, Invitrogen).

12. Powerpack.

13. Growth factors for stimulation of cells, such as VEGF, EGF, bFGF, HGF (Sigma Aldrich).

14. Phorbol 12-myristate 13-acetate (PMA) (Sigma Aldrich).

2.10. Relative Quantitation of Matrix Metalloproteinases in Zymogram

Use a GelPro analyzer 3.1 (Media Cybernetics Europe) and a digital camera for relative quantitation of matrix metalloproteinases in zymogram.

3. Methods

3.1. Cell Culture and Maintenance

The culture of normal primary and early-passage human ECs requires specialised culture media. Careful culture techniques are necessary to ensure cell survival and maintenance of expression of key proteins. Normal primary ECs have a finite life span. In practice, this means that, depending on the cell type, they can be cultured *in vitro* for a maximum of 10–15 passages (*see* **Note 4**) before becoming senescent, at which point the cells become aetiolated and stop proliferating.

1. Thaw media supplements rapidly at 37°C and add to basal media using a sterile Gilson tip. Pre-warm the mixture to 37°C.

2. Add the requisite amount of medium to the tissue culture flask, such as 5 mL per flask for T25, 15 mL per flask for T75.

3. Thaw a vial of cryopreserved ECs rapidly at 37°C and wipe the vial with ethanol to ensure sterility. Calculate the number of cells required to seed the minimum density of cells in the appropriate tissue culture flask. This will depend on the cell line. Microvascular ECs require a minimum of 5×10^3 cells/cm^2; however, large-vessel ECs can be seeded at half this density (*see* **Note 4**).

4. Add cells to the flask and place in an incubator at 37°C set at 5% CO_2 in air. The procedure should take no more than 5–8 min to ensure maximum cell viability.

5. Change medium after 24 h to remove cryoprotectant.

6. Feed the cells with fresh medium every other day.

7. Harvest or subculture cells when they reach 60–80% confluence.

3.2. Subculture

1. Aspirate medium from the flasks.

2. Rinse the cell layer gently with 5 mL PBS/EDTA.

3. Add 1 mL trypsin/EDTA or TrypLE Express to the cell layer ensuring that the entire surface of the cell sheet monolayer is covered. Pour off the excess fluid, leaving a thin layer covering the cells, and return the flask to the incubator.

4. After about 2 min, examine the cells microscopically to monitor detachment. If the cells appear to be lifting off the surface, tap the flask gently to complete the process. If not, return and monitor at 30- to 60-s intervals until this occurs (*see* **Note 4**).

5. Add 5 mL of fresh medium and resuspend the cells carefully by aspirating the cells using a disposable pipet and transferring to a suitable size vial.

6. To determine the number of viable cells per millilitre, remove 20 µL cell suspension from the vial, dilute with 20 µL trypan blue, and count the viable (dye-excluding) cells using a haemocytometer.

7. Reseed flasks at the minimum density as above (*see* 3.1.3).

3.3. Cell Proliferation Assays

3.3.1. Screening Compounds for Growth Inhibitory Activity

Most current assays for measuring cell proliferation are based on the reduction of the yellow tetrazolium salt MTT (3-(4,5-dimethylthiazolyl-2)-2,5-diphenyltetrazolium bromide) by the action of dehydrogenase enzymes from metabolically active cells. The resulting intracellular purple formazan can be solubilised and quantified by reading the absorbance of the product on a spectrophotometer. An alternative cheaper metabolic assay is based on a ready-to-use buffered alkaline phosphatase substrate containing pNPP. Prior to use, the substrate appears as a solution that is colourless to pale yellow. After incubation with cellular-derived alkaline phosphatase, the compound turns a bright yellow, and this colour intensity can be measured on a spectrophotometer. One final alternative method to assess cell survival that may be used is the sulforhodamine-B (SRB) assay. SRB is a water-soluble dye that binds to the basic amino acids of the cellular proteins. The colorimetric measurement of the bound dye provides an estimate of the total protein mass that is related to the cell number. While this method is the suggested preferred technique for high-through-put screening, it is reportedly less sensitive for the measurement of ECs, which, compared with tumor cells, have relatively low protein levels (69).

1. Remove growth medium from a T75 flask of ECs.

2. Rinse with 5 mL of PBS/EDTA solution.

3. Add 3 mL TrypLe Express and pour off excess. When cells have detached, add 10 mL EC basal medium and resuspend cells gently until a single-cell suspension is produced.

4. Count viable cells using a haemocytometer and adjust cell concentration to 2×10^4 cells/mL.

5. Using a multichannel pipet, dispense a 100-µL volume of cells into each well of a 96-well plate with the exception of the outside wells. To the outside wells, add 100 µL of distilled water.

6. Make doubling dilutions of the agent of interest in a replicate plate before transferring to the cell plate as in the following steps (see **Note 5**).

7. Add 100 µL tissue culture medium to the wells of a 96-well plate except the outside wells and the first column.

8. Add 200 µL of the agent at its maximum concentration to each of three wells in the first empty column of wells. Repeat this for any further compounds to be tested by adding the agent to the next group of three wells.

9. Fit a multidispensing pipet with tips and set to 100 µL. Collect 100 µL of compound from the first row, add to the second row, mix, and take 100 µL on to the next row. Repeat the procedure to the penultimate row. Leave the last row as a negative control.

10. Remove the medium from the plate containing the cells, transfer the drug dilutions across, and dispense into the wells using the multipipet.

11. Incubate the plates at 37°C and 5% CO_2 for 4 d or until control wells appear just confluent.

12. Remove the medium by gently tapping the inverted plate on an absorbent paper towel.

13. Add 100 µL of paranitrophenylphosphate (3 mg/mL in 0.1*M* sodium acetate containing 0.1% Triton X-100) and incubate for 2 h at 37°C.

14. Stop the reaction by adding 50 µL 1*M* NaOH.

15. Measure absorbance at 405 nm in a spectrophotometer.

3.4. Chemotaxis

3.4.1. Cell Labelling with CellTracker Fluorescent Dye

1. Add CMFDA to cell culture flask medium containing growth medium to give 5 µ*M* final concentration of dye (*see* **Note 6**).

2. Leave cells at 37°C for a minimum of 45 min. Remove medium and wash once with PBS. Apply trypsin to cells as described (*see* **Subheading 3.2, steps 1–6** to remove cells from the flask and use in chemotaxis assay).

3.5. Screening Compounds for Effects on Migration in a FluoroBlok Transwell Assay (see Fig. 9.1)

1. Remove growth medium from a labelled T75 flask of ECs.

2. Rinse with 5 mL of PBS/EDTA solution.

3. Add 3 mL TrypLe Express and pour off excess.

4. When cells have detached, add 10 mL EC basal medium and resuspend gently until a single-cell suspension is produced.

5. Count viable cells using a haemocytometer and adjust cell concentration to 4×10^5 cells/mL.

6. Add 150 µL of the diluted cell suspension to the top chamber of the Transwell insert placed in a companion plate. If the experiment involves an investigation of single growth factor-mediated chemotaxis, then the use of a fibronectin-coated Transwell is recommended (*see* **Note 7**). Otherwise, if FCS is used as a chemoattractant, use an uncoated Fluoro Blok insert.

7. Dilute compound to be screened to double the concentration desired. Add 150 µL to the cells in the top chamber of the insert to give a final volume of 300 µL (*see* **Note 8**).

8. Add 150 µL serum-free EC basal medium (containing drug vehicle at the highest concentration that is on the compound plate) to one or more inserts, which will then serve as an untreated cell controls.

9. To the bottom chamber of each insert, add 800 µL of serum-free medium (*see* **Subheading 3.5, step 8**) containing both the chemoattractant and compound at the same concentration as the upper chamber. Include in the experiment at least one insert that contains cells in the top chamber but does not contain any chemoattractant. Unless a specific attractant is required, HUVECs will migrate well to 5% FCS.

10. Add 150 µL of the suspension of labelled cells to one well of a 24-well plate. Add 150 µl of growth medium. This is a

control to confirm that the cells are viable for the duration of the experiment and can be counted separately at the end of the migration period.

11. Incubate the plates at 37°C in a humidified atmosphere of 5% CO_2 in air for up to 6 h for FCS and up to 16 h for single chemoattractants such as VEGF.

12. View the cells that migrate in real time and photograph under ultraviolet (UV) illumination by selecting the U-MWB filter set cube and the ×10 objective on the microscope. Monitor the cells migrating onto the lower surface of the membrane every 2 h (*see* **Note 9**).

13. After terminating the assay, either count the cells immediately or fix by removing the medium in the lower chamber and replacing it with 4% paraformaldehyde in PBS.

14. Acquire images of the migrated cells using the CCD camera at ×10 magnification in at least three random representative areas under each of the membranes.

15. Count migrated cells as described in **Subheading 3.9.**

3.6. Invasion Assay

Invasion assays are performed as above, but the observation time needs to be increased as the cells also have to pass through the Matrigel barrier above the filter (e.g., allow about 18 h for HUVEC invasion assays).

Coating of the upper surface of the FluoroBlok inserts with a matrix barrier may be performed in the laboratory for the sake of economy but can be problematic. Matrigel solutions solidify rapidly at room temperature and have a tendency to form a meniscus on the membranes. This can lead to uneven invasion of cells through the gel. To avoid these problems, it may be preferable to buy the pre-coated inserts from Becton Dickinson marketed as the BD BioCoat Tumor Invasion System.

Re-hydrate the inserts as follows:

1. Remove the package from –20°C storage and allow it to come to room temperature.

2. Add 0.5 mL warm (37°C) PBS to the interior of the insert wells.

3. Allow the plate to rehydrate for 2 h at 37°C in non-CO_2 environment.

4. After rehydration, carefully remove the medium from the insert wells without disturbing the layer of BD Matrigel matrix on the membrane.

5. Pre-label the cells with CMFDA CellTracker as described in **Subheading 3.4, steps 1–2** and seed them into the upper chamber (*see* **Note 10**).

6. Proceed as described for chemotaxis assays (**Subheading 3.3**) but with a longer observation period.

7. Observe the underside of the membranes under UV illumination (*see* **Note 11**).

8. Acquire images of the migrated cells using the CCD camera at ×40 magnification in three random representative areas under each of the membranes.

9. Count migrated cells as described in **Subheading 3.9**.

3.7. Haptotaxis Assays

1. Harvest HUVECs from flasks as described in **Subheading 3.1** and plate in 24-well plates at 5×10^4 cells/well. Incubate at 37°C in a humidified incubator at 5% CO_2 in air. Allow cells to become 90% confluent.

2. Prepare cell culture medium with appropriate inhibitors and controls and keep at 37°C in a heated water bath.

3. Using the sterile tip on a Gilson pipet, carefully score the cell monolayers vertically down the centre of each well. Wash the surface once with PBS to remove detached cells.

4. Add fresh growth medium containing inhibitors or an equivalent volume of vehicle as a negative control. To ensure that wound closure is due solely to migration, inhibit cell proliferation by adding mitomycin C at 2.5 µg/mL.

5. Monitor closure every hour by acquiring duplicate images of the wound in each well under phase contrast from time 0 (to measure the initial wound width) for up to 24 h. This procedure is simplified by the use of a mechanised stage and image acquisition software (Image-Pro Plus).

6. Express results as percentage change in wound width per time point. In Image-Pro Plus, the images acquired can also be compressed and then displayed as a movie file. Running the movie of untreated cultures against the treated wells will provide qualitative illustrations of the effects that inhibitors are having on wound closure (*see* **Note 12**).

3.8. Tubular Differentiation

1. Place Matrigel on ice in an insulated container at 4°C and leave overnight to thaw out completely.

2. Place endothelial growth medium at 4°C.

3. Place any tips or plates that will be coming into contact with Matrigel at 4°C until cooled sufficiently.

4. All subsequent procedures should be performed in a tissue culture hood.

5. Mix Matrigel by gentle pipetting.

6. Add 300 µL of Matrigel to each well of a 24-well plate.

7. Place plate at 37°C for at least 30 min to gel.

8. Label ECs with 5 µ*M* CMFDA CellTracker Green dye for 1 h.

9. Remove growth medium from labelled T75 flask of HUVECs and rinse cells with 5 mL of PBS/EDTA solution.

10. Add 1 mL TrypLe Express, remove excess, and incubate briefly at 37°C to release cells.

11. When cells have detached, add 10 mL EC basal medium and resuspend gently until a single-cell suspension is produced.

12. Count cells using a haemocytometer and adjust cell concentration with fresh medium to 6×10^4 cells/mL.

13. Add 500 µL of cell suspension to each Matrigel-coated well. Include an uncoated well as a negative control.

14. Incubate plate for about 6 h at 37°C in a humidified incubator at 5% CO_2 in air (*see* **Note 13**).

15. Acquire images of tubule formation using a cooled CCD camera on an inverted fluorescence microscope using a U-MWB filter set cube and ×10 objective.

16. Use Image-Pro Plus to analyse tubule formation as described in **Subheading 3.10.**

3.9. Quantitation of Migration and Invasion

1. Open image file in Image-Pro.

2. Select Measure, Count/Size from the pull-down menu on the control bar at the top of the screen.

3. In the Count/Size window select Automatic Bright Objects, select the Options button, and ensure that the options selected read outline style: Outline, Label Style None, and Four connect.

4. Perform a count. This initial cell count will include all bright objects in the image, including the pores in the membrane. Filter these out as in the following steps.

5. Select Measurements in the Count/Size window, highlight the Area default setting, and then select the Edit range option.

6. Drag the vertical bar slowly to the right until the pores are excluded from the final count. Click the Filter Objects option.

7. The final cell count, along with any statistics, may be transferred by DDE (dynamic data exchange) to Excel as in the following steps.

8. Click View on the top toolbar and select Statistics. Select File and then DDE to Excel.

3.10. Quantitation of Tubularisation (see Fig. 9.3)

1. If the images are acquired in colour, it will be necessary to select the colour channel that contains the highest contrast of image (i.e., red, green, or blue). If the cells are stained with CMFDA CellTracker Green, then this channel is obviously green.

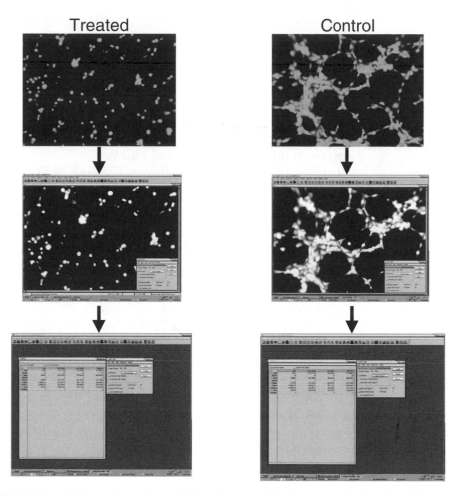

Fig. 9.3. Endothelial cell tubularisation and quantitation using Image-Pro Plus (area, branching, branch length, etc). (*See Color Plates*)

2. Select Process, Colour Channel, and Extract from the menu bar. Click the G box in Generate Channel and then OK.

3. Select Measure, Count/size, and Measure (select Measurements).

4. There is an option within this menu to select the parameters that are used to discriminate the images from each other. We find that the best option to use is Area (polygon). This choice may be cell type dependent, and optimal discrimination should be established by selecting a number of parameters and seeking the greatest ratio between test and control images.

3.11. Zymography   The zymography technique involves the electrophoresis of secreted protease enzymes through discontinuous polyacrylamide gels containing enzyme substrate (either type III gelatin or B-casein). After electrophoresis, removal of sodium dodecyl sulphate (SDS) from the gel by washing in 2.5% Triton X-100 solution allows enzymes to renature and degrade the protein substrate. Staining of the gel with a protein dye such as colloidal blue or Coomassie blue allows the proteolytic activity to be detected as clear bands of lysis against a blue background, **for example,** screening inhibitors for effects on matrix metalloproteinase production in **ECs** grown on different substrates (*see* **Fig. 9.4**).

1. Remove growth medium from a T75 flask of ECs.

2. Rinse with 5 mL of PBS/EDTA solution.

3. Add 3 mL TrypLe Express and pour off excess. When cells have detached, add 10 mL EC basal medium and resuspend cells gently until a single-cell suspension is produced.

4. Count viable cells using a haemocytometer and adjust cell concentration to 2×10^4 cells/mL.

5. Using a multichannel pipet, dispense a 100-µL volume of cells into each well of a 96-well plate with the exception of the outside wells. Repeat this procedure with plates that have been coated with extracellular matrix proteins (*see* **Note 14**). Add 100 µL of distilled water to the outside wells.

6. After 24 h, incubate the cells in serum-free medium for an additional 24 h.

7. Make dilutions of the agent of interest in a replicate plate in serum-free medium before transferring to the cell plate.

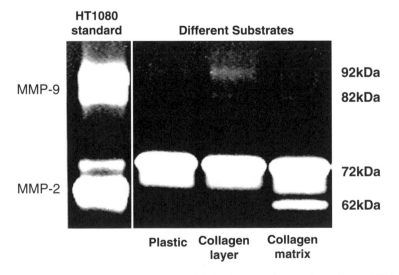

Fig. 9.4. Zymogram showing *matrix metalloproteinase* 2 (MMP2) latent and active forms (72 and 62 kDa) and MMP9 latent and active forms (92 and 82 kDa). Supernatants of PMA-stimulated HT1080 cells can be used as a standard since this contains both latent and active forms of MMP2 and MMP9 and serves in effect as a molecular weight marker. Note that endothelial cells grown on plastic often express little MMP9 and that activated forms (of both enzymes) are more readily detected on physiological substrates such as collagen matrix.

8. Remove the medium from the plate containing the cells, transfer the drug dilutions across, and dispense into the wells using the multipipet.

9. Incubate the plates at 37°C at 5% CO_2 in air for 2 h.

10. Stimulate the cells with 20 ng VEGF.

11. Remove conditioned medium from plate at 24 or 48 h.

12. Place samples on ice and mix 1:1 with 2X Tris-glycine SDS sample buffer and incubate at room temperature for 10 min.

13. Place a 10% Novex Tris-glycine SDS gel containing gelatin in an XCell II Minicell electrophoresis tank as described in the manufacturer's instructions. Fill tank with Tris-glycine SDS running buffer to recommended level.

14. Add samples to the wells in the gel. Include MMP2/MMP9 standards. Alternatively use PMA-stimulated HT1080 conditioned medium (*see* **Note 15**).

15. Connect electrophoresis equipment to powerpack and run at 125 V for 90 min.

16. Dilute 1X zymogram renaturing buffer and 10X developing buffer 1:9 with deionised water.

17. Remove gel from tank and incubate with renaturing buffer for 30 min with gentle agitation.

18. Remove buffer, replace with developing buffer, and repeat incubation with agitation.

19. Remove buffer, replace with fresh developing buffer, and incubate for at least 4 h or preferably overnight.

20. Stain with either colloidal blue stain (Invitrogen) or Coomassie blue (40% v/v methanol, 10% v/v acetic acid containing 0.5% w/v Coomassie blue for 1 h with gentle agitation) (*see* **Note 16**).

21. Destain using either distilled water for the colloidal blue stain (7 h) or destain solution (30% w/v methanol, 10% v/v acetic acid replaced every 4 h) for the Coomassie blue stained gels.

22. Place the gel on a light box and acquire an image of the clear areas of lysis using a digital camera.

23. Analyse the areas of proteolysis in Gel Pro as in the following steps.

24. Open the image file, select 1D gels and Show toolbar. Click the Lane option.

25. A window will pop up with Extract Intensity. Accept this option, and the gel image will appear as black and white.

26. Click Find Lanes and adjust the bars that appear over the lanes to reduce the area of interest to just the areas of lysis and click OK.

27. On the 1D Gel window, select Mass/IOD option and then select IOD.

28. The numbers that appear will give an absolute integrated intensity measurement of the bands. By comparing these figures, it is possible to obtain relative quantitation of the areas of lysis in different lanes.

4. Notes

1. ECs can be obtained from many different species. In the past, bovine and porcine aortic ECs were commonly used (70, 71), but human ECs are now commercially available. ECs can be obtained from large and small veins and arteries, from capillaries, or from specialized vascular areas such as the umbilical vein of newborns, blood vessels in specific organs (e.g., ovary, lung), or from solid tumors (7, 72–75). Most commonly, they are derived from human umbilical veins from pooled donors, and for the purpose of most assays this is the most reliable and economical source. However, it should be recognised that neoangiogenesis generally originates from microvascular (rather than macrovascular) ECs, and it is advisable to perform key studies in such cells, for example, commercially available human dermal microvascular cells (HDMECs) (ZHC-2226 TCS Cellworks). In addition, human lymphatic ECs are now also supplied and can be used in much the same way as vascular ECs, although requiring specific growth medium (SC-2500 TCS Cellworks). Both of these EC lines and appropriate media for their propagation are also available from Cambrex Bio Science Nottingham (http://www.cambrex.com).

2. It is of the utmost importance to maintain cell lines free of any contaminating species of mycoplasma, and this should be regularly monitored. Many methods are available to do this; it may be by direct culture or by indirect assay. Indirect methods include DNA staining, biochemical detection, nucleic acid hybridization, immunoassays, and polymerase chain reaction (PCR) DNA amplification. Indirect methods have the advantage of detecting those mycoplasma species that are not easily cultivated. Assay methods can typically detect several, if not all, of the most common cell culture contaminants (*Mycoplasma hyorhinis, M. arginini, M. orale,*

M. fermentans, M. salivarium, and *Acholesplasma laidlawii*) as well as some of the more unusual species.

3. Trypsin/EDTA treatment is the standard method for cell detachment. A prolonged exposure of cells to trypsin, however, can result in a loss of viability. We find that TrypLE Express treatment is significantly less damaging to cells and results in fewer cell clumps.

4. Subculture the cells when they reach 60–80% confluence (if the recommended seeding densities are used, this usually takes 7–10 d, depending on the doubling time). Generally, the cells will remain viable for at least 10 passages before approaching senescence. It is important to ensure that the cells are still proliferating well if they are to be used, and if there is evidence of aetiolation, the cells should be discarded. When subculturing ECs, it is preferable to avoid pelleting them by centrifugation as they are very easily damaged in the process.

5. Before setting up the assay to evaluate inhibitors of proliferation, it is always advisable to ensure that the assay will give linear measurements for either colour change technique by the construction of a standard curve. This is done by seeding the cells at a range of densities that reflect those that may occur during the course of the assay. A 96-well plate should be seeded in triplicate at the appropriate range of cell-seeding densities and allowed to adhere. The substrate should be the added and the assay run as previously described. A graph drawn of cell-seeding density against mean absorbance should appear linear.

6. One of the advantages of using fluorescent dyes such as CMFDA is that it is possible to stain different cell lines (or cells treated with different agents) with alternative dyes and monitor these mixed-cell populations in co-cultures. For example, differentially labelled ECs and tumor cells can be co-migrated to examine two-way modulating effects; cells pre-treated with drugs or small interfering RNA can be mixed with controls and separately counted and more. There are a variety of similar cell dyes available from Molecular Probes that possess differing absorption and emission maxima that correspond to red, orange, and blue; details of these compounds are available from the Web site http:// probes.invitrogen.com/.The selection of the appropriate filter sets will enable the visualisation of these mixed-cell populations under the microscope's UV light source. Quantum dots are also increasingly being considered for these types of applications as they can provide high-intensity luminescent signals (76), and luciferase-mediated luminescence is also a further alternative (77–79). The latter will require

transient transfection of a luciferase gene (e.g., from the firefly *Photinus pyralis*). Similarly, green fluorescent protein from *Aequoria victoria* or *Renilla* spp. could be used to generate fluorescence, although all transfection procedures are most commonly used in immortal cells rather than normal cells with a finite life span.

7. In the majority of cases, cells will migrate to defined growth factors more readily when a substrate that promotes adhesion is present; indeed, in some cases it is not possible to obtain any migration to a soluble ligand if a natural substrate is lacking. The appropriate substrate and optimal dilution for coating will need to be determined empirically for each cell line and ligand pair. A range of substrates is commercially available (e.g., collagen I, collagen IV, laminin, fibronectin), and coating procedures are supplied with each product data sheet. FCS works well as a chemoattractant because it not only contains a plethora of potent chemoattractant factors but also contains high levels of proteins such as fibronectin, which coat the membrane naturally. In most cases, the rate and extent of migration to a defined ligand will be less than when using FCS. On the other hand, if inhibitors under test are targeting a specific signalling pathway, for instance, chemotaxis to FCS may only be partially inhibited as other mechanisms will contribute.

8. In some instances, the cells may require pre-incubation with the inhibitory compound prior to stimulation with a chemoattractant. If this is necessary, it may be convenient to add the inhibitor to a flask of cells for a predetermined length of time before seeding. Fresh compound should be added once the cells are in the chamber to ensure continued activity. However, if the compound acts rapidly, both the cell suspension and compound can be added simultaneously to the top chamber and medium and compound to the bottom chamber. The addition of the chemoattractant to the lower chamber is then delayed until initiation of chemomigration is desired. If this approach is used, the length of time necessary for inhibition of the appropriate target (and its duration) should be established in pilot studies.

9. ECs will generally migrate to 5% FCS in adequate numbers within 6 h, and performing a cell count at intervals may not be necessary if an endpoint assay is sufficient.

10. The cells may also be labelled with the lipophilic dye DiI (see http://probes.invitrogen.com/) according to the manufacturer's instructions.

11. It is important to observe the lower surface of the membrane under UV illumination before capturing images of the invading

cells. If there is an uneven coating of Matrigel, which will sometimes arise due to problems in manufacture or manual coating, the cells may tend to preferentially migrate around the edge of the membrane instead of uniformly across its surface. Alternatively, if there is a meniscus effect, such that the Matrigel is thicker at the circumference, then migration will preferentially occur in the centre where the layer is thinnest. The use of Matrigel in chemoinvasion has been discussed in several publications *(39, 40, 80)*.

12. Dynamic assays (as opposed to endpoint-only assays) will also indicate if inhibition is sustained throughout the observation period or if any recovery is taking place. This may be useful to determine the stability of test compounds.

13. For the sake of economy, Matrigel may be diluted up to 1:1 with ice-cold serum-free medium prior to plating, and/or the volume used may be reduced to 200ui. Although ECs can normally be cultured on Matrigel for up to 24 h, tubularisation usually reaches an optimum level (before partially degrading) at less than 10 h. There are some reports that 24–48 h after plating on Matrigel, viability of ECs is decreased *(81)*.

14. A thin layer of collagen I is prepared by diluting rat tail collagen type I to 50 µg/mL with 0.02*M* acetic acid and 50 µL to each well. Incubate at room temperature for 1 h. Remove solution and wash each well with PBS. A collagen matrix may be prepared by diluting the collagen to 3 mg/mL. Add 100 µL of a mixture containing serum-free growth medium, 0.1*M* sodium bicarbonate buffer, 3 mg/mL rat tail collagen type I, and distilled water (2:1:4:2) to each well and allow to gel at 37°C for 30 min.

15. If a semi-quantitative analysis of the zymography samples is to be performed, the protein content of the samples loaded in each lane will need to be standardised. For generating control supernatant containing high levels of gelatinases, serum starve an 80% confluent 75-cm³ flask of HT1080 cells overnight. Add fresh serum-free growth medium containing 20 ng/mL PMA. Leave for a further 24 h and then collect the conditioned medium. A sample of this medium run on the zymogram will show both active and latent forms of MMP2 and MMP9. These bands will appear at 62 and 72 kDa (MMP2) and at 82 and 92 kDa (MMP9).

16. Either method of staining the gels is suitable, although a fivefold higher sensitivity may be obtained using the colloidal blue technique. If the gels are to be stored, incubate them in 100 mL of 2% glycerol for 30 min, then dry overnight using a gel drier system (Gibco-Invitrogen).

706   References

1. Folkman, J. (1971) Tumor angiogenesis: therapeutic implications. *N Engl J Med* 285, 1182–1186.

2. Alitalo, K., Tammela, T., Petrova, T. V. (2005) Lymphangiogenesis in development and human disease. *Nature* 438, 946–953.

3. Benelli, R., Lorusso, G., Albini, A., Noonan, D. M. (2006) Cytokines and chemokines as regulators of angiogenesis in health and disease. *Curr Pharm Des* 12, 3101–3115.

4. Carmeliet, P. (2005) Angiogenesis in life, disease and medicine. *Nature* 438, 932–936.

5. Folkman, J. (2007). Is angiogenesis an organizing principle in biology and medicine? *J Pediatr Surg* 42, 1–11.

6. Luttun, A. and Carmeliet, P. (2004).Angiogenesis and lymphangiogenesis: highlights of the past year. *Curr Opin Hematol* 11, 262–271.

7. Shaked, Y., Bertolini, F., Man, S., (2005) Genetic heterogeneity of the vasculogenic phenotype parallels angiogenesis; implications for cellular surrogate marker analysis of antiangiogenesis. *Cancer Cell* 7, 101–111.

8. Tammela, T., Enholm, B., Alitalo, K., Paavonen, K. (2005) The biology of vascular endothelial growth factors. *Cardiovasc Res*, 65, 550–563.

9. Tammela, T., Petrova, T. V., Alitalo, K. (2005) Molecular lymphangiogenesis: new players. *Trends Cell Biol* 15, 434–441.

10. Wissmann, C., Detmar, M. (2006) Pathways targeting tumor lymphangiogenesis. *Clin Cancer Res* 12, 6865–6868.

11. Bisacchi,D.,Benelli,R.,Vanzetto,C.,Ferrari,N., Tosetti, F., Albini, A. (2003) Anti-angiogenesis and angioprevention: mechanisms, problems and perspectives. *Cancer Detect Prev* 27, 229–238.

12. Carmeliet, P. (2005) VEGF as a key mediator of angiogenesis in cancer. *Oncology* 69 Suppl 3, 4–10.

13. Ferrara, N., Kerbel, R. S. (2005) Angiogenesis as a therapeutic target. *Nature* 438, 967–974.

14. Ferretti, C., Bruni, L., Dangles-Marie, V., Pecking, A. P., Bellet, D. (2007) Molecular circuits shared by placental and cancer cells, and their implications in the proliferative, invasive and migratory capacities of trophoblasts. *Hum Reprod Update* 13, 121–141.

15. Folkman, J. (2003) Angiogenesis inhibitors: a new class of drugs. *Cancer Biol Ther* 2, S127–S133.

16. He, Y., Karpanen, T., Alitalo, K. (2004) Role of lymphangiogenic factors in tumor metastasis. *Biochim Biophys Acta* 1654, 3–12.

17. Hirakawa, S., Brown, L. F., Kodama, S., Paavonen, K., Alitalo, K., Detmar, M. (2007) VEGF-C-induced lymphangiogenesis in sentinel lymph nodes promotes tumor metastasis to distant sites. *Blood* 109, 1010–1017.

18. Kerbel, R. S. (2004) Antiangiogenic drugs and current strategies for the treatment of lung cancer. *Semin Oncol* 31, 54–60.

19. Munoz, R., Shaked, Y., Bertolini, F., Emmenegger, U., Man, S., Kerbel, R. S. (2005) Anti-angiogenic treatment of breast cancer using metronomic low-dose chemotherapy. *Breast* 14, 466–479.

20. Naumov, G. N., Akslen, L. A., Folkman, J. (2006) Role of angiogenesis in human tumor dormancy: animal models of the angiogenic switch. *Cell Cycle* 5, 1779–1787.

21. Saharinen, P., Tammela, T., Karkkainen, M. J., Alitalo, K. (2004) Lymphatic vasculature: development, molecular regulation and role in tumor metastasis and inflammation. *Trends Immunol*, 25, 387–395.

22. Sanz, L., Alvarez-Vallina, L. (2005) Antibody-based antiangiogenic cancer therapy. *Expert Opin Ther Targets* 9, 1235–1245.

23. Shields, J. D., Emmett, M. S., Dunn, D. B., et al. (2007) Chemokine-mediated migration of melanoma cells towards lymphatics— a mechanism contributing to metastasis. *Oncogene* 26, 2997–3005.

24. Sridhar, S. S., Shepherd, F. A. (2003) Targeting angiogenesis: a review of angiogenesis inhibitors in the treatment of lung cancer. *Lung Cancer* 42 Suppl 1, S81–S91.

25. Verhoef, C., de Wilt, J. H., Verheul, H. M. (2006) Angiogenesis inhibitors: perspectives for medical, surgical and radiation oncology. *Curr Pharm Des* 12, 2623–2630.

26. Kerbel, R. S. (2006) Antiangiogenic therapy: a universal chemosensitization strategy for cancer? *Science*, 312, 1171–1175.

27. Kerbel, R. S., Kamen, B. A. (2004) The anti-angiogenic basis of metronomic chemotherapy. *Nat Rev Cancer* 4, 423–436.

28. Lavie, G., Mandel, M., Hazan, S., et al. (2005) Anti-angiogenic activities of hypericin

in vivo: potential for ophthalmologic applications. *Angiogenesis*, 8, 35–42.

29. Griggs, J., Skepper, J. N., Smith, G. A., Brindle, K. M., Metcalfe, J. C., Hesketh, R. (2002) Inhibition of proliferative retinopathy by the anti-vascular agent combretastatin-A4. *Am J Pathol* 160, 1097–1103.

30. Stellmach, V., Crawford, S. E., Zhou, W., Bouck, N. (2001) Prevention of ischemia-induced retinopathy by the natural ocular antiangiogenic agent pigment epithelium-derived factor. *Proc Natl Acad Sci U S A* 98, 2593–2597.

31. Tong, Y., Zhang, X., Zhao, W., (2004) Anti-angiogenic effects of Shiraiachrome A, a compound isolated from a Chinese folk medicine used to treat rheumatoid arthritis. *Eur J Pharmacol* 494, 101–109.

32. Dupont, E., Savard, P. E., Jourdain, C., (1998) Antiangiogenic properties of a novel shark cartilage extract: potential role in the treatment of psoriasis. *J Cutan Med Surg* 2, 146–152.

33. Sauder, D. N., Dekoven, J., Champagne, P., Croteau, D., Dupont, E. (2002) Neovastat (AE-941), an inhibitor of angiogenesis: randomized phase I/II clinical trial results in patients with plaque psoriasis. *J Am Acad Dermatol* 47, 535–541.

34. Griffin, R. J., Molema, G., Dings, R. P. (2006) Angiogenesis treatment, new concepts on the horizon. *Angiogenesis* 9, 67–72.

35. Saaristo, A., Tammela, T., Farkkila, A., (2006) Vascular endothelial growth factor-C accelerates diabetic wound healing. *Am J Pathol* 169, 1080–1087.

36. Detillieux, K. A., Cattini, P. A., Kardami, E. (2004) Beyond angiogenesis: the cardioprotective potential of fibroblast growth factor-2. *Can J Physiol Pharmacol* 82, 1044–1052.

37. Levy, A. P., Levy, N. S., Loscalzo, J., et al. (1995). Regulation of vascular endothelial growth factor in cardiac myocytes. *Circ Res* 76, 758–766.

38. Eccles, S. A. (2004) Parallels in invasion and angiogenesis provide pivotal points for therapeutic intervention. *Int J Dev Biol* 48, 583 598.

39. Albini, A., Benelli, R., Noonan, D. M., Brigati, C. (2004) The "chemoinvasion assay": a tool to study tumor and endothelial cell invasion of basement membranes. *Int J Dev Biol* 48, 563–571.

40. Benelli, R., Albini, A. (1999) *In vitro* models of angiogenesis: the use of Matrigel. *Int J Biol Markers*, 14, 243–246.

41. Mastyugin, V., McWhinnie, E., Labow, M., Buxton, F. (2004) A quantitative high-throughput endothelial cell migration assay. *J Biomol Screen* 9, 712–718.

42. Montanez, E., Casaroli-Marano, R. P., Vilaro, S., Pagan, R. (2002) Comparative study of tube assembly in three-dimensional collagen matrix and on Matrigel coats. *Angiogenesis* 5, 167–172.

43. Oberringer, M., Meins, C., Bubel, M., Pohlemann, T. (2007) A new *in vitro* wound model based on the co-culture of human dermal microvascular endothelial cells and human dermal fibroblasts. *Biol Cell* 99, 197–207.

44. Schneider, M., Tjwa, M., Carmeliet, P. (2005) A surrogate marker to monitor angiogenesis at last. *Cancer Cell* 7, 3–4.

45. Brkovic, A., Pelletier, M., Girard, D., Sirois, M. G. (2007) Angiopoietin chemotactic activities on neutrophils are regulated by PI-3K activation. *J Leukoc Biol* 81, 1093–1101.

46. Chang, L. K., Garcia-Cardena, G., Farnebo, F., (2004) Dose-dependent response of FGF-2 for lymphangiogenesis. *Proc Natl Acad Sci U S A* 101, 11658–11663.

47. Gualandris, A., Lopez Conejo, T., Giunciuglio, D., (1997) Urokinase-type plasminogen activator overexpression enhances the invasive capacity of endothelial cells. *Microvasc Res* 53, 254–260.

48. Rak, J., Kerbel, R. S. (1998) Basic fibroblast growth factor and the complexity of tumour angiogenesis. *Expert Opin Investig Drugs* 7, 797–801.

49. Rosenkilde, M. M., Schwartz, T. W. (2004) The chemokine system—a major regulator of angiogenesis in health and disease. *APMIS* 112, 481–495.

50. Strieter, R. M., Burdick, M. D., Mestas, J., Gomperts, B., Keane, M. P., Belperio, J. A. (2006) Cancer CXC chemokine networks and tumour angiogenesis. *Eur J Cancer* 42, 768–778.

51. Alghisi, G. C., Ruegg, C. (2006) Vascular integrins in tumor angiogenesis: mediators and therapeutic targets. *Endothelium* 13, 113–135.

52. Cai, W., Chen, X. (2006) Anti-angiogenic cancer therapy based on integrin alphavbeta3 antagonism. *Anticancer Agents Med Chem* 6, 407–428.

53. Mettouchi, A., Meneguzzi, G. (2006) Distinct roles of beta1 integrins during angiogenesis. *Eur J Cell Biol* 85, 243–247.

54. Serini, G., Valdembri, D., Bussolino, F. (2006) Integrins and angiogenesis: a sticky business. *Exp Cell Res* 312, 651–658.

55. Lakka, S. S., Gondi, C. S., Rao, J. S. (2005) Proteases and glioma angiogenesis. *Brain Pathol* 15, 327–341.

56. Rundhaug, J. E. (2005) Matrix metalloproteinases and angiogenesis. *J Cell Mol Med* 9, 267–285.

57. van Hinsbergh, V. W., Engelse, M. A., Quax, P. H. (2006) Pericellular proteases in angiogenesis and vasculogenesis. *Arterioscler Thromb Vasc Biol* 26, 716–728.

58. Rabbani, S. A., Mazar, A. P. (2001) The role of the plasminogen activation system in angiogenesis and metastasis. *Surg Oncol Clin N Am* 10, 393–415, x.

59. Stefansson, S., McMahon, G. A., Petitclerc, E., Lawrence, D. A. (2003) Plasminogen activator inhibitor-1 in tumor growth, angiogenesis and vascular remodeling. *Curr Pharm Des* 9, 1545–1564.

60. Chantrain, C. F., Henriet, P., Jodele, S., et al. (2006) Mechanisms of pericyte recruitment in tumour angiogenesis: a new role for metalloproteinases. *Eur J Cancer* 42, 310–318.

61. Kiselyov, A., Balakin, K. V., Tkachenko, S. E. (2007) VEGF/VEGFR signalling as a target for inhibiting angiogenesis. *Expert Opin Investig Drugs* 16, 83–107.

62. Liu, C. C., Shen, Z., Kung, H. F., Lin, M. C. (2006) Cancer gene therapy targeting angiogenesis: an updated review. *World J Gastroenterol* 12, 6941–6948.

63. O'Dwyer, P. J. (2006) The present and future of angiogenesis-directed treatments of colorectal cancer. *Oncologist* 11, 992–998.

64. Telliez, A., Furman, C., Pommery, N., Henichart, J. P. (2006) Mechanisms leading to COX-2 expression and COX-2 induced tumorigenesis: topical therapeutic strategies targeting COX-2 expression and activity. *Anticancer Agents Med Chem* 6, 187–208.

65. Trachsel, E., Neri, D. (2006) Antibodies for angiogenesis inhibition, vascular targeting and endothelial cell transcytosis. *Adv Drug Deliv Rev* 58, 735–754.

66. Wanebo, H. J., Argiris, A., Bergsland, E., Agarwala, S., Rugo, H. (2006) Targeting growth factors and angiogenesis; using small molecules in malignancy. *Cancer Metastasis Rev*, 25, 279–292.

67. Eccles, S. A., Box, C. (2005) Court W Cell migration assays and their application in cancer drug discovery. *Biotech Annu Rev* 11, 391–421

68. Sanderson, S., Valenti, M., Gowan, S., et al. (2006) Benzoquinone ansamycin heat shock protein 90 inhibitors modulate multiple functions required for tumor angiogenesis. *Mol Cancer Ther* 5, 522–532.

69. Connolly, D. T., Knight, M. B., Harakas, N. K., Wittwer, A. J., Feder, J. (1986) Determination of the number of endothelial cells in culture using an acid phosphatase assay. *Anal Biochem* 152, 136–140.

70. Jackson, S. J., Venema, R. C. (2006) Quercetin inhibits eNOS, microtubule polymerization, and mitotic progression in bovine aortic endothelial cells. *J Nutr* 136, 1178–1184.

71. Michaelis, U. R., Fisslthaler, B., Barbosa-Sicard, E., Falck, J. R., Fleming, I., Busse, R. (2005) Cytochrome P450 epoxygenases 2C8 and 2C9 are implicated in hypoxia-induced endothelial cell migration and angiogenesis. *J Cell Sci* 118, 5489–5498.

72. Alessandri, G., Chirivi, R. G., Castellani, P., Nicolo, G., Giavazzi, R., Zardi, L. (1998) Isolation and characterization of human tumor-derived capillary endothelial cells: role of oncofetal fibronectin. *Lab Invest* 78, 127–128.

73. Garrafa, E., Alessandri, G., Benetti, A., (2006) Isolation and characterization of lymphatic microvascular endothelial cells from human tonsils. *J Cell Physiol* 207, 107–113.

74. Invernici, G., Ponti, D., Corsini, E., et al (2005) Human microvascular endothelial cells from different fetal organs demonstrate organ-specific CAM expression. *Exp Cell Res* 308, 273–282.

75. Park, H. J., Zhang, Y., Georgescu, S. P., Johnson, K. L., Kong, D., Galper, J. B. (2006) Human umbilical vein endothelial cells and human dermal microvascular endothelial cells offer new insights into the relationship between lipid metabolism and angiogenesis. *Stem Cell Rev* 2, 93–102.

76. Gu, W., Pellegrino, T., Parak, W. J., et al. (2007) Measuring cell motility using quantum dot probes. *Methods Mol Biol* 374, 125–132.

77. Gildea, J. J., Harding, M. A., Gulding, K. M., Theodorescu, D. (2000) Transmembrane motility assay of transiently transfected cells by fluorescent cell counting and luciferase measurement. *Biotechniques* 29, 81–86.

78. Spessotto, P., Giacomello, E., Perri, R. (2002) Improving fluorescence-based assays for the *in vitro* analysis of cell adhe-

sion and migration. *Mol Biotechnol* 20, 285–304.

79. Spessotto, P., Giacomello, E., Perris, R. (2000) Fluorescence assays to study cell adhesion and migration in vitro. *Methods Mol Biol* 139, 321–343.

80. Albini, A. (1998) Tumor and endothelial cell invasion of basement membranes. The Matrigel chemoinvasion assay as a tool for dissecting molecular mechanisms. *Pathol Oncol Res* 4, 230–241.

81. Ranta, V., Mikkola, T., Ylikorkala, O., Viinikka, L., Orpana, A. (1998) Reduced viability of human vascular endothelial cells cultured on Matrigel. *J Cell Physiol*, 176, 92–98.

Chapter 10

Tube Formation: An *In Vitro* Matrigel Angiogenesis Assay

M. Lourdes Ponce

Abstract

Neovascularization plays a role in several pathological conditions, including tumor growth, arthritis, and choroidal neovascularization. Investigators from different fields can choose from several available angiogenesis assays according to their specific needs. This chapter describes an easy-to-perform assay that is based on the differentiation of endothelial cells and the formation of tube-like structures on an extracellular matrix, Matrigel. The assay can be used to screen compounds for angiogenic activity or to determine if it has an effect on angiogenesis, depending on the conditions chosen. It is a quick assay, easy to set up, and highly reproducible. It can be used to test one or two samples, or it can quickly be scaled up to screen hundreds of compounds. The flexibility that this assay provides makes it a good first choice to test if a compound or a series of compounds may play a role in angiogenesis.

Key words: Angiogenesis, endothelia, HUVEC, Matrigel, tube assay.

1. Introduction

Angiogenesis, the formation of new blood vessels from preexisting ones, can be studied by using many different methods that are available to the investigator. Choosing the most appropriate assay can be based on a number of factors, such as reagent availability, equipment, time it will take to perform the assay, cost, and skills needed. Two *in vivo* models that are widely accepted are the corneal implant assay and the chick choriallantoic assay (CAM). However, both assays are difficult to perform since the corneal assay requires special equipment and a skilled person (usually a trained ophthalmologist) to implant the beads in the eyes of animals. The CAM assay does not require highly technical skills, but it is a labor-intensive assay because it requires a large number

S. Martin and C. Murray (eds.), *Methods in Molecular Biology, Angiogenesis Protocols, Second edition, Vol. 467*
© Humana Press, a part of Springer Science+Business Media, LLC, 2009
Book doi: 10.1007/978-1-59745-241-0_10

of samples due to its intrinsic variability and to gradual loss of viable samples.

In our laboratory, we have developed a quick and highly reliable method for testing numerous compounds for angiogenic or antiangiogenic activity. The method is based on the differentiation of endothelial cells on a basement membrane matrix, Matrigel, derived from the Engelbreth-Holm-Swarm tumor *(1)*. Endothelial cells from human umbilical cords as well as from other sources differentiate and form capillary-like structures on Matrigel in the presence of 10% bovine calf serum (BCS) and 1 mg/mL of endothelial cell growth supplement (ECGS) *(2)*, which is a mixture of both acidic and basic fibroblast growth factor (**Fig. 10.1C**). The formation of tube-like vessels under these conditions can be used to assess compounds that either inhibit or stimulate angiogenesis. In the assay, substances that affect angiogenesis, such as the laminin-1 peptide containing the IKVAV sequence, disturb the formation of capillary-like structures, and the distinctive morphological cell characteristics resulting are indicative of the potential activity of the compound (**Fig.10.1D**). However, it cannot be determined from the morphological characteristics whether a compound is angiogenic or antiangiogenic.

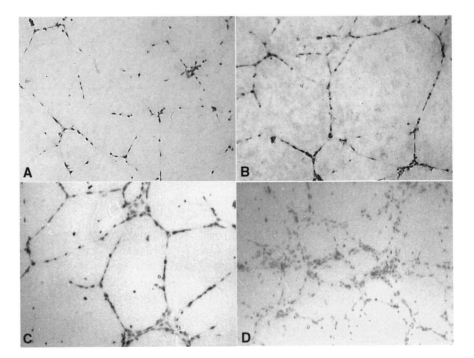

Fig. 10.1. Endothelial cells on Matrigel. Human umbilical vein endothelial cells (HUVECs) were plated on Matrigel in the presence of HUVEC media containing either 5% (**A** and **B**) or 10% **bovine calf serum** (BCS) (**C** and **D**). Cells incubated with 25 ng/mL of hepatocyte growth factor (HGF) formed tubes in serum-reduced media (**B**) as compared to their control (**A**). Cells that were plated with 0.1 mg/mL of a laminin-1 peptide that contained the IKVAV sequence did not form complete tube-like structures (**D**) as compared to their negative control devoid of peptide (original magnification ×5) (**C**).

Tubes do not form when both the serum and the ECGS concentration are reduced (**Fig. 10.1A**). This allows one to determine whether a substance is angiogenic. Under these circumstances, human umbilical vein endothelial cells (HUVECs) do not completely differentiate; instead, they form incomplete, short, tube-like structures. In the presence of angiogenic compounds such as fibroblast growth factor or hepatocyte growth factor (HGF), the endothelial cells differentiate into well-defined tube-like structures (**Fig. 10.1B**). An additional advantage of these assays is that they can be scaled down, and 48- or even 96-well plates can be used if many compounds need to be tested or their quantities are limited. The use of either one or both of these assays can help to quickly and efficiently identify compounds with angiogenic or antiangiogenic activity. Using both of these assays, a number of new compounds that affect angiogenesis have been identified including HGF, haptoglobin, estrogen, chemokine IP-10, and numerous laminin-1 peptides, among others *(3–6)*. It should be noted that the activity in all cases should be confirmed by an additional assay.

2. Materials

1. Matrigel (Sigma-Aldrich, St. Louis, MO, or BD Biosciences, Franklin Lake, NJ, or Trevigen, Gaithersburg, MD).
2. 24-well plates.
3. Pipetmen and sterile tips.
4. HUVECs prepared from umbilical veins or commercially obtained from Vec Technologics (Rensselaer, NY) or Cambrex (East Rutherford, NJ) (see Note 1).
5. RPMI 1640 media (1 L) (Gibco, Frederick, MD).
6. Table centrifuge.
7. 50-mL conical tubes.
8. Hematocytometer.
9. Trypsin-EDTA (ethylenediaminetetraacetic acid) solution: 0.05% trypsin, 0.53 mM EDTA in Hank's balanced solution (Gibco).
10. Incubator with a 95% air/5% CO_2 mixture.
11. HUVEC Medium: 500 mL of RPMI 1640, 100 mL of BCS, defined and supplemented (HyClone Laboratories, Logan, UT), 0.1 g of endothelial cell growth supplement (BD Biosciences), 2500 U of heparin sodium (Fisher Scientific, Fair Lawn, NJ), 5 mL of fungizone solution (Gibco), 1.25 mg

gentamicin, 5 mL of penicillin/streptomycin solution (10,000 U/mL of penicillin G sodium and 10 mg/mL streptomycin sulfate in 0.85% saline; Gibco), Diff-Quik fixative and solution II (Dade-Behring, Newark, DE, or VWR, West Chester, PA).

3. Methods Testing for Angiogenic Compounds

1. Allow the Matrigel to thaw at 4°C; once thawed, keep on ice at all times since it solidifies at room temperature.

2. Under sterile conditions, coat 24-well plates with 300 µL per well of Matrigel without introducing air bubbles. Let the plates sit at room temperature for at least 15 min to allow gelling of the Matrigel (*see* **Note 2**).

3. In the meantime, rinse a confluent 150-mm dish or equivalent containing between 3 and 4×10^6 endothelial cells with 5 mL of RPMI 1640 medium.

4. Add 5 mL of trypsin-EDTA solution and incubate until the cells detach from the plate (approximately 5 min).

5. Place the cells in a 50-mL conical sterile tube and add 5 mL of HUVEC medium to neutralize the trypsin.

6. Centrifuge the tube on a tabletop centrifuge for 5 min at 50g; remove the supernatant, tap the pellet gently, and suspend the cell pellet in 2 mL of HUVEC medium.

7. Set aside a 0.2-mL cell aliquot as control.

8. Count a 10-mL aliquot in a hemocytometer or use a Coulter Counter to determine the total number of cells in the tube.

9. Adjust the volume with HUVEC medium (containing 20% BCS) to have 2×10^6 cells/mL (*see* **Note 3**).

10. To reduce simultaneously the concentration of serum to 5% and the number of cells to 0.5×10^6/mL, add an amount of RPMI 1640 medium equal to three times the volume present in the tube.

11. Plate 192 µL of the cell mixture, which contains 48,000 cells, into each Matrigel-coated well.

12. Add the test compound in a total volume of 8 µL.

13. Controls that will contain 10% serum are similarly prepared from the aliquot that was set aside, except that 1×10^6 cells/mL (instead of 2×10^6 cells/mL) are suspended in HUVEC medium (20% serum) and are subsequently diluted to double their volume with RPMI 1640 medium.

14. Gently swirl the dishes to evenly mix the test compound with the cells.

15. Incubate dishes overnight at 37°C in a 5% CO_2/95% air incubator.

16. The next day, the medium is gently aspirated from each well and is incubated with approximately 200 μL of Diff-Quik fixative for 30 s.

17. The fixative is aspirated, and the cells are stained for 2 min with solution II that has been diluted 1:1 with water (*see* **Note 4**).

18. The tube structures are observed under a microscope, and pictures can be taken (*see* **Note 5**).

3.1. Screening Compounds that Influence Angiogenesis

This method to screen compounds that influence angiogenesis is a variation of the assay described in **Subheading 3.1.** and can be used for screening compounds that either induce or inhibit angiogenesis. The protocol should be followed in an identical manner up to **step 8.** In **step 9**, the cells should be suspended in HUVEC medium containing 20% BCS to yield 1×10^6 cells/mL. The cells are subsequently diluted with an equal volume of RPMI 1640 medium lacking serum to reduce by half the amount of serum and the number of cells per milliliter. The rest of the assay is carried out identically as described. Controls should include cells containing 5% and 10% serum. The method can be easily scaled down to screen multiple samples (*see* **Note 6**). The angiogenic or antiangiogenic activity of a sample can also be determined with this assay (*see* **Note 7**). It is important to note that some Matrigel lots may allow HUVECs to form tubes even at low serum concentrations (*see* **Note 8**).

4. Notes

1. HUVECs should not be older than six passages since they are primary cell cultures and tend to lose their differentiated characteristics. They should always be handled with gloves since they are from human origin.

2. Several concentrations of each compound should be tested at least in triplicate. Include six wells for controls (three positive and three negative).

3. The amount of HUVEC medium in which the cells are suspended in **step 9** will vary from experiment to experiment according to the total number of cells harvested from each plate.

4. Staining of the Matrigel with Diff-Quik solution II for longer than 2 min or without previous dilution will excessively stain the Matrigel, making it difficult to observe the tubular structures.

5. Dishes containing stained cells can be stored wrapped with plastic for a week or longer at 4°C. Pictures should be taken within a few days since the Matrigel tends to dry.

6. To perform assays on 48-well dishes, wells are coated with 200 mL of Matrigel, and 24,000 cells are plated in 150 mL of medium. The 96-well plates are coated with 100 mL of Matrigel and plated with 14,000 cells in 100 mL of medium.

7. Unknown samples should be tested using both serum concentrations.

8. HUVECs can form tubes even in the absence of serum in some commercially available batches of Matrigel. It is recommended to do a control experiment using low serum concentrations every time a new Matrigel lot is used.

References

1. Kubota, Y., Kleinman, H. K., Martin, G. R., Lawley, T. J. (1988) Role of laminin and basement membrane in the morphological differentiation of human endothelial cells into capillary-like structures. *J Cell Biol* 107, 1589–1598.

2. Grant, D. S., Kinsella, J. L., Fridman, R., et al.(1992) Interaction of endothelial cells with a laminin A chain peptide (SIKVAV) *in vitro* and induction of angiogenic behavior *in vivo*. *J Cell Physiol* 153, 614–625.

3. Grant, D. S., Kleinman, H. K., Goldberg, I. D., et al. (1993) Scatter factor induces blood vessel formation *in vivo*. *Proc Natl Acad Sci U S A* 90, 1937–1941.

4. Cid, M. C., Grant, D. S., Hoffman, G. S., Auerbach, R., Fauci, A. S., Kleinman, H. K. (1993) Identification of haptoglobin as an angiogenic factor in sera from patients with systemic vasculitis. *J Clin Invest* 91, 977–985.

5. Ponce, M. L., Nomizu, M., Delgado, M. C.,et al. (1999) Identification of endothelial cell binding sites on the laminin gamma 1 chain. *Circ Res* 84, 688–694.

6. Malinda, K. M., Nomizu, M., Chung, M., et al. (1999) Identification of laminin alpha1 and beta1 chain peptides active for endothelial cell adhesion, tube formation, and aortic sprouting. *FASEB J* 13, 53–62.

Chapter 11

Three-Dimensional *In Vitro* Angiogenesis in the Rat Aortic Ring Model

David C. West and Mike F. Burbridge

Abstract

Angiogenesis is a complex sequential process involving endothelial activation, basement membrane degradation, endothelial sprouting from the parent vessel, invasion of the extracellular matrix, endothelial proliferation, vessel elongation, branching, anastomosis, increases in vessel diameter, basement membrane formation, pericyte acquisition, and remodelling. Most *in vitro* angiogenesis assays are two-dimensional and measure only one facet of this process, generally endothelial proliferation, migration, or tube formation. The two-dimensional nature of the assays also ignores the differences in endothelial phenotype seen in three-dimensional models and *in vivo*. The *in vitro* serum-free three-dimensional rat aortic model closely approximates the complexities of angiogenesis *in vivo*, from endothelial activation to pericyte acquisition and remodelling, and most of these can be quantified by image analysis, immunohistochemistry, and biochemical analysis. It is easily manipulated using molecular biological intervention or exogenous inhibitors and activators in a relatively controlled system.

Key words: Angiogenesis, aorta, image analysis, *in vitro* model.

1. Introduction

Angiogenesis is an essential component for normal development, tissue repair, and the oestrous cycle, as well as a wide variety of inflammatory and pathological processes, including tumor growth and dissemination, diabetic retinopathy, rheumatoid arthritis, and psoriasis. Consequently, the 37 years since the isolation of tumor angiogenesis factor (TAF), in 1971, have seen extensive research into the regulation of neovascularisation, particularly in tumors, with an emphasis on the detection of angiogenic and antiang-iogenic factors. Thus, assays have commonly been developed to

S. Martin and C. Murray (eds.), *Methods in Molecular Biology, Angiogenesis Protocols, Second edition, Vol. 467*
© Humana Press, a part of Springer Science + Business Media, LLC 2009
DOI: 10.1007/978-1-59745-241-0_11

screen for modulation of vascular growth *in vivo* and the stimulation, or inhibition, of endothelial cell proliferation, migration, and tube formation, a measure of differentiation, in simple two-dimensional culture systems *(1, 2)*. However, animal models are extremely complex, difficult to interpret for routine screening, and relatively expensive. While the two-dimensional culture systems study only a limited number of processes involved in the complex angiogenic process, ignoring endothelial activation and sprouting from the parent vessel, invasion, alignment, branching, anastomosis, vessel diameter, basement membrane formation, and remodelling. Thus, in recent years, several methods for studying angiogenesis have been introduced in an attempt to investigate and quantify angiogenesis in three-dimensional collagen or fibrin matrices *in vitro (3–9)*.

Montesano et al. *(3, 4)* and Schor et al. *(7)* examined endothelial invasion of collagen and fibrin gels. In both systems, the cells invade as "vessels" in the presence of angiogenic stimuli, with invasion apparently dependent on a proteolytic environment. This more closely resembles the *in vivo* situation compared with the two-dimensional and Boyden chamber migration assays.

Others have examined the invasive spread of vessels formed by endothelial cells cultured on microcarrier beads embedded in a collagen or fibrin matrix *(10, 11)*. Madri et al. *(8)* found that capillary endothelial cells cultured in a three-dimensional collagen matrix retain their endothelial properties to a greater extent and respond differently to cytokines compared with two-dimensional systems. Other phenotypic differences have also been reported *(12–14)*.

Probably the *in vitro* model most closely approximating the *in vivo* process is that developed by Nicosia and Ottinetti *(6)*. In this method, rat aortic rings are embedded in a plasma clot, collagen, or fibrin gel and cultured in an optimised defined serum-free growth medium for microvascular endothelial cells *(15)*. A complex network of branching and anastomosing microvessels develops from the endothelial cells of the aortic intima *(16)*, interspersed with solitary fibroblasts, thus providing a closer approximation to the complexities of the *in vivo* angiogenic process (**Fig. 11.1**). This is in response to endogenous growth factors and proteolytic enzymes released on resection of the aorta, giving essentially a controlled environment. No other *in vitro* system exhibits all the processes observed *in vivo* (except, of course, blood flow), and potential angiogenic stimulators and inhibitors may be applied throughout the entire length of culture or over defined periods to observe their impact at various stages of the angiogenic response.

Recently, this model has been employed to examine the effect of matrix constituents, cytokines, RGD peptides (peptides containing an arginine-glycine-asparate amnio acid sequence), antiangiogenic compounds, oxidised low-density lipoproteins (ox-LDLs), and neutralising antibodies on vessel growth and has been further developed into a "quantitative," serum-free matrix system *(6,16–23)*. Using

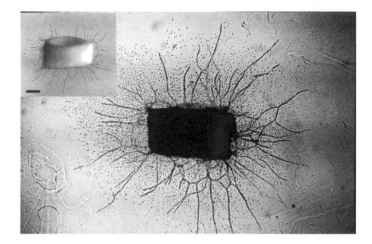

Fig. 11.1. A 5-d-old aortic ring culture stimulated with angiogenic hyaluronan oligosaccharides. Endothelial cells are seen arranged as microvessels and fibroblasts as individual cells. The microvessels show evidence of branching and anastomosis. *Insert*: A 6-d-old culture of a rat aortic ring embedded in collagen and cultured in an optimised serum-free growth medium for microvascular endothelial cells (MCDB131). Endothelial cells are seen arranged as microvessels and fibroblasts as individual cells (bar 500 µm). Again, there is microvessel branching and anastomosis.

this quantitative, serum-free, *in vitro* angiogenesis model, we have examined the effect of pH on vessel growth *(24)*, elucidated the importance of vascular endothelial growth factor in the metastatic propensity of prostatic cancer cell lines *(25)*, investigated the role of metalloproteinases in this model *(26)*, and employed it as a general screening method for antiangiogenic compounds *(25)*.

Several recent variations on the model have employed rabbit aorta *(23)*, human placental vessel *(27, 28)*, porcine carotid artery *(29)*, mouse aorta *(30–33)*, rat inferior vena cava *(34)*, mouse embryonic kidney *(35)*, rat myocardium *(36)*, and human tumor tissue *(37, 38)*. However, few of these employ a purely serum-free system.

2. Materials

2.1. Preparation of Agarose Culture Wells

1. Agarose (type VII, low gelling temperature, cell culture tested; Sigma) (*see* **Note 1**).
2. 100-mm diameter cell culture-treated Petri dishes (Costar).
3. Cutters, or cork borers, 10- and 17-mm diameter.
4. 100-mm diameter bacteriological Petri dishes (Falcon).

2.2. Type 1 Collagen Gel Preparation

1. Collagen R: 2 mg/mL rat-tail collagen in 0.5 *M* acetic acid (Serva) (*see* **Note 2**).
2. 10X Eagle's minimum essential medium (10X MEM; Gibco), containing phenol red pH indicator.

3. 1.4% w/v sodium bicarbonate solution: Diluted from 7.5% solution (Sigma).

4. Sterile aqueous $1M$ NaOH.

2.3. Fibrin Gel Preparation

1. Sterile bovine fibrinogen (culture tested; Sigma) (*see* **Note 3**).

2. Sterile bovine thrombin (Sigma) (*see* **Note 3**).

3. Eagle's MEM (Gibco).

2.4. Preparation of Rat Aortic Segments

1. Male Fischer-344 rats, 8–12 wk old, sacrificed by intraperitoneal phenobarbital or by CO_2 inhalation (*see* **Note 4**).

2. Microdissection (iridectomy) scissors and forceps.

3. Syringe (1 mL) fitted with a 23-gauge needle.

4. Ice-cold serum-free culture medium such as MEM (Gibco) or a balanced salt solution such as Dulbecco's phosphate-buffered saline.

5. Dissecting microscope (preferably in a laminar flow cabinet).

6. Scalpel blade.

7. Petri dish (cell culture treated).

2.5. Culture

1. Serum-free culture medium, MCDB131 (Gibco), supplemented with glutamine (2 mM), penicillin (100 U/mL), streptomycin (100 µg/mL), and 25 mM NaHCO$_3$ (*see* **Notes 5** and **6**).

2. Humidified incubator with a 5% CO_2/air atmosphere at 37°C.

2.6. Quantification

1. Inverted microscope, such as Olympus IMT-2, with non-phase-contrast optics (*see* **Note 7**).

2. Camera and monitor or image analyser (*see* **Notes 8** and **9**).

2.7. Histology

1. Lab-Tek Permanox 8-well chamber slides (Nunc International) (*see* **Note 10**).

2. Fluorescein-labelled acetylated low-density lipoprotein (DiI-Ac-LDL; Biomedical Technologies).

3. Polyclonal sheep antibody to rat von Willebrand factor (Cedarlane) (*see* **Note 11**).

4. Fluorescence microscope, such as the Olympus IMT-2.

3. Methods

3.1. Preparation of Agarose Culture Wells

1. Prepare a 1.5% w/v sterile aqueous solution of agarose, then autoclave to dissolve and sterilise.

2. Pour 30 mL each into 100-mm cell culture-treated Petri dishes and allow to gel at room temperature in sterile conditions 30 min (*see* **Notes 12** and **13**).

3. Punch concentric circles in the agarose (up to 15 per dish) using cork borers of 10- and 17-mm diameter.

4. Remove central portions and with a bent spatula transfer the agarose rings to 100-mm diameter bacteriological Petri dishes (*see* **Note 14**). Using bacteriological plastic here improves adherence of agarose to the base of the dish so that wells do not become detached during culture.

3.2. Preparation of Rat Aortic Segments

1. Remove thoracic aortas rapidly and, using fine microdissection forceps, place immediately in ice-cold culture medium or balanced salt solution (*see* **Note 15**).

2. Flush aorta gently with ice-cold culture medium (from abdominal end) using a 1-mL syringe fitted with a 23-gauge needle until medium runs clear and aorta is free of clotted blood (1–2 mL should suffice) (*see* **Note 16**).

3. Under a dissecting microscope, carefully remove fibroadipose tissue and colateral vessels with fine microdissection scissors (iridectomy scissors) (*see* **Note 17**).

4. Using a scalpel blade, cut rings 1 mm long. Approximately 30 may be obtained from each aorta (*see* **Note 18**).

5. Transfer rings to 10 mL of fresh ice-cold medium in a sterile culture tube (*see* **Note 19**).

6. Allow rings to settle to bottom of tube, pour off medium, and replace with fresh. Repeat five times to ensure adequate rinsing of rings. Swirl tube to resuspend rings and pour medium and rings into a Petri dish on ice.

7. With sterile microdissection forceps, touch each ring gently onto a clean Petri dish (culture treated) to remove excess medium and place in culture wells such that lumen is oriented horizontally. Do not allow rings to dry out during this step. Prepare fresh collagen or fibrin solution as described and gently fill each well with 200 µL to cover each aortic ring.

3.3. Preparation of Type 1 Collagen Gels and Embedding of Aortic Rings

1. Working under sterile conditions, mix, on ice, 7.5 volumes of 2 mg/mL collagen solution with 1 volume of 10X MEM (containing phenol red pH indicator), 1.5 volumes of 1.4% w/v $NaHCO_3$, and a predetermined volume of $1M$ NaOH (approximately 0.1 volumes) to adjust to pH 7.4.

2. Mix gently using a magnetic stirrer, avoiding the creation of air bubbles and allowing time for the pH to equilibrate (*see* **Notes 20** and **21**).

3. Add 200 µL collagen solution to coat the bottom of each agarose well and allow to gel in a humidified incubator to avoid dehydration.

4. With sterile microdissection forceps, touch each ring gently onto a clean Petri dish (culture treated) to remove excess medium and place in culture wells (four per 100-mm dish) such that lumen is oriented horizontally. Do not allow rings to dry out during this step.

5. Prepare fresh collagen as described and gently fill each well with 200 µL to cover each aortic ring.

6. Add 30 mL of MCDB131 medium to each Petri dish and incubate at 37°C in a humidified incubator in 5% CO_2/air.

3.4. Preparation of Fibrin Gels and Embedding of Aortic Rings

1. Prepare fibrinogen solutions at 1.5 mg/mL by dissolving lyophilised cell culture-tested bovine fibrinogen in MEM (*see* **Note 22**).

2. Filter the solution through a 0.4µm sterile filter to remove remaining clumps of fibrinogen molecules, which would interfere with the uniform polymerisation of the gel.

3. Rapidly vortex 1 volume of this fibrinogen solution with 0.02 volumes of a 50 U/mL sterile solution of thrombin and use within 10 s (*see* **Note 23**).

4. Add 200 µL collagen solution to coat the bottom of each agarose well and allow to gel in a humidified incubator to avoid dehydration.

5. With sterile microdissection forceps, touch each ring gently onto a clean Petri dish (culture treated) to remove excess medium and place in culture wells (four per 100-mm dish) such that lumen is oriented horizontally. Do not allow rings to dry out during this step.

6. Prepare fresh fibrin as described and gently fill each well with 200 µL to cover each aortic ring.

7. Add 30 mL of MCDB131 medium containing 300 µg/mL ε-amino-n-caproic acid (*see* **Note 6**) to each Petri dish and incubate at 37°C in a humidified incubator in 5% CO_2/air.

3.5. Culture and Response to Test Materials

In the standard procedure, culture wells of 400 µL are formed by agarose rings placed in Petri dishes (four per dish). These are filled with gels of extracellular matrix components such as fibrin or collagen in which are embedded short segments of rat aorta (**Fig. 11.2**). Each 100-mm Petri dish receives 30 mL medium and is incubated at 37°C in a humidified 5% CO_2/air atmosphere. Cultures may be maintained for a period of up to 2 wk before excessive breakdown of the collagen or fibrin matrix occurs. It is not necessary to replace the medium during culture as the large volume of medium, relative to the small tissue fragments, is

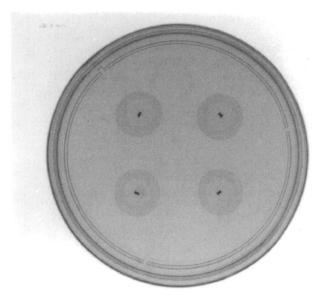

Fig. 11.2. Culture wells are formed by agarose rings placed in a Petri dish. Each well contains a 1-mm segment of rat aorta embedded in a collagen or fibrin gel.

sufficient to avoid depletion of medium components throughout the culture period. Indeed, if medium is replaced, regression of microvessels (*see* **Section 3.5.1**) is accelerated, probably due to removal of endogenous growth factors. In experiments in which medium replenishment is necessary to remove substances applied only for short periods of culture, the medium of control dishes should also be replaced at the same time. Cultures may be observed over a period of time or at a defined time point depending on the information required.

3.5.1. The Angiogenic Response

Individual fibroblasts are seen to grow out of the aortic explants within 2 d of culture, forming an expanding "carpet" of individual cells (**Fig. 11.3A**). By d 3, endothelial cells begin to migrate into the matrix as microvessels, clearly visible by d 4 (**Fig. 11.3B**). Growth continues throughout the first week of culture, initially in terms of an increase in number and length of microvessels, then as a lengthening and thickening of existing vessels (**Fig. 11.3C**). In the second week of culture, regression of the microvessels is observed (**Fig. 11.3D**), probably due to a depletion of growth factors and other soluble components and the reduction in integrity of the surrounding matrix.

3.5.2 Intra- and Interaorta Variation

The spontaneous angiogenic response of each aortic ring may vary twofold. Mean values of all rings from 2 rats should be very similar, even if rats of different age (6 wk to 6 mo) are compared.

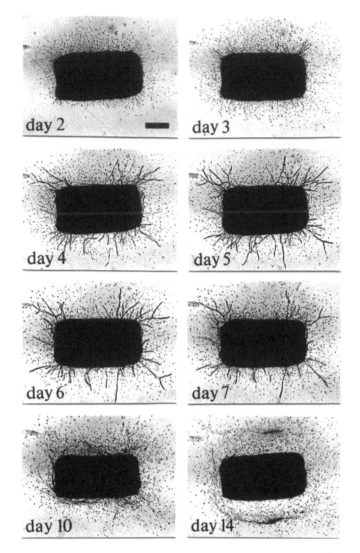

Fig. 11.3. The spontaneous angiogenic response of a rat aortic ring embedded in collagen. Culture was maintained for 2 wk without medium change. See text for details (bar 500 μm).

It should be borne in mind, however, that response to stimulators and inhibitors of aortic rings from rats of different ages will not necessarily be the same.

3.5.3. Interpretation of the Angiogenic Response

In view of the complexity of the vascular network observed in such a system, it is important to be able to fully quantify the many parameters of microvessel outgrowth if a comprehensive description of the process and its modification are required. Criteria should be defined to enable valid interpretation of the angiogenic response and its modification by exogenous stimulators and inhibitors.

The most obvious quantification would simply be a count of the number of microvessels. However, this parameter may not nec-

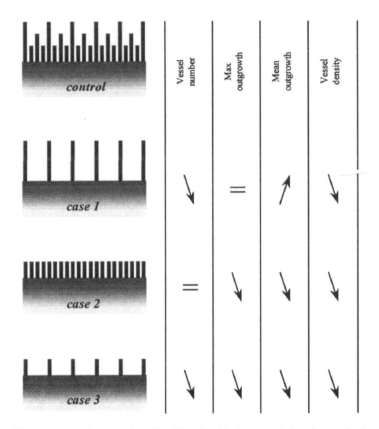

Fig. 11.4. Three possible cases of angiogenesis inhibition. *Top:* Graphical representation of an aortic ring and its endothelial outgrowth. *Below*: Three possible cases of inhibition. The variations in four parameters of microvessel outgrowth for each case are shown on the right.

essarily be sufficient to detect a variation in angiogenic response. In **Fig. 11.4**, the aortic ring is represented along with its vascular outgrowth and, below, three possible cases of inhibition. In case 1, there is an obvious reduction in number of vessels. However, in case 2, while inhibition is evident, a vessel count would lead to similar results as in the control. In case 3, an accumulation of cases 1 and 2, vessel number is again reduced. Another parameter would be a measure of distance of outgrowth of the microvascular network—either that of maximum outgrowth (i.e., the distance migrated by the most distal vessel tip) or of mean outgrowth, taking into account all vessel tips. While in cases 2 and 3 both maximal and mean outgrowths obviously decrease, in case 1 maximal outgrowth is unchanged, and mean outgrowth, due to the absence of intermediary length vessels, actually *increases*. However, only a measure of vascular density (*see* **Section 3.6**) would give an indication of inhibition in all three cases. In other words, while in case 3 all four parameters would indicate inhibition, in case 2 only outgrowth (maximal or mean) and density are reduced and vessel number is unchanged, and in case 1 only two parameters would indicate a reduction in the angiogenic response. The measurement of the degree of angiogenic stimulation is simply an inverse situation, and a measure of vascular density would again be most reliable. Other parameters, such as branching, individual vessel length, or total vessel length, could also be measured, and it is important to bear in mind all these parameters and their possible modification when evaluating the angiogenic response. Microvascular growth curves showing the evolution of the spontaneous angiogenic response in terms of different parameters are presented in **Fig. 11.5.**

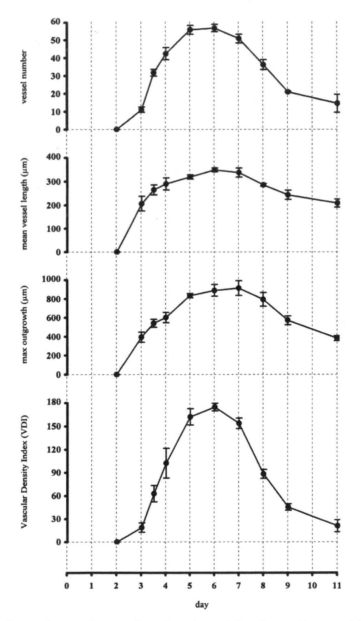

Fig. 11.5. Microvascular growth curves for serum-free culture of rat aorta in collagen gel in terms of different parameters. Vessel growth is expressed in terms of vessel number, mean vessel length, maximum distance of outgrowth, and the vascular density index (VDI) as described in the text. Each point represents the mean value of four explants for a representative experiment. Error bars indicate standard error of the mean.

Angiogenesis in this model closely approximates that observed in wound healing, in which a rapid growth rate of vessels and migration of fibroblasts is followed by slowing in growth rate as microvessels mature and the wound is healed. However, if we consider vessel growth in the case of pathological situations such as tumor growth and diabetic retinopathy, a notion of growth rate rather than that of maximal density would be more applicable. It is thus important to consider whether a measure of any of these parameters at any one time point or rather a comparison of kinetic profiles would be more relevant to the subject of investigation. Microvascular growth curves for stimulation by vascular endothelial growth factor are shown in **Fig. 11.6**. It is apparent that percentage stimulation calculated for this growth factor would depend both on the parameter and on the day chosen for measurement.

3.6. Quantification: Image Analysis

Examine cultures under an inverted microscope with non-phase-contrast optics. For optimal contrast and depth of field, reduce the aperture as far as possible. Transfer images to a monitor screen for image analysis.

3.6.1. Semiautomated Image Analysis

If a detailed analysis of microvascular outgrowth is required, an accurate reproduction of the microvessel network is acquired manually before analysis is performed (*see* **Note 24**).

1. Using suitable software, trace the outline of each vessel on the monitor screen.

2. Transfer the coordinates of these traced vessels to graphic software (such as Microsoft Excel) to reproduce the pattern of microvessel outgrowth in the form of a silhouette (**Fig. 11.7A,B**).

3. A number of parameters may be automatically measured, including vessel number, mean vessel length, maximum and mean distance of outgrowth, and number of branchings.

4. An imaginary grid placed over the microvessel network provides a histogram of intersections at increasing distances from the ring and can be considered to give an "outgrowth profile" for each culture (**Fig. 11.7C,D**).

5. The vascular density index, or VDI, is defined as the total number of vessel intersections with a grid with lines at intervals of 100 µm (for greater precision, lines at intervals of 50 µm are used; the corresponding VDI is then calculated as half the number of intersections). The time required for the analysis of each aortic ring is between 2 and 5 min, depending on the extent of outgrowth.

3.6.2. Automated Image Analysis

When using the model for high-throughput screening of angiogenic or antiangiogenic molecules, a less-precise analysis of cultures may be considered sufficient. Automated image analysis techniques may readily be used to isolate the vessel patterns from the surrounding fibroblasts and to quantify angiogenesis in terms of area occupied by these vessels and extent of outgrowth

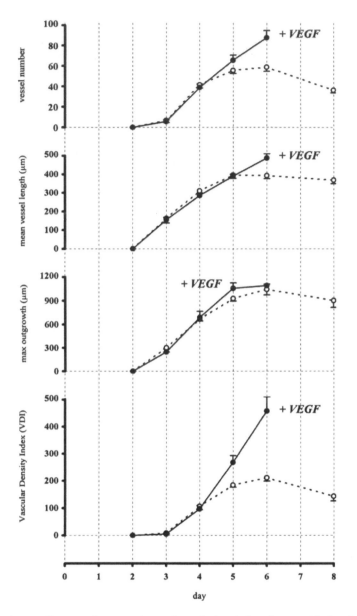

Fig. 11.6. Microvascular growth curves for serum-free culture of rat aorta in collagen gel in the absence (– – – –) and presence (———) of 10 ng/mL vascular endothelial growth factor (VEGF). Vessel growth is expressed in terms of vessel number, mean vessel length, maximum distance of outgrowth, and the vascular density index (VDI) as described in the text. Each point represents the mean value of four explants for a representative experiment. Error bars indicate standard error of the mean.

(**Fig. 11.7E**). An approximation to the number and distance of migration of fibroblasts may also be obtained by such techniques (*see* **Note 25**). Use a black-and-white camera for optimal image definition. Care should be taken when using such automated image analysis techniques since artefacts in the captured image (such as incomplete removal of fibroadipose tissue or bubbles or

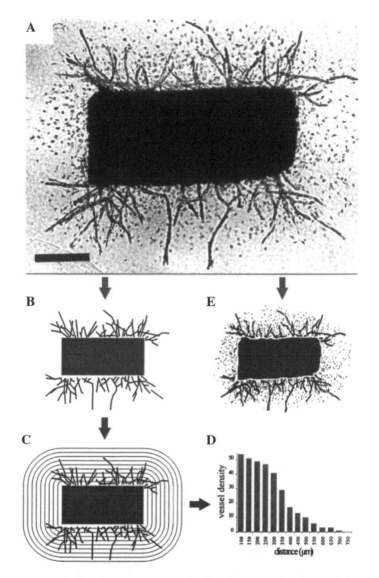

Fig. 11.7. Angiogenesis quantification. (**A**) Photomicrograph of aortic ring d 6 (bar 500 µm). (**B**)–(**D**) Semiautomated image analysis. (**E**) Automated image analysis. See text for details.

other defects in the collagen or fibrin gels) may significantly alter results if not carefully monitored. Quantification is also highly sensitive to illumination level during image capture.

Fibroblast colonisation can also be quantified to give an assessment of the number of single fibroblast cells in the matrix and their distance of migration (**Fig. 11.8**).

A modification of this automatic method prelabels the endothelial cells Dil-Ac-LDL and examines the cultures by fluorescent microscopy *(39, 40)* (*see* **Note 19**).

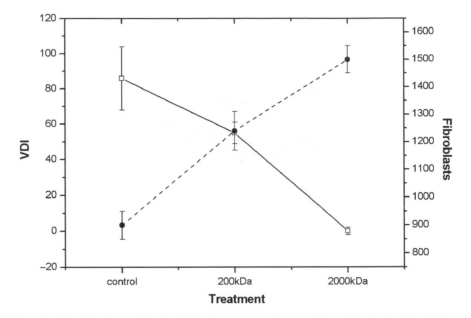

Fig. 11.8. A graph of the size-dependent effect of hyaluronan on angiogenesis (vascular density index, VDI), ∘ — ∘, and fibroblast numbers, •– –•, determined by image analysis. Angiogenesis is inhibited by hyaluronan in a size-dependent manner, while fibroblast numbers increase with the size of hyaluronan.

3.7. Variations on the Standard Technique

3.7.1. Repressed Cultures

The spontaneous outgrowth observed in the aortic cultures is stimulated by the endogenous growth factors released by the aortic wall on resection. When cultured in basal media such as MEM, unsuited to serum-free growth of endothelial cells, this outgrowth is absent. Replacement of this medium on d 5 by MCDB131 leads to minimal outgrowth. However, the addition of certain growth factors to these repressed cultures will cause marked stimulation, suggesting that cells remain viable under such conditions and that the reduced outgrowth is due rather to the removal of released endogenous growth factors along with the MEM. This technique thus provides a means by which to investigate the intrinsic effect of growth factors, bearing in mind that growth factors may work in synergy and alone may lead to minimal response. **Figure 11.9** shows stimulation by vascular endothelial growth factor of such a repressed culture.

3.7.2. Multiwell cultures

As described, this model is useful for the long-term maintenance and observation of aortic ring cultures. However, it may be necessary to adapt the model to situations in which smaller volumes of media are required, as is the case when using angiogenic stimulators or inhibitors with costs that would prohibit dilutions into volumes of 30 mL.

3.7.2.1. 12-/24-Well Culture

Cultures may be carried out in 12- or 24-well culture plates with minor modifications.

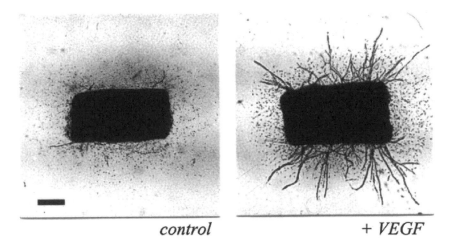

control   *+ VEGF*

Fig. 11.9. Repressed cultures. Stimulation with 10 ng/mL vascular endothelial growth factor (VEGF) (photo d 9, bar 500 μm). See text for details.

1. Embed the aortic segments in the collagen or fibrin gels as outlined for the standard method.
2. Carefully remove the agarose rings from the gels with the bent spatula.
3. Transfer the collagen or fibrin gels, again with the bent spatula, to wells of culture plates filled with medium (e.g., 3 mL per well for a 12-well plate).

Outgrowth in these cultures is reduced as compared to cultures in 100-mm Petri dishes, due partially to an alteration in mechanical forces on the collagen matrix (removal of the supporting agarose enables retraction of the collagen gel) and possibly to a modification in the medium environment due to the reduced culture volume (agarose-free cultures in Petri dishes lead to outgrowth superior to that in 12-well plates). Angiogenesis in terms of VDI is half-maximal on d 5 as opposed to d 4 for the standard Petri dish culture. This difference in outgrowth should be borne in mind when comparing results. The action of stimulators and inhibitors of angiogenesis may conceivably be altered to a certain degree by differences in growth kinetics, especially when the stability of the added compound is limited.

However, the results are of course no less valid than in Petri dish culture, and such multiwell culture is useful for high-throughput screening using automated image analysis. With four rats, around 100 potential antiangiogenic compounds may be screened in one such experiment (which should be repeated for confirmation).

3.7.2.2. 96-Well Culture

If volumes of medium available are exceedingly small, it is possible to culture directly in collagen or fibrin gels in 96-well plates (two layers of 50 μL each/well). However, the small volume of medium is rapidly spent (the ratio of medium to gel is also very low), and outgrowth is less satisfactory. For analysis, plates should be immersed in a medium bath to enable observation of cultures without image interference due to meniscus within each well.

3.8. Histological Examination

Although it is possible to fix and dehydrate cultures for paraffin embedding, massive shrinkage of the gels during the various dehydration and embedding steps leads to distorted arrangement of fibroblasts and endothelial cells, rendering interpretation and extrapolation to the three-dimensional cultures difficult. Moreover, due to the high water content of the gels, frozen sections are not feasible.

Specific whole-cell or antigen marking using immunohistochemical or other techniques may be performed on intact cultures. To enable rapid penetration and washing out of antibodies and other markers, small fragments rather than rings of rat aorta are embedded in thin films of gel.

3.8.1. Culture Wells

1. Culture wells are formed by the silicone gasket of eight-well Lab-Tek Permanox chamber slides after removal of the upper chambers. The use of Permanox plastic ensures minimal autofluorescence if using fluorescent markers.

2. Attach each slide to the base of a Petri dish with a drop of cyanoacrylate gel (Loctite Super Glue).

3.8.2. Aortic Fragments

1. Prepare small fragments of 0.25–0.5 mm² by first cutting each aorta in half lengthwise with a scalpel blade, then cutting each half into strips. Chop each strip into small lengths to produce square fragments.

2. Wash as for rings, remove excess medium, and place in Chamber Slide wells (5–10 fragments per well).

3. Carefully add 50 μL collagen or fibrin to each well without disturbing fragments and place in a humidified incubator to gel before covering with medium (**Fig. 11.10A**) (*see* **Note 26**).

3.8.3. Fluorescent Staining of Viable Endothelial Cells

Endothelial cells within these thin-film cultures may be stained by their selective uptake of DiI-Ac-LDL.

1. Carefully remove aortic fragments.

2. Incubate living cultures for 4 h at 37°C with a solution of DiI-Ac-LDL (10 μg/mL) in MEM medium, followed by three washings of 1 h each in fresh medium.

3. Examine under a fluorescence microscope set up to view fluorescein labelling. Vessels are clearly labelled, whereas uptake by individual fibroblasts is minimal (**Fig. 11.10B**).

3.8.4. Antibody Labelling of Fixed Cultures

1. If required, thin-gel cultures from which fragments have been carefully removed may be fixed by either 4% paraformaldehyde or ethanol.

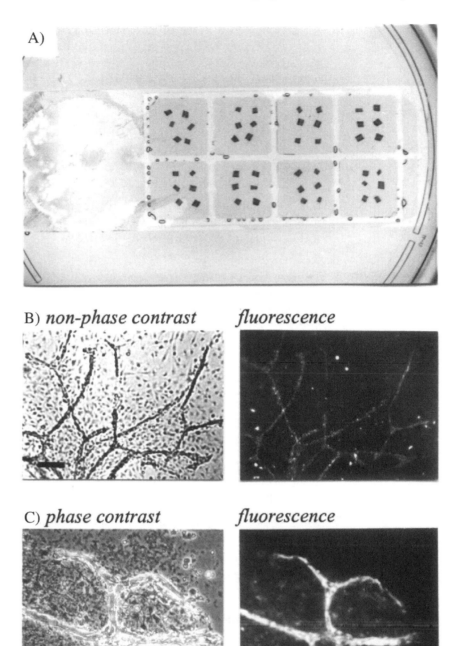

Fig. 11.10. Fluorescent staining of endothelial cells. (**A**) Aortic fragments are cultured in wells formed by the silicone gasket of Lab-Tek chamber slides. (**B**) Fluorescein-labelled acetylated low-density lipoprotein (DiI-Ac-LDL) uptake (bar 200 µm). (**C**) Fluorescent marking for von Willebrand factor (bar 100 µm): Cultures are fixed in ethanol 15 min at room temperature and allowed to air-dry. Nonspecific binding sites are blocked 30 min in 10% foetal calf serum. Preparations are then incubated 2 h with 10 µg/mL polyclonal sheep antibody to rat von Willebrand factor (Cedarlane), 1 h with biotin-conjugated rabbit antisheep immunoglobulin G (Rockland), and 1 h with ExtrAvidin-R-phycoerythrin conjugate (Sigma).

2. Air-dry the gels to leave preparations of a thickness of a few tens of microns, to which may be applied standard immunohistological staining techniques. A typical culture labelled for von Willebrand factor is shown in **Fig. 11.10C**.

3.8.5. Parallel Extraction of Messenger RNA and Protein from Rat Aortic Ring Cultures

1. Carefully remove aortic rings with forceps and snap freeze the collagen or fibrin gels.

2. Add three volumes of Trizol-LS (Invitrogen) containing 0.7% v/v 2-mercaptoethanol to each sample and homogenise in a power homogeniser.

3. Phase separation is performed according to the manufacturer's instructions. The total RNA is recovered from the aqueous phase and further purified using the RNaid kit (Qbiogene, Illkirch, France).

4. The interface and lower organic phase are mixed with ethanol to precipitate DNA per the manufacturer's instructions, and the protein is isolated by isopropanol precipitation as set out in the Trizol-LS protocol.

5. Polymerase chain reaction (PCR) and Western blotting can be performed by standard techniques (see **ref**. *26*).

4. Notes

1. Type VII is ideal for its melting and solidifying temperatures and rigidity.

2. Sterile type I collagen solutions may be prepared from rat tails, although this is a tedious and time-consuming procedure and may lead to large variations between experiments when using different batches. Commercially prepared solutions of rat-tail collagen in acetic acid are now readily available and will generally lead to more satisfactory and consistent results (e.g., Collagen R, 2 mg/mL in 0.5M acetic acid; Serva). Other commercially available sources of collagen (e.g., bovine skin) have been found to give gels of a less-rigid consistency. For rat-tail collagen, a concentration of 1.5 mg/mL is optimal; lower concentrations lead to excessively labile gels, whereas higher concentrations may lead to a reduction in outgrowth from the aortic explants.

3. Fibrinogen and thrombin may be of various origins; human or bovine preparations give similar results and are relatively inexpensive.

4. The rat is the chosen species for these experiments since it provides rings of a practical size and since rings from mouse aortas have been found not to give rise to satisfactory vessel

outgrowth, possibly due to a reduced local concentration of growth factors.

5. MCDB131 is a culture medium optimised for the culture of microvascular endothelial cells and enables growth in serum-free conditions (see **ref**. *15*). This commercially available medium contains a concentration of 14 mM NaHCO$_3$, which will equilibrate at pH 7.4 in atmospheric CO$_2$ concentration of 2%. If conditions of 5% CO$_2$ are used, pH may be corrected by adding NaHCO$_3$ to a final concentration of 25 mM or by adding NaOH until equilibrium is reached at pH 7.4, the net result in terms of NaHCO$_3$ concentration and buffering capacity being the same *(41)*. Maintenance of pH 7.4 is of utmost importance for optimal growth.

6. If fibrin gels are used, 300 µg/mL ε-amino-n-caproic acid should be added to culture medium to prevent spontaneous fibrinolysis.

7. For optimal contrast and depth of field, reduce the aperture as far as possible.

8. Use a black-and-white camera for optimal image resolution, especially when automated image analysis is required.

9. The image analyser used in our laboratory is the Biocom Visiolab™ 2000 (Biocom SA, Les Ulis, France), using specialised software developed specifically for analysis of the three-dimensional aortic ring cultures.

10. The use of Permanox plastic ensures minimal autofluorescence when using fluorescent markers.

11. In view of the large species heterogeneity of von Willebrand factor, antibodies raised to the human antigen may be unsuccessful, even though species cross-reactivity is claimed.

12. Cell culture-treated dishes are preferred to ensure reduced adherence of agarose for easier manipulation.

13. At this stage, agarose may be stored at 4°C for up to 3 d; any longer, agarose tends to dehydrate.

14. Using bacteriological plastic here improves adherence of agarose to the base of the dish so that wells do not become detached during culture.

15. The rat aortic endothelium is exceptionally well retained to the intima, and unlike that of larger species, the formation of clotted blood will not lead to a stripping off of endothelial cells if aortas are rinsed within 15 min of removal from the animal.

16. If more than one aorta is used in an experiment, this stage should be completed for each aorta before moving on to the next.

17. It is not necessary to keep Petri dishes on ice during dissection procedures under the dissecting microscope. Frequent replenishment of ice-cold medium is sufficient to maintain

integrity of the aorta and its cellular components. Total procedure may last up to 4 h with no detrimental effects on the subsequent spontaneous angiogenic response.

18. The optimal length of aortic rings is found to be 1 mm; shorter lengths are more difficult to handle and position correctly and give rise to a lower level of outgrowth, probably due to a lower local concentration of growth factors. Longer lengths, although giving satisfactory outgrowth, would obviously lead to a lower yield per aorta.

19. While all preceding steps may be performed on the workbench in a culture laboratory, from here on perform work under a laminar flow hood. At this stage, the aortic segments may be pretreated to fluorescently label the endothelial cells *(39, 40)* or transfected with adenovirus small interfering RNA (siRNA) constructs, nude siRNA, or other constructs *(31)*.

20. The polymerisation of the collagen gel will vary with pH.

21. Prepared collagen solutions may be kept on ice up to 30 min; at room temperature, they will gel rapidly (3 min).

22. As for collagen, a concentration of 1.5 mg/mL is found to be optimal.

23. Take care to work at pH 7.4 for satisfactory gel polymerisation.

24. The time required for the analysis of each aortic ring is between 2 and 5 min, depending on the extent of outgrowth.

25. Care should be taken when using such automated image analysis techniques since artefacts in the captured image (such as incomplete removal of fibroadipose tissue or bubbles or other defects in the collagen or fibrin gels) may significantly alter results if not carefully monitored. Quantification is also highly sensitive to illumination level during image capture.

26. This method is similar to the "thin prep rat aortic ring assay" published by Zhu and Nicosia *(42)*, in which they used anti-α-smooth muscle actin to stain for vessel-associated pericytes.

Acknowledgements

The support of the BBSRC is acknowledged by D. C. W.

References

1. Auerbach, R., Lewis, R., Shinners, B., Kubai, L., Akhtar, N. (2003) Angiogenesis assays: a critical overview. *Clin Chem* 49, 32–40.

2. Staton, C. A., Stribbling, S. M., Tazzyman, S., Hughes, R., Brown, N. J., Lewis, C. E. (2004) Current methods for assaying angiogenesis *in vitro* and *in vivo*. *Int J Exp Pathol* 85, 233–248.

3. Montesano, R., Orci, L. (1985) Tumor-promoting phorbol esters induce angiogenesis *in vitro*. *Lab Invest* 42, 469–477.

4. Montesano, R., Pepper, M. S., Orci, L. (1990) Angiogenesis *in vitro*: morphogenetic and invasive properties of endothelial cells. *News Physiol Sci* 5, 75–79.

5. Madri, J. A., Pratt, B. M., Tucker, A. M. (1988) Phenotypic modulation of endothelial cells by transforming growth factor beta depends upon the composition and organization of the extracellular matrix. *J Cell Biol* 106, 1375–1384.

6. Nicosia, R. F., Ottinetti, A. (1990) Growth of microvessels in serum-free matrix culture of rat aorta—a quantitative assay of angiogenesis *in vitro*. *Lab Invest* 63, 115–122.

7. Schor, A. M., Ellis, I., Schor, S. L. (1999) Collagen gel assay for angiogenesis: induction of endothelial cell sprouting, in (Murray, J. C., ed.), *Methods in Molecular Medicine—Angiogenesis: Reviews and Protocols*, pp. 145–162. Humana Press, Totowa, NJ.

8. Madri, J. A., Merwin, J. R., Bell, L., et al. (1992) Interactions of matrix components and soluble factors in vascular responses to injury, in (Simionescu, N., Simionescu, M., eds), *Endothelial Cell Dysfunctions*, pp. 11–30. Plenum Press, New York.

9. Bishop, E. T., Bell, G. T., Bloor, S., Broom, I. J., Hendry, N. F. K., Wheatley, D. N. (1999) An *in vitro* model of angiogenesis: basic features. *Angiogenesis* 3, 335–344.

10. Nehls, V., Herrmann, R., Huhnken, M. (1998) Guided migration as a novel mechanism of capillary network remodelling is regulated by basic fibroblast growth factor. *Histochem Cell Biol* 109, 319–329.

11. Nakatsu, M. N., Sainson, R. C., Aoto, J. N., et al. (2003) Angiogenic sprouting and capillary formation modelled by human umbilical vein endothelial cells (HUVEC) in fibrin gels: the role of fibroblasts and angiopoietin. *Microvasc Res* 66, 102–112.

12. Pröls, F., Loser, B., Marx, M. (1998) Differential expression of osteopontin, PC4, and CEC5, a novel mRNA species, during *in vitro* angiogenesis. *Exp Cell Res* 239, 1–10.

13. Aitkenhead, M., Wang, S. J., Nakatsu, M. N., Mestas, J., Heard, C., Hughes, C. C. (2002) Identification of endothelial cell genes expressed in an *in vitro* model of angiogenesis: induction of ESM-1, (β)ig-h3, and NrCAM. *Microvasc Res* 63, 159–171.

14. Bell, S. E., Mavila, A., Salazar, R., et al. (2001) Differential gene expression during capillary morphogenesis in 3D collagen matrices: regulated expression of genes involved in basement membrane matrix assembly, cell cycle progression, cellular differentiation and G-protein signalling. *J Cell Sci* 114, 2755–2773.

15. Knedler, A., Ham, R. G. (1987). Optimized medium for clonal growth of human microvascular endothelial cells with minimal serum. *In Vitro Cell Dev Biol* 23, 481–491.

16. Nicosia, R. F., Bonanno, F., Villaschi, S. (1992) Large-vessel endothelium switches to a microvascular phenotype during angiogenesis in collagen gel culture of rat aorta. *Artherosclerosis* 95, 191–199.

17. Nicosia, R. F., Bonanno, F. (1991) Inhibition of angiogenesis *in vitro* by arg-gly-asp-containing synthetic peptide. *Am J Pathol* 138, 829–833.

18. Nicosia, R. F., Nicosia, S. V., Smith, M. (1994) Vascular endothelial growth factor, platelet-derived growth factor and insulin-like growth factor-1 promote rat aortic angiogenesis *in vitro*. *Am J Pathol* 145, 1023–1029.

19. Nicosia, R. F., Lin, Y. J., Hazelton, D., Qian, X. H. (1997) Endogenous regulation of angiogenesis in the rat aorta model—role of vascular endothelial growth factor. *Am J Pathol* 151, 1379–1386.

20. Derringer, K. A., Linden, R. W. A. (1998) Enhanced angiogenesis induced by diffusible angiogenic growth factors released from human dental pulp explants of orthodontically moved teeth. *Eur J Orthod* 20, 357–367.

21. Wakabayashi, T., Kageyama, R., Naruse, N., et al. (1997) Borrelidin is an angiogenesis inhibitor; disruption of angiogenic capillary vessels in a rat aorta matrix culture model. *J Antibiot* 50, 671–676.

22. Bocci, G., Danesi, R., Benelli, U., et al. (1998) Inhibitory effect of suramin in rat models of angiogenesis *in vitro* and *in vivo*. *Cancer Chemother Pharmacol* 43, 205–212.

23. Chen, C. H., Cartwright, J., Li, Z., et al. (1997) Inhibitory effects of hypercholes-

terolemia and Ox-LDL on angiogenesis-like endothelial growth in rabbit aortic explants—essential role of basic fibroblast growth factor. *Arterioscl Thromb Vasc Biol* 17, 1303–1312.

24. Burbridge, M. F., West, D. C., Atassi, G., Tucker, G. C. (1999) The effect of pH on angiogenesis *in vitro*. *Angiogenesis* 3, 281–288.

25. Burbridge, M. F. (2000) The rat aortic ring model of angiogenesis *in vitro* as an assay for angiogenic modulators. The role of the matrix metalloproteinases in vessel formation. Ph.D. thesis, University of Liverpool.

26. Burbridge, M. F., Coge, F., Galizzi, J.-P., Boutin, J. A., West, D. C., Tucker, G. C. (2002) The role of the matrix metalloproteinases during *in vitro* vessel formation. *Angiogenesis* 5, 215–226.

27. Brown, K. J., Maynes, S. F., Bezos, A., Maguire, D. J., Ford, M. D., Parish, C. R. (1996) A novel *in vitro* assay for human angiogenesis. *Lab Invest* 75, 539–555.

28. Jung, S. P., Siegrist, B., Wade, M. R., Anthony, C. T., Woltering, E. A. (2001) Inhibition of human angiogenesis with heparin and hydrocortisone. *Angiogenesis* 4, 175–186.

29. Stiffey-Wilusz, J., Boice, J. R., Ronan, J., Fletcher, A. M., Anderson, M. S. (2001) An *ex vivo* angiogenesis assay utilizing commercial porcine carotid artery: modification of the rat aortic ring assay. *Angiogenesis* 4, 3–9.

30. Li, Q., Olson, B. R. (2004) Increased angiogenic response in aortic explants of collagen XVIII/ endostatin-null mice. *Am J Pathol* 165, 415–424.

31. Masson, V., Devy, L., Grignet-Debrus, C., et al. (2002) Mouse aortic ring: a new approach of molecular genetics of angiogenesis. *Biol Protocol Online* 4, 24–31.

32. Chun, T.-H. , Sabeh, F., Ota, I, et al. (2004) MT1-MMP-dependent neovessel formation within the confines of the three-dimensional extracellular matrix. *J Cell Biol* 167, 757–767.

33. Zhu, W.-H. Iurlaro, M., MacIntyre, A., Fogel, E., Nicosia, R. F. (2003) The mouse aorta model: Influence of genetic background and aging on bFGF- and VEGF-induced angiogenic sprouting. *Angiogenesis* 6, 193–199.

34. Nicosia, R. F., Zhu, W.-H., Fogel, E., Howson, K. M., Aplin, A. C. (2005) A new *ex vivo* model to study venous angiogenesis and arterio-venous anastamosis formation. *J Vasc Res* 42, 111–119.

35. Antes, L. M., Villar, M. M., Decker, S., Nicosia, F. F., Kujubu, D. A. (1998) A serum-free *in vitro* model of renal development. *Am J Physiol* 274, F1150–F1160.

36. Kiefer, F. N., Munk, V. C., Humar, R., Dieterle, T., Landmann, L., Battegay, E. J. (2004) A versatile *in vitro* assay for investigating angiogenesis of the heart. *Exp Cell Res* 300, 272–282.

37. Gulec, S. A., Eugene, A., Woltering, M. D. (2003) A new *in vitro* assay for human tumor angiogenesis: three-dimensional human tumor angiogenesis assay. *Ann Surg Oncol* 11, 99–104.

38. Woltering, M. D., Lewis, J. M., Maxwell, P. J., et al. (2003) Development of a novel *in vitro* human tissue-based angiogenesis assay to evaluate the effect of antiangiogenic drugs. *Ann Surg* 237, 790–800.

39. Blacher, S., Devy, L., Noel, A., Foidart, J.-M. (2003) Quantification of angiogenesis on the rat aortic ring assay. *Image Anal Stereol* 22, 43–48.

40. Blacher, S., Devy, L., Burbridge, M. F., et al. (2001) Improved quatification of angiogenesis in the rat aortic ring assay. *Angiogenesis* 4, 133–142.

41. Freshney, R. I. (1987) In *Culture of Animal Cells. A Manual of Basic Technique*. pp. 71–103. Liss, Inc., New York.

42. Zhu, W.-H., Nicosia, R. F. (2002) The thin prep rat aortic ring assay: a modified method for the characterisation of angiogenesis in whole mounts. *Angiogenesis* 5, 81–86.

Chapter 12

Static and Dynamic Assays of Cell Adhesion Relevant to the Vasculature

Lynn M. Butler, Helen M. McGettrick, and Gerard B. Nash

Abstract

Methods are described for analysing adhesion of isolated cells (such as leucocytes, tumor cells, or precursor cells) to purified adhesion receptors or cultured endothelial cells. "Static" assays (in which cells are allowed to settle on the adhesive substrates) and flow-based assays (in which cells are perfused over the substrates) are compared. Direct observations of the time course of adhesion and migration can be made when purified proteins or endothelial cells are cultured in plates, after cells are allowed to settle onto them for a desired period. In the flow-based assay, cells are perfused through coated glass capillaries or flow channels incorporating coated plates. Again, direct video-microscopic observations are made. In this assay, various stages of capture, immobilisation and migration can be followed. In general, the static systems have higher throughput and greatest ease of use, but yield less-detailed information, while the flow-based assay is most difficult to set up but is most physiologically relevant if one is interested in the dynamics of adhesion in the vasculature.

Key words: Adhesion, blood vessel, endothelial cells, flow, leucocyte, tumor cells.

1. Introduction

Adhesion of several types of cells in the vascular system may be relevant to the process of angiogenesis: leucocytes, platelets, metastatic tumor cells, endothelial cell progenitors, and stem cells. We have recently described methods for studying dynamics of leucocyte migration (1) and attachment of flowing platelets (2) on endothelial monolayers. Here, we concentrate on "generic" methods, both static and with imposed flow, for studying cell adhesion to purified proteins or to cultured endothelial cells. These may be adapted for a variety of adhesion proteins or endothelial

S. Martin and C. Murray (eds.), *Methods in Molecular Biology, Angiogenesis Protocols, Second edition, Vol. 467*
© Humana Press, a part of Springer Science + Business Media, LLC 2009
DOI: 10.1007/978-1-59745-241-0_12

stimuli or for different adherent cells, relevant to specific scenarios chosen for investigation.

In general, flow-based assays are required to evaluate capture processes occurring in the vasculature, where cells are initially travelling rapidly on a scale of the receptors used. Microscopic observation of such models allows the different steps in the adhesion process to be dissected (e.g., capture, rolling, activation-dependent stabilisation, and migration observed for leucocytes) *(3–5)*. While these stages have been well described for flowing leucocytes, their existence for other cell types remains unclear and worthy of examination. Static assays are easier to set up and utilise, but results relating to the roles of specific receptors may be ambiguous when several are available, and one cannot be certain whether an interaction described could occur under more realistic flow conditions in the vasculature. Specificity in assays (static or flow based) can be improved by choosing individual adhesion receptors to coat surfaces, while endothelial cells may be used to generate more "physiological" data, provided one realises that the definition of the operative adhesive mechanisms may require subsequent detailed evaluation. Either type of surface can be used to assess motility of adherent cells, with endothelial monolayers offering the potential to study transendothelial diapedesis.

In this chapter, we describe static assays on purified protein and on cultured human umbilical vein endothelial cells (HUVECs). The proteins most relevant for these assays would be basement membrane constituents such as collagen, laminin, or fibronectin. HUVECs are widely used primary cells capable of expressing a range of adhesion molecules. Alternatively, endothelial cell lines or endothelial cells with derivation that is described elsewhere in this book might also be used. In the following description of flow-based assays, relevant purified proteins are those adapted for "capture," such as E-selectin, P-selectin, or vascular cell adhesion molecule 1 (VCAM-1). Endothelial cells would typically be stimulated with agonist such as inflammatory cytokines (tumor necrosis factor-α [TNF-α], or interleukin 1 [IL-1]), so that such capture receptors are expressed *(4, 6)*.

2. Materials

All reagents are from Sigma (Sigma-Aldrich, Gillingham, UK) unless stated otherwise.

2.1 Culture of Endothelial Cells

1. Medium 199 with glutamine (M199; Gibco) supplemented with gentamycin sulphate (35 µg/mL), human epidermal growth factor (10 ng/mL; Sigma E9644), and foetal calf

Color Plates

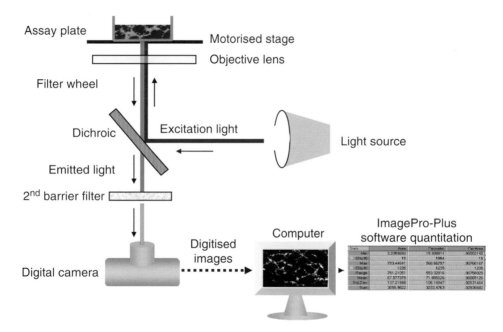

Fig. 9.2. Diagram of the image analysis setup used to quantify cell migration, invasion, haptotaxis, and tubular differentiation. The example shown is the tubular differentiation.

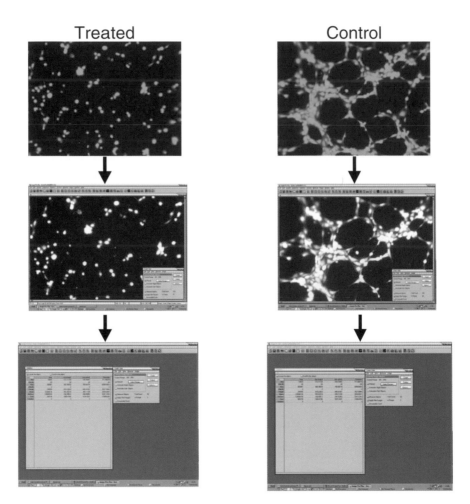

Fig. 9.3. Endothelial cell tubularisation and quantitation using Image-Pro Plus (area, branching, branch length, etc).

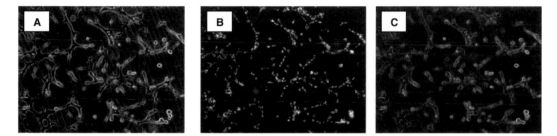

Fig. 14.3. *In vitro* angiogenesis. The angiogenesis of wound healing can be quantified using the *in vitro* fabricated layer-on-layered three-dimensional (3D) skin construct. Capillary endothelial cells are prelabeled by incubating monolayer cultures with 50–100 µg/mL tetramethylrhodamine-dextran sulfate (10,000 MW at 37°C, Molecular Probes). Prelabeled endothelial cells can then be trypsinized and plated within or on the 3D matrix for *in vitro* angiogenesis analyses. Shown are phase contrast (**A**), rhodamine-fluorescence (**B**), and merged, pseudocolored composite image (**C**; *green* phase contrast, *red* rhodamine dextran).

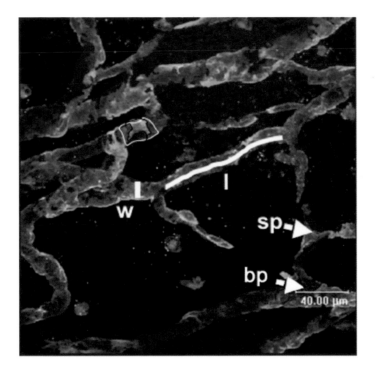

Fig. 15.2. Measurement of microvascular parameters on stained mesenteries. *Red* isolectin-stained vessels, *green* NG2-stained pericytes. Parameters of width (measured line by w), length (measured line by l), sprouts (sp), branches (bp), and pericyte coverage (delineated green divided by delineated red areas by pericyte). Note that the pericytes can be seen right up to the tips of the sprouts (grey arrowhead).

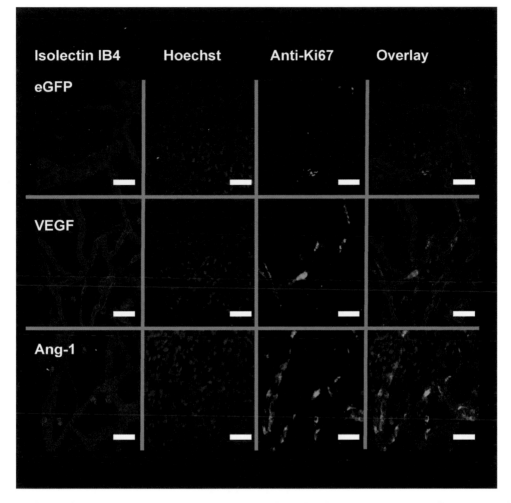

Fig. 15.4. Growth factor expression stimulates endothelial cell (EC) proliferation. Fluorescent staining of mesenteries after injection of adenovirus-enhanced green fluorescent protein (Ad-eGFP) (control) or Ad-growth factors. Isolectin IB4-TRITC (red) stains endothelial cells ECs, Hoechst 33324 (blue) stains all mesenteric nuclei and antibodies to Ki67-AF488 (green) to detect proliferating cells. Overlaying of the stack images is used to calculate the number of proliferating endothelial cells (PECs). Images are triple stained with TRITC-streptavidin and biotinylated GSL (Griffonia Simplicifolia Lectin, Ec, red) lectin IB4 (EC, red). Alexa Fluor 488-labeled goat antimouse immunoglobulin G and mouse monoclonal anti-Ki67 antibody (proliferating cells, green) and overlay. The PECs can be distinguished from other cells by their position within the vessel wall. White arrow demonstrates a PEC. *VFGF* vascular endothelial growth factor. Scale bar 40 μm

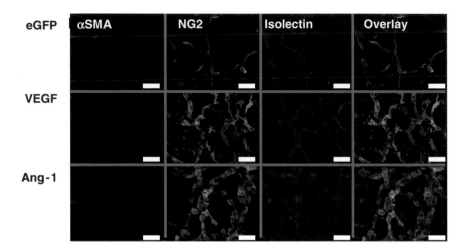

Fig. 15.7. Growth factor expression stimulates endothelial cell (EC) proliferation. Fluorescent staining of mesenteries after injection of adenovirus-enhanced green fluorescent protein (Ad-EGFP) (control) or Ad-growth factors. Isolectin IB4-TRITC (red) stains ECs, Hoechst 33324 (blue) stains all mesenteric nuclei and antibodies to Ki67-AF488 (green) to detect proliferating cells. Overlaying of the stack images is used to calculate the number of proliferating endothelial cells (PECs). Images are triple stained with TRITC-streptavidin and biotinylated GSL lectin IB4 (ECs, red). Alexa Fluor 488-labeled goat antimouse immunoglobulin G and mouse monoclonal anti-Ki67 antibody (proliferating cells, green) and overlay. The PECs can be distinguished from other cells by their position within the vessel wall. White arrow demonstrates a PEC. Scale bar 40 μm. *SMA* smooth muscle area

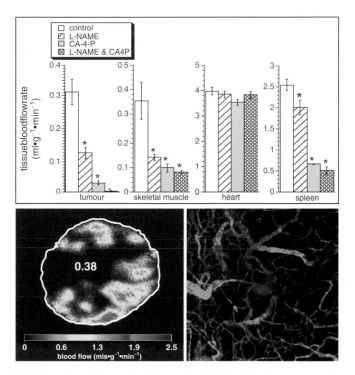

Fig. 16.2. Estimation of tissue blood flow rate **F** in the P22 rat sarcoma and several normal rat tissues using the tissue equilibration method with ^{125}I-IAP (**a**) and ^{14}C-IAP (**b**). Results in panel (**a**) were obtained by calculating C_{tiss} from gamma counts of large tissue samples. Data, from left to right, shows the effects of saline, the vascular disrupting agent combretastatin A4-P, the nitric oxide synthase inhibitor, nitro-L-arginine (L-NNA) and tumor the combination of the two. Asterices represent a significant difference from the control group. The image in panel (**b**) was obtained from an untreated P22 tumor by calculating multiple values for C_{tiss} from quantitative autoradiography of tumor sections. The mean **F** is 0.8 mL·g^{-1}·min^{-1}. The image in panel (**c**) illustrates the vascular networks in the P22 tumor obtained by multiphoton fluorescence microscopy.

serum (20% v/v heat inactivated). Adding hydrocortisone (1 μg/mL from 10 mg/mL stock in ethanol) improves growth if going beyond first passage.

2. Phosphate-buffered saline (PBS) with 1 mM Ca^{2+} and 0.5 mM Mg^{2+} (Gibco, Invitrogen, Paisley, UK).

3. Bovine skin gelatine (type B, 2% solution, culture tested).

4. Collagenase (type IA) stored at −20°C at 10 mg/mL in PBS. Thaw and dilute to 1 mg/mL with M199 for use.

5. Autoclaved cannulae and plastic electrical ties.

6. Ethylenediaminetetraacetic acid (EDTA) solution (0.02%, culture tested).

7. Trypsin (culture tested).

8. 70% (v/v) ethanol or industrial methylated spirits.

9. TNF-α and iIL-1β, stored in aliquots at 80°C.

2.2. Purified Proteins (see Note 1)

Purified proteins are all dissolved in PBS and stored at −20°C unless stated otherwise.

1. Basement membrane or extracellular matrix proteins: Collagen type IV (human placenta); fibronectin (human foreskin fibroblasts); laminin type 1 (human placenta); human laminin type 10 (human placenta) (Chemicon).

2. Capture receptors for flowing cells, such as recombinant human VCAM-1 (R&D Systems, Abingdon, UK); recombinant human P-selectin stop protein (R&D Systems).

3. Bovine serum albumin (BSA) (diluted in PBS from 7.5% w/v culture-tested solution, typically to 1% w/v).

2.3. Surfaces for Protein Coating or Endothelial Culture

1. Multiwell plates (Falcon, tissue culture treated; Becton Dickinson Labware, Franklin Lakes, NJ); we typically use six-well plates, but other formats may be used (*see* **Subheadings 3.1.1.** and **3.2.3.**).

2. Chamber slides: Plastic chamber with cover, mounted on glass slide (Lab-tek, Nalge Nunc International, Naperville, IL).

3. Microslides: Glass capillaries with rectangular cross section (0.3 × 3 mm; 50 mm length) (Vitro Dynamics, Rockaway, NJ, USA; available through Camlab, Cambridge, UK).

4. Aminopropyltriethoxysilane (APES) (4% v/v) in acetone with mole-cular sieve (BDH Laboratory Supplies, Poole, UK) added to ensure anhydrous.

5. A special culture dish for use with microslides (**Fig. 12.1**) constructed by fusing either three or six glass tubing side arms into the wall of a Pyrex glass Petri dish 100 mm in diameter (Fisons Scientific Equipment, Leicester, UK.) and made to order by the Glassblowing Workshop of the School of Chemistry, University of Birmingham, United Kingdom.

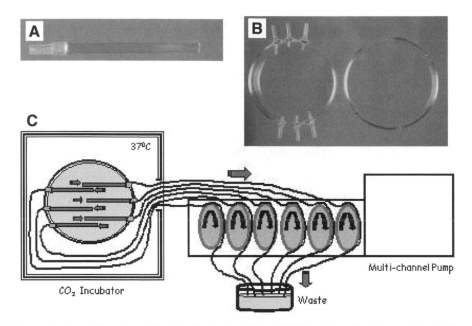

Fig. 12.1. Apparatus for culture of endothelial cells in microslides. **(A)** Microslide with silicone rubber tubing adapter attached. **(B)** Glass culture dish with six ports fused into the wall 90-mm diameter, ~50 mL volume of culture. (C) Schematic diagram of apparatus for culture of endothelial cells in six microslides. The glass culture dish with six microslides attached is placed in a CO_2 incubator. Tubing attached to each outlet passes through a port in the incubator wall. Perfusion is supplied for 30 s of each hour by a multichannel roller pump that pumps to waste.

2.4. Flow-Based Adhesion Assay

1. Flow system: Syringe pump with smooth flow (e.g., PHD2000 infusion/withdrawal, Harvard Apparatus, South Natick, MA, USA). Electronic three-way microvalve with zero dead volume (LFYA1226032H, Lee Products, Gerrards Cross, Buckinghamshire, UK) with 12-V direct current power supply for valve. Silicon rubber tubing, internal diameter/external diameter (ID/OD) of 1/3 and 2/4 mm, respectively (Fisher Scientific, Loughborough, UK). Scotch double-sided adhesive tape, about 1 cm wide (3M, Bracknell, Berkshire, UK). Three-way stopcocks (BOC Ohmeda AB, Helsinborg, Sweden). Sterile, disposable syringes (2-, 5-, 10-mL Becton Dickinson, Oxford, UK) and glass 50-mL syringe for pump (Popper Micromate; Popper and Sons, New York, NY, USA).

2. For chamber slides: Custom-made flow channel assembly, made to order by Wolfson Applied Technology or by suitable workshop (**Fig. 12.2**). This includes a gasket, cut to the size of the chamber slide, with a slot that forms the flow channel when the assembly is clamped together (**Fig. 12.2**). The gasket can be cut fresh each time using lengths of Parafilm sealing film and a scalpel (typical depth ~ 130 μm) or cut from silicone rubber sheet (e.g., 250-μm thick; ESCO, Bibby Sterilin., Staffordshire, UK) and reused. In either case, the

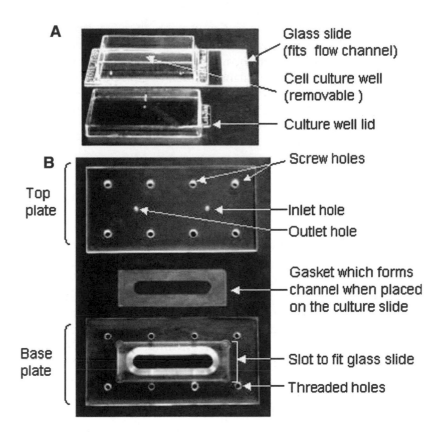

Fig. 12.2. Apparatus for flow-based adhesion assay using chamber slides. **(A)** Photograph of chamber slide incorporated in flow channel. **(B)** Photograph of parallel plate flow channel to hold chamber slide.

depth of the channel should be checked after the system is clamped together by focussing alternately on the upper and lower surfaces of the channel on a microscope using a calibrated microscope focus control.

3. Using coated, 35-mm diameter culture dishes, a commercially available flow channel is available from GlycoTech (www.glycotech.com). We have no experience of this system, but it has been used in flow-based adhesion assays by others (e.g., **ref.** 7).

4. Video microscope: Microscope with heated stage or, preferably, stage and attached flow apparatus enclosed in a temperature-controlled chamber at 37°C and phase contrast optics. Fluorescence capability is desirable for some variants of assay. Video camera (e.g., analogue Cohu 4912 monochrome camera with remote gain control), monitor, and video recorder (e.g., time lapse; Panasonic AG-6730) or digital camera for direct capture to computer.

5. Image analysis: Computer with video capture card (if recording to video) and specialist software for counting cells, measuring motion, and so on. There is a range of commercial packages available, as well as image analysis software available free over the Internet (NIH Image, http://rsb.info.nih.gov/nih-image/). We currently use Image Pro software (DataCell Limited, Finchampstead, UK).

3. Methods

This chapter aims to provide detailed protocols for adhesion assays that can be used for various cell types (such as leucocytes, stem cells, or tumor cell lines). Cell isolation protocols are given in the chapters in Section II of this volume or in previous works on leucocyte adhesion (1). Isolation of HUVECs is described. We primarily use microscopy to directly examine the adhesive behaviour of different leucocyte subsets to purified proteins or endothelial cells (but *see* **Note 2**). For flow-based assays, various systems can be used. We have used glass capillaries (microslides) coated with either purified proteins or cultured endothelial cells (8). Microslides are economical (content only ~ 50 µL) and easy to use with purified proteins, but endothelial cells require a specialised culture system (*see* **Subheading 3.2.5.**). For endothelial cells, it may be easier initially to culture in commercially available chamber slides and incorporate them in a custom-made flow channel (**Fig. 12.2**) or to use 35-mm dishes and the GlycoTech system (*see* **Subheading 2.4.3.**). Flow systems have been used mainly with leucocytes, and they will need to be optimised (e.g., with respect to flow rates, cell concentrations, etc.) for other cell types. In addition, pretreatment of the endothelial cells (e.g., with cytokines) may be necessary to stimulate them to present adhesion receptors for the cells under study.

3.1. Preparing Surfaces Coated with Purified Protein for Adhesion Assays

3.1.1. Coating Multiwell Plates (see **Note 3**) with Purified Proteins

1. Make up the purified protein (e.g., a basement membrane or extracellular matrix constituent) to the desired concentration (*see* **Note 4**).

2. Add 1 mL to each well in a six-well format or proportionately less in smaller-well formats.

3. Incubate for 2 h or overnight at 37°C.

4. Aspirate off residual protein solution and add 2mL 1% (w/v) BSA to each well for 2 h to block nonspecific binding sites.

3.1.2. Coating Microslides with Purified Proteins

3.1.2.1. Pretreatment with APES

1. Immerse microslides in nitric acid (50% v/v in distilled water) for 24 h (e.g., in batch of 100–300).

2. Wash thoroughly in beaker using running tap water and rinse through with deionised distilled water.

3. Blot water on tissue paper and dry microslides at 37°C .

4. Place in polystyrene tubes and rinse twice with anhydrous cetone by gently inverting the tubes for 30 s.

5. Immerse in a freshly prepared solution of APES (4% v/v in anhydrous acetone) for 1 min, ensuring all capillaries are filled (*see* **Note 5**).

6. Remove microslides from the APES and blot out onto tissue, ensuring all capillaries are emptied.

7. Reinsert into a fresh aliquot of the APES solution for a further 1 min.

8. Remove APES by blotting and rinse the microslides once with anhydrous acetone, followed by three washes with deionised distilled water, and then dry at 37°C. Between each change, care must be taken to remove all the liquid from the microslides.

9. Attach a short length of 2-mm ID silicon rubber tubing to APES-coated microslide. The tubing assists in handling and filling of microslides and is required for adaptation to the endothelial culture system.

10. Autoclave the microslides at 121°C for 11 min and store aseptically indefinitely.

3.1.2.2. Coating with Purified Proteins

1. Using a pipettor inserted in the silicon tubing adaptor, aspirate about 50 µL of adhesion receptor at desired concentration into a microslide (*see* **Note 6**).

2. Incubate at 37°C for 2 h.

3. Aspirate about 100 µL of PBS with 1% w/v bovine albumin into a microslide as above and incubate at 37°C for 2 h to block nonspecific binding sites.

3.2. Culture of Endothelial Cells on Chosen Surfaces

There are various methods for culture of endothelial cells from different sources, and for the novice, it is probably best to start by buying cells and media from commercial suppliers. Our current method for isolating and culturing HUVECs is given, adapted from Cooke et al. (8), who also described methods for coating microslides (*see* **Subheading 3.1.2.1.**) and culturing HUVECs in them (*see* **Subheading 3.2.5.**).

3.2.1. Isolation and Primary Culture of HUVECs

1. Place the cord on paper towelling in a tray and spray liberally with the 70% ethanol. Choose sections about 3–4 in. that do

not have any clamp damage. Each 3- or 4-in. piece of cord equates to a 25-cm^2 flask of primary cells.

2. Locate the two arteries and one vein at one end of the cord.

3. Cannulate the vein and secure the cannula with an electrical tie.

4. Carefully wash through the vein with PBS using a syringe and blow air through to remove the PBS.

5. Cannulate the opposite end of the vein and tie off.

6. Inject collagenase (~10 mL per 3–4 in.) into vein until both cannulae bulbs have the mixture in them.

7. Place the cord into an incubator for 15 min at 37°C.

8. Remove from the incubator and tighten the ties. Massage the cord for about 1 min.

9. Flush the cord through using a syringe and 10 mL PBS into a 50-mL centrifuge tube.

10. Push air through to remove any PBS; repeat this twice more (3 × 10 mL).

11. Centrifuge at 400g for 5 min. Discard supernatant.

12. Resuspend the cells in about 1 mL of culture medium and mix well with pipet.

13. Make up to 4 mL in complete medium.

14. Add cell suspension to a 25-cm^2 culture flask.

15. Culture at 37°C in a 5% CO$_2$ incubator.

16. Change medium after 2 h, again the next day, and every 2 d thereafter. Cells should be confluent in 3–7 d.

3.2.2. Dispersal of Endothelial Monolayers for Seeding New Surfaces

1. Rinse a flask containing a confluent primary monolayer of HUVECs with 2 mL EDTA solution.

2. Add 2 mL of trypsin solution and 1 mL of EDTA for 1–2 min at room temperature until the cells become detached. Tap on bench to loosen.

3. Add 8 mL of culture medium to the flask to neutralise the trypsin and remove the resulting suspension and centrifuge at 400g for 5 min.

4. Remove supernatant and resuspend the cell pellet in 0.5 mL of culture medium and disperse by sucking them in and out of a 1-mL pipet tip.

5. Make up to desired volume of culture medium for seeding onto the chosen surface.

*3.2.3. Seeding HUVECs in Multiwell Plates (see **Note 3**)*

1. Add 1mL of 1% gelatine (in PBS) to each well in a six-well format or proportionately less in smaller-well formats for 15 min.

2. Trypsinise a single flask of HUVECs as in **Subheading 3.2.2.**

3. Make up to 8 mL with culture medium (*see* **Note** 7) and add 2 mL of HUVEC suspension to each of four wells or proportionately less in smaller- well formats.

4. Culture at 37°C in a 5% CO_2 incubator.

5. Replace the medium 24 h later and culture for 1–3 d (*see* **Note** 7).

6. If required, add cytokine stimulant, such as TNF (100 U/ mL) or IL-1 (5×10^{-11} g/mL), to cultures for desired period before assay (e.g., typically 4–24 h when studying neutrophils or lymphocytes).

3.2.4. Seeding HUVECs in Chamber Slides

1. Trypsinise a single flask of HUVECs as in **Subheading 3.2.1.**

2. Make up to 6 mL with culture medium (*see* **Note** 7).

3. Remove chamber cover and add 2 mL to each Lab-Tek chamber slide (Nalge Nunc International, Naperville, IL) (**Fig. 12.2**).

4. Replace plastic cover.

5. Use the tray in which the chamber slides come as an incubation rack and incubate at 37°C in a 5% CO_2 incubator.

6. Replace the medium after 24 h and culture for 1–3 d. (*see* **Subheading 3.2.3, point 6.**)

7. Add cytokine stimulant (as suggested above) to culture for desired period before migration assay.

3.2.5. Culturing HUVECs in Microslides (see **Note** 8)

1. Prepare the special culture dish (**Subheading 2.3.5.**) by attaching a length of silicon rubber tubing (40 cm long) onto each external arm.

2. Autoclave the dish with tubing attached at 121°C for 11 min before use.

3. Draw in 1% gelatine (in PBS) using a pipettor with tip inserted into the adaptor tubing on microslides. Allow to coat for 30 min. Microslides hold about 50 µL.

4. Wash microslides with PBS followed by air to remove excess gelatine using a pipettor with tip inserted into the adaptor tubing.

5. Trypsinise a single flask of HUVECs as in **Subheading 3.2.2.**

6. Resuspend cell pellet to about 400 µL with culture medium and transfer to one corner of a tilted 35-mm Petri dish.

7. Aspirate about 50 µL of cell suspension into each of the six microslides using a pipettor inserted in the silicon tubing adaptor.

8. Place a sterile, glass microscope slide inside a 100-mm Petri dish, rest the filled microslides across it (to keep them horizontal), and incubate in the dish at 37°C for 1 h.

9. Add 50 mL of culture medium to the special culture dish, prime the tubing, and clamp ends.

10. Connect the microslides aseptically to the internal side arms of the special culture dish, via adaptor tubing, using sterile forceps.

11. Placed the dish in a humidified CO_2 incubator and pass the silicon tubing through a service port located in the incubator wall (manufactured to order; e.g., either model GA2000, LEEC, Nottingham, UK, or Nuaire DH, Triple Red, Thame, UK).

12. Attach the tubing to individual channels of a multichannel roller pump (e.g., Watson Marlow 500 series pump with 308MC pumpheads; Watson-Marlow Bredel Pumps, Falmouth, UK), itself linked to a timed power supply (e.g., RS Components, Corby, UK).

13. Pump medium through each microslide to waste (*see* **Fig. 12.1** for layout) at a flow rate of about 0.2 mL/min for 30 s each hour to change medium contained in the microslides.

14. The original seeding is designed to yield confluent monolayers within 24 h.

15. After 24 h, treat HUVECs with TNF (100 U/mL) or IL-1 (5 × 10^-11 g/mL) if desired. Cytokines can be added to the dish to treat all microslides equally. Or, detach the microslides with tubing adaptor and place them in a separate disposable plastic culture dish. Aspirate differently diluted cytokines into the separate microslides as desired and repeat aspiration at hourly intervals.

3.3. Static Cell Adhesion Assays

3.3.1. Cell Adhesion to Purified Proteins or Endothelial Cell Monolayers (see **Note 2**)

1. Prewarm the microscope and (PBSA) PBS with 0.15% BSA wash buffer to 37°C.

2. Pretreat the endothelial cells with cytokine if desired.

3. Rinse the surface of the plate/monolayer (six-well format) with 2 mL PBSA, using a plastic pipette to remove any residual protein.

4. Add 2 mL of cell suspension to each well.

5. Leave to settle for desired time (*see* **Note 9**).

6. Aspirate off the cellular suspension and gently rinse twice with PBS.

7. Add 2 mL PBSA and view the well under phase contrast video microscope with an objective magnification of ×20.

8. Make video/digital recordings immediately after the wash stage, choosing at least five different fields at random and record them for 5 s each to allow counting of adherent cells. If migratory behaviour is of interest, choose one field at random and record continuously for 5 min (for leucocytes) or longer for analysis of movement (*see* **Note 10**).

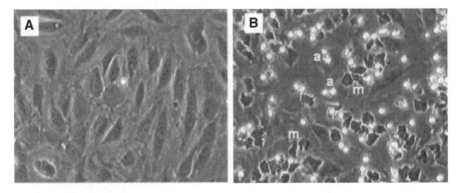

Fig. 12.3. Phase contrast micrographs of confluent monolayers of (A) untreated HUVECs or (B) HUVECs treated with 100 U/mL TNF after completion of a neutrophil adhesion assay. Neutrophils were settled for 5 min on HUVECs, and the non-adherent cells were washed off and allowed to migrate for a further 15 min. (B) Phase-bright neutrophils are adherent to the surface of the HUVECs (a) and phase-dark spread cells (m) are migrated underneath the monolayer.

3.3.2. Data Analysis

1. Make video/digital recordings of a microscope stage micrometer to calibrate the size of the field observed on the monitor and to calibrate scale of the image analysis system.

2. To measure cell adhesion: Count all cells visible in each video field (*see*, e.g., **Fig. 12.3**) (*see* **Note 11**). Take the average for the fields and convert to number per square millimeter using known dimensions of field. Multiply this by the area of the wells (9.6 cm² for six wells) and divide by the number of cells added. Multiply by 100 to obtain percentage of cells adherent.

3. To measure cell migration: Take images at intervals from the prolonged video/digital sequence into a program such as Image Pro in a sequence (e.g., at 1-min intervals for leucocytes). Outline cells using a pointer and record the positions of their centroids. Calculate the distances and directions of cell migration in each interval. Calculate the average migration velocity over time and the direction of migration as required.

4. To measure cell transmigration in endothelial cell assays (*see* **Note 11**): Express the count of phase-dark cells in each video field as a proportion of the total number of adherent cells. Take the average for the repeated fields. For leucocytes, phase-dark cells are under the monolayer. This can be verified, for example, by removing the endothelial layer with trypsin to leave the transmigrated cells *(5)*.

3.4. Flow-Based Assay of Cell Adhesion

3.4.1. Setting Up the Flow Assay

1. Assemble flow system shown in **Fig. 12.4** but without chamber slide or microslide attached. The electronic valve has a common output and two inputs, from "Wash reservoir" and "Sample reservoir," that can be selected by turning the electronic valve on or off.

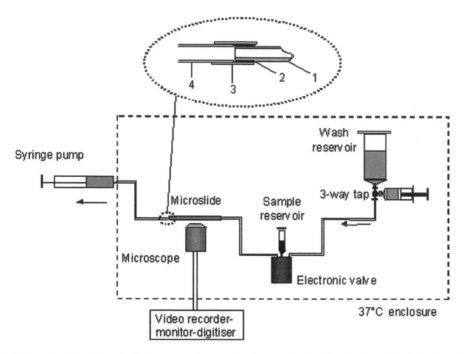

Fig. 12.4. Schematic of flow-based adhesion assay flow system. In expanded section: 1 microslide, 2 double-sided tape, 3 silicon rubber tubing with ID/OD 2/4 mm, 4 silicon rubber tubing with ID/OD 1/2 mm.

2. Fill wash reservoir with PBSA and rinse through all tubing, valves, and connectors with PBSA, ensuring bubbles are displaced (e.g., using syringe attached to three-way tap for positive ejection). Fill sample reservoir with PBSA and rinse through valve and attached tubing. Prime downstream syringe and tubing with PBSA and load into syringe pump. All tubing must be liquid filled.

3.4.2. Connecting the Chamber Slides (Endothelial Cell Coated)

1. Remove the medium from the chamber slide.

2. To detach slide from the chamber, grip one end of slide and gently squeeze both ends of chamber toward the centre, lifting the chamber as its sealing gasket releases.

3. To remove the gasket: Insert the tip of a thin-blade spatula or similar tool under gasket at one corner. Without stretching or tearing the gasket, smoothly lift it away from the slide (*see* **Note 12**).

4. Use dampened cotton bud to remove any gasket residue.

5. Put the glass slide in the bottom Perspex plate of the flow channel and place a precut gasket over the slide. Put the top Perspex plate onto the gasket (**Fig. 12.2**).

6. Screw the top and bottom Perspex plates together evenly, tightening diagonally opposite screws alternately, a bit at a time.

7. Put the tubing into the inlet and outlet holes in the top Perspex plate of flow chamber.

3.4.3. Connecting the Microslides (Purified Protein or Endothelial Cell Coated)

1. Glue a coated microslide across the middle of a glass microscope slide using two spots of cyanoacrylate adhesive (Superglue Locktite UK, Welwyn Garden City, UK) applied to the edges of the slide. Discard tubing adaptor used for filling.

2. Wrap double-sided adhesive tape around each end of the microslide without obstructing lumen.

3. Connect microslide to silicon rubber tubing by pushing over each of the taped ends. Start at the upstream (sample) end to avoid injection of air. Squeeze the 2-mm ID silicone rubber tubing to flatten and ease over rectangular end of microslide, one corner at a time.

3.4.4. Perfusing Cellular Suspension and Recording Behaviour

1. Place chamber slide or microslide onto microscope stage and start flow by turning on syringe pump in withdrawal mode, with electronic valve and three-way tap in position to allow delivery of PBSA from wash reservoir.

2. Wash out culture medium and observe surface using phase contrast microscopy.

3. Adjust flow rate to that required for assay (*see* **Note 13**).

4. Perfusion of leucocytes through microslides in our group is typically carried out at a flow rate (Q = 0.37 mL/min) equivalent to a wall shear rate of 140 s^{-1} (which is comparable to that found in postcapillary venules) and wall shear stress of 0.1 Pa (= 1 dyn/cm^2).

5. Load isolated cells into sample reservoir (**Fig. 12.4**) and allow to warm for 5 min.

6. Switch the electronic valve so cell suspension is drawn through the microslide.

7. Deliver timed bolus (e.g., 4 min). Typically, flowing cells will be visible after about 30 s, the time required to displace dead volume in valve and tubing.

8. Switch electronic valve so that PBSA from wash reservoir is perfused. Again, 30–60 s will be required before all cells have been washed through the chamber slide or microslide.

9. Video recordings can be made as desired during inflow and washout of cells. Typically, a series of fields should be recorded along the centre line of the microslide during inflow (e.g., six fields recorded for 10 s each during the last minute of the bolus) for offline analysis of the behaviour (e.g., rolling or stationary adhesion) of the cells. Another series should be made after 1-min washout (when the bolus is complete) for analysis of the number of adherent cells. Fields are recorded over a desired period or at later times (e.g., after a further 5 and 10 min) to assess cellular movement or transmigration through the monolayer when endothelial cells are used. All analyses are made offline.

10. If a defined timing protocol is developed, digital images or sequences of digital images could be recorded instead of video images. The continual recording of the latter gives flexibility in analysis.

3.4.5. Analysis of Cell Behaviour from Video Recordings

1. Calibrate size of video fields and image analysis system as in **Subheading 3.3.2.**

2. Digitise a sequence of images at 1-s intervals from recordings made at desired times.

3. Count cells present on a stop-frame video field at the start of a sequence. Repeat and average the counts for the series of sequences recorded (e.g., after washout of nonadherent cells). Convert this to count of adherent cells/square millimeter.

4. Divide this by the number of cells perfused (in units of 10^6 cells) to obtain number adherent/square millimeter/10^6 perfused. The number perfused is simply calculated by multiplying the concentration of the suspension (typically 10^6/mL) by the flow rate (e.g., 0.37 mL/min) by the duration of the bolus (e.g., 4 min). This normalisation allows correction for changes in conditions (bolus duration, cell concentration, flow rate) between experiments and effectively calculates an efficiency of adhesion.

5. The count includes cells that are rolling (circular phase-bright cells tumbling slowly at ~ 1–10 µm/s over the surface) or stabley adherent on the endothelial surface (phase-bright cells typically with distorted outline and migrating slowly on the surface at < 10 µm/min) or transmigrated cells (phase-dark spread cells migrating under the HUVECs at > 10 µm/min). Nonadherent cells will only be visible as blurred streaks.

6. To assist in obtaining data for proportion of adherent cells behaving in the different ways, play the digitised sequence as a loop. Observing cells in turn, it is easy to classify them as rolling, stabley adherent, or transmigrated.

7. Repeat the analysis at different times (e.g., after 1, 5, or 10 min of washout) to quantify the progress of migration through the endothelium or any changes in behaviour, such as transmigration through the endothelial monolayer.

8. To measure rolling velocity, mark the leading edges of a series of cells to be followed and move to a second captured frame. Re-mark the leading edges and record the distance moved. Repeat through the 10-s sequence. This will yield data for position versus time. Velocity for each cell can be averaged over the observation time and estimates of variation in velocity made if desired.

4. Notes

1. The proteins suggested are typical of those used in static or flow-based assays for leucocytes. Previous studies of leucocyte adhesion have used laminin type I from Engelbreth-Horm-Swarm sarcoma (EHS Sarcoma), but this is not strictly appropriate as it is only found in the prenatal human basement membrane. The alternative laminin mixture, containing predominantly LN10 along with LN1, LN2, LN6 and LN8, may be more physiologically relevant.

2. Microscopy gives unequivocal, direct evaluation of cell adhesion and allows the state of the cells and surface and uniformity of adhesion to be checked. There are alternative methods; for example, a radioactive or fluorescent label or a dye can be preloaded into the target cells. The label is released on lysis of the cells adherent to the surface and quantitated by measurement of radioactivity, fluorimetry, or densitometry, respectively. There can be problems of label leakage and uptake by the endothelial cells. Other methods include total lysis of adherent cells and measurement of total protein or analysis of a released enzyme (e.g., myeloperoxidase from neutrophils) or cell-specific marker. All of these approaches should be calibrated using known numbers of cells to obtain quantitative results.

3. Six-well plates are useful for adhesion assays as the large-well format allows for efficient washing and provides a large, optically clear, area to view under the microscope. However, if limited purified proteins or cells are available, a smaller-well format can be used. The 12- or 24-well plates give reasonable washing and viewing. Others have used 96-well plates, but visual quantification of adhesion is problematic (due to the limited area and poor optical properties). In our experience, the problem with well-based assays lies in difficulty in achieving efficient washing and the tendency of cells to collect around the edges of the well. This is worse with the smaller wells, and higher nonspecific background adhesion occurs, which decreases sensitivity. Direct visual observation alleviates this problem to a degree but is again better in larger wells. With the smaller-well plates, washing can be improved using a swinging bucket plate centrifuge, with a sealed plate, to "spin" cells off the surface before analysis.

4. The main constituents of the endothelial basement membrane are laminin, collagen type IV, and fibronectin, which we have used at coating concentrations ranging between 0.02 and 20 µg/mL. A 0.2 µg/mL laminin solution deposited protein onto the surface equivalent to that found in the basement

membrane deposited from HUVECs that had been cultured for 20 d (measured by enzyme-linked immunosorbent assay) *(9)*. The same was true for a collagen solution of around 2 μg/mL. It is advisable to titrate proteins for particular cells under study.

5. Successful coating with APES requires that the reagent be anhydrous. We buy small volumes adequate to coat a batch of microslides, use a fresh bottle for each batch, and discard any unused reagent. It is important to efficiently remove all liquid from each microslide between changes and to ensure all bubbles are displaced on refilling with agents.

6. In general, microslides are coated with receptors involved in capture of flowing cells, such as P- or E-selectin or VCAM-1 (–), for which concentrations of about 1 μg/mL have been effective. Leucocytes typically form unstable rolling attachments on these receptors. In our hands, titration of P-selectin has shown that as the concentration used increases, attachment quickly saturates, but velocity at which cells roll decreases steadily. If capture receptors are combined with other receptors for firm adhesion, then cells may become stabley attached and migrate. For leucocytes, this typically requires addition of an activating agent, such as a chemokine, to activate integrin receptors.

7. One confluent 25-cm^2 flask of HUVECs, resuspended in 6 mL, will seed three wells (2 mL per well) to produce a confluent monolayer within 24 h. Alternatively, one 25-cm^2 flask can be resuspended in 8 mL and used to seed four wells, which will be confluent in 2–3 d. We have found no difference in neutrophil adhesion or migration on 1- or 3-d cultures *(9)*. Chamber slides can be seeded as if equivalent to a single well in a six-well plate.

8. The microslide culture system is one we have developed for culture of endothelial cells under flow as well as for testing of adhesion and migration *(10)*. The culture dishes can have three ports or six ports. Three-port dishes are ideal for setting up different cytokine treatments or culture regimes; six-port dishes are ideal when those parameters are fixed, but the endothelial cells or perfused cells are different. A system is available commercially (GlycoTech, Rockville, MD) based on culture of endothelial cells in 35-mm diameter culture dishes. An adaptor and gasket are inserted on top of the cells and held in place by vacuum. The device allows perfusion of fluid (with or without suspended cells of interest) over a central region of the endothelial surface.

9. When studying neutrophil adhesion, we found 5 min to be an appropriate settling time. This short duration and the

effective wash procedure in six-well plates result in little or no background neutrophil adhesion on albumin-coated proteins or unstimulated HUVEC controls (*see* **Fig. 12.3A**). However, the settling time needs to be tested depending on the type of cell studied and may need to be longer for poorly adherent cells.

10. We have found that 5–10 min are sufficient to visualise neutrophils migrating on proteins or over, through, and underneath endothelial monolayers. However, the rate of migration varies between cell types. Longer recordings maybe required to allow analysis of migration over hours rather than minutes (e.g., for endothelial cells or fibroblasts).

11. Adherent leucocytes can be classified into two groups: phase-bright cells adherent to the upper surface of the endothelial cells and phase-dark, spread cells that are transmigrated under the endothelial monolayer (*see* **Fig. 12.3B**). Leuco-cyte adhesion levels should not change between the two sets of recordings—only the proportion that have transmigrated.

12. Take care with this procedure so the endothelial monolayer is not disrupted.

13. The flow rate (Q) required to give a desired wall shear rate (γ_w in s^{-1}) or wall shear stress (τ_w in Pa) is calculated from the internal width (w) and internal depth (h) of the micro-slide (or flow channel) and the viscosity (n) of the flowing medium using the formulas $\gamma = (6.Q)/(w.h^2)$ and $\tau = n.\gamma$. For microslides, since w = 3 mm, h = 0.3 mm, n = 0.7 mPa.s for simple cell suspension buffers at 37°C, this can be manipulated to give Q (mL/min) = 0.0027 . γ_w (s^{-1}) or Q (mL/min) = 3.95 . τ_w (Pa). When using flow channels that are formed by slots cut in the gasket, the length and width of the slot and the thickness of the gasket (i.e., depth) are used in the calculation.

Acknowledgement

Development of adhesion and migration assays in our laboratory was supported by grants from the British Heart Foundation.

Reference

1. McGettrick, H. M., Butler, L. M. and Nash, G. B. (2005) Analysis of leukocyte migration through monolayers of cultured endothelial cells. *Methods Mol Biol.* 370: 37–54

2. Nash, G. B. (2004) Adhesion between platelets and leukocytes or endothelial cells. *Methods Mol. Biol.* 272: 199–214.

3. Springer, T. A. (1995) Traffic signals on endothelium for lymphocyte recirculation and leukocyte emigration. *Ann Rev Physiol.* 57: 827–72.

4. Bahra, P., Rainger, G. E., Wautier, J. L., Luu, N.-T. and Nash, G. B. (1998) Each step during transendothelial migration of

flowing neutrophils is regulated by the stimulatory concentration of tumour necrosis factor-alpha. *Cell Ad Commun*. 6: 491–501.

5. Luu, N. T., Rainger, G. E. and Nash, G. B. (1999) Kinetics of the different steps during neutrophil migration through cultured endothelial monolayers treated with tumour necrosis factor-alpha. *J Vasc Res*. 36: 477–85.

6. Smith, C. W., Kishimoto, T. K., Abbassi, O., Hughes, B., Rothlein, R., McIntire, L. V., Butcher, E., Anderson, D. C. and Abbass, O. (1991) Chemotactic factors regulate lectin adhesion molecule 1 (LECAM-1)-dependent neutrophil adhesion to cytokine-stimulated endothelial cells in vitro. *J. Clin. Invest*. 87: 609–618.

7. Cinamon, G., Shinder, V. and Alon, R. (2001) Shear forces promote lymphocyte migration across vascular endothelium bearing apical chemokines. *Nature Immunol*. 2: 515–522.

8. Cooke, B. M., Usami, S., Perry, I. and Nash, G. B. (1993) A simplified method for culture of endothelial cells and analysis of adhesion of blood cells under conditions of flow. *Microvasc Res*. 45: 33–45.

9. Butler, L. M., Rainger, G. E., Rahman, M. and Nash, G. B. (2005) Prolonged culture of endothelial cells and deposition of basement membrane modify the recruitment of neutrophils. *Exp Cell Res*. 310: 22–32.

10. Sheikh, S., Gale, Z., Rainger, G. E. and Nash, G. B. (2004) Methods for exposing multiple cultures of endothelial cells to different fluid shear stresses and to cytokines, for subsequent analysis of inflammatory function. *J. Immunol. Methods*. 288: 35–46.

Chapter 13

Genetic Manipulation of Corneal Endothelial Cells: Transfection and Viral Transduction

Eckart Bertelmann

Abstract

The corneal endothelium plays a key role in the physiology of the cornea, maintaining its transparency by regulating corneal hydration. Moreover, corneal endothelial cells play the central role in irreversible corneal graft rejection as human corneal endothelial cells are predominantly postmitotic, and destroyed cells cannot be replaced. Therefore, gene transfer to the corneal endothelium to modify the corneal immune response for prophylaxis of corneal endothelial rejection has become a fast-developing research field. An addition pivotal advantage of gene transfer to the cornea is the possibility of *ex vivo* transfection during organ culturing, minimizing the risk of systemic spread of the vector or the transgene expression. A wide variety of vectors has been found suitable for gene transfer to the corneal endothelium, and therapeutic efficacy has been demonstrated in some experimental models of corneal disease. However, the transfection efficiency varies widely among the different vectors, and the optimal transfection efficiency to provoke a desired effect is still unclear. Moreover, it certainly depends on the biological function of the chosen transgene (cytokine, growth factor, etc.). As a consequence, relatively few studies have been able to demonstrate significant prolongation of corneal allograft survival after gene transfer to the endothelium, and the ideal transfer strategy has not been found. In contrast, different transfer strategies compete today, each with its special advantages and disadvantages. Physical, viral, and nonviral techniques have been used to transfer transgenes into endothelial cells. In the introduction of this chapter, a short overview of the different gene transfer strategies for endothelial cells is given; the materials and methods sections describe in detail the most widely used viral gene transfer technique (adenoviral) and an important nonviral alternative technique (liposomal transfection) to endothelial cells.

Key words: Adeno-associated viral transduction, adenoviral transduction, ballistic gene transfer, hybrid vectors, liposomal transfection, retro- and lentiviral transduction.

S. Martin and C. Murray (eds.), *Methods in Molecular Biology, Angiogenesis Protocols, Second edition, Vol. 467*
© Humana Press, a part of Springer Science+ Business Media, LLC 2009
DOI: 10.1007/978-1-59745-241-0_13

1. Introduction

The possibilities of gene transfer into corneal endothelial cells were described in a multitude of articles *(1–6)*. Potential therapeutic genes can be transferred by different strategies into the corneal endothelium. On the one hand, physical techniques like the application of naked DNA, electroporation-mediated gene transfer, and ballistic gene transfer by means of a "gene gun" have been used for transgene introduction into endothelial cells; on the other hand, a variety of viral and nonviral vectors and vehicles such as liposomes were used successfully for transfection and transduction of endothelial cells.

1.1. Physical Gene Transfer Methods

1.1.1. Transfer of Naked DNA

The transfer of a "naked" transgene containing plasmid has the advantage of inducing a localised target cell transfection without associated inflammation. The disadvantages include predominantly low transfection efficiency and short-term transgene expression. Moreover, the endothelium localised on the posterior face of the cornea is hard to reach by this technique. Using syringe pressure sufficient to cause stromal hydration, effective transfer of a plasmid encoding a reporter gene to the murine cornea but not the endothelium was achieved by manual intrastromal injection *(7)*. The reporter gene expression could be demonstrated for up to 10 d afterward. Injection of naked DNA encoding an interleukin 1 receptor antagonist into injured murine corneas resulted in transgene expression for up to 2 wk *(8)*.

1.1.2. Electroporation and Iontophoresis

Gene delivery by electroporation refers to transduction with plasmid DNA after the application of a high field strength electric pulse to the target tissue, sufficient to create pores in the membranes of cells within that tissue. Iontophoresis is different from electroporation in that it makes use of an electric current to carry ions, in this context charged plasmid DNA molecules, into target tissues *(9)*. These methods seem to be more efficient than application of naked DNA. Iontophoresis has been used to carry DNA of up to 8000 kDa across human sclera *in vitro (10)*. Electroporation can be similarly effective in epithelial cells and keratocytes. Rat keratocytes could be transduced *in vivo* with a reporter gene following insertion of plasmid DNA into a corneal stromal pocket and the application of an electric pulse to the corneal epithelium *(11)*. The expression time of the transgene was found to be 4–10 d in this study.

The quantity of DNA administered to the eye by electroporation and iontophoresis has ranged from approximately 2.5 ng to 50 mg. Transgene expression has generally been transient *(9)*.

1.1.3. Ballistic Gene Transfer

Ballistic gene transfer is accomplished by firing plasmid DNA adsorbed to gold or tungsten microparticles into a target tissue by means of a commercially available gene gun, typically helium driven. Although it has been used with some success for transduction of the corneal epithelium *in vitro* and *in vivo* (12), the technique has been unsuitable for gene transfer to the endothelium *in vitro* (13). Bombardment of the corneal surface can cause epithelial cell damage, which may be reduced by using an adaptor to control the distance between the gene gun and the cornea and by reducing the mass of microparticles used (9). However, ballistic gene transfer to the cornea can induce an inflammatory response of the recipient organism. It was shown that direct bombardment of ovine corneas from the endothelial surface at pressures ranging from approximately 700 to 2000 kPa results in relatively poor rates of transfection and can cause extensive mechanical damage of the endothelial cells (13). In summary, ballistic gene transfer to the endothelium (in contrast to other corneal layers) seems to remain associated with several method-dependent problems.

1.2. Viral Vectors

Replication-deficient adenovirus, adeno-associated virus (AAV), herpes simplex virus (HSV), retrovirus, and other lentiviruses have been used for transfection of corneal endothelial cells.

1.2.1. Adenoviral Vectors

Adenoviruses are the most frequently used carriers for the transfer of genetic information in human gene therapy trials. They consist of an icosahedric capsid 70 nm in diameter harbouring a double-stranded, linear DNA of approximately 36 kb. More than 50 serotypes belonging to different subgroups are known (14).

The eye is a potential target tissue for adenoviruses since infection of the eye may result in severe keratoconjunctivitis. However, this disease will not occur with adenoviral gene therapy vectors since essential genes for virus replication are deleted. In addition, it is of interest that gene-modified adenovirus affects predominantly the corneal endothelium, whereas the epithelium is almost not susceptible for adenovirus-mediated gene transfer. The deletion of viral genes important for DNA replication and transcription (E1a, E1b) opened the possibility to use these viruses as vectors for gene therapy. Up to 8 kb of foreign DNA can be efficiently inserted into the adenovirus vector. In addition to their large packaging capacity, adenovirus vectors have several advantages that make them superior to other gene therapy vectors. They can be propagated *in vitro* at high titres (up to 10^{11} infectious virions/mL), which is a prerequisite for *in vivo* experiments, and the expression levels of the therapeutic gene are generally very high. On the other hand, adenovirus vectors induce a cascade of immune reactions against the vector itself (15) and against the transduced cells, which finally lead to transient expression of the desired transgene. In addition, repeated application of adenovirus vectors is difficult to achieve due to the generation of neutralizing antibodies after the first injection of the adenovirus vector. Despite the generation of immune responses against adenovirus particles, numerous studies have been performed using adenoviral vectors in gene therapy trials, mainly in treating patients with end-stage tumors.

1.2.2. Adeno-Associated Viral Vectors

Adeno-associated virus (AAV) is an integrative DNA virus that can produce long-term transgene expression when used as a gene therapy vector (16). AAV is more difficult to grow to high titre than adenovirus and has a relatively small capacity for exogenous complementary DNA (cDNA). Recombinant AAV has achieved transfection rates of approximately 2% of human and rabbit corneal endothelial cells *in vitro* over a period of 3–4 wk (17). However, when AAV encoding a reporter gene was injected into the anterior chamber of rabbits *in vivo* and an inflammatory stimulus was applied in the form of an intraocular injection of endotoxin, approximately 90% of corneal endothelial cells expressed the reporter gene (18). Expression levels were maintained for 2 wk and were reestablished after a second injection of endotoxin.

1.2.3. Herpes Simplex Virus Vectors

Herpes simplex virus (HSV) transduction has the advantages of stable lifelong infection of the target cells. Moreover, HSV can infect postmitotic cells.

On the other hand, the transfection efficiency for endothelial cells varies widely. It ranges from occasional cells in sheep (19) to 5% of rabbit and human cells (17). Moreover, a viral toxicity for the target cells has been found.

1.2.4. Retroviral and Lentiviral Vectors

Replication-deficient retrovirus and lentivirus are integrative vectors that can produce sustained transgene expression in target cells (16). Both can be pseudotyped to ensure a wide host range. Retroviral vectors are unable to infect postmitotic cells efficiently, whereas lentiviral vectors will transduce both mitotically active and inactive cells. The former are thus theoretically suitable for infection of corneal epithelial cells and keratocytes but are inappropriate for infecting corneal endothelial cells. Lentiviral vectors should be suitable for infection of all corneal cell types.

A human immunodeficiency virus type 1 (HIV-1)-based lentiviral vector encoding a reporter gene has been found to infect human corneal endothelial cells, keratocytes, and epithelial cells for up to 60 d in extended organ culture (20), and injection of a similar lentiviral vector into the anterior chamber of neonatal mice also resulted in efficient and stable infection of corneal endothelial cells (21). HIV-based lentiviral vectors suffer from the perception, if not necessarily the reality, that they may contain replication-competent virions.

1.3. Nonviral Vectors

Nonviral vectors include a heterogeneous group of molecules and chemical structures. A part of these compounds consists of lipids or dendrimers, whereas other vehicles take advantage of the cell-attacking properties of integrins or antibodies. They can uptake relatively large amounts of DNA; another advantage of these vehicles is the relatively safety of these structures as no potential infective particles are involved. The disadvantages include an often low transfection efficiency and a variable duration of the transgene expression. The introduction of several hybrid vectors could partly improve transfection efficiency.

The technique of liposomal transfection of target cells is based on the reports of Felgner *(22)*. Liposomal transfection of corneal endothelial cells has been extensively described in reports from 2001 and 2005 *(14, 23)*. Highest transfection rates are achieved by cationic liposomes that transfer the DNA, probably by endocytosis rather than membrane fusion into the target cell *(24, 25)*. The toxicity of liposomes for the cellular membrane is variable, with a multitude of compounds not toxic for the endothelial cell membrane in adequate concentrations. The cell toxicity in many target cells can be easily measured by an acid phosphatase assay *(26, 27)*.

Liposomal transfection can obtain transfection efficiencies of 5–10% of endothelial cells *(14, 28)*. Liposomes can be topically applied to the ocular surface; direct intraocular injection (intracameral) is also possible. Intracameral injection of liposomes interestingly can lead to transfection of ciliary body epithelial cells and cells on the corneal surface (epithelium), whereas the "closer" endothelium is not affected *(29)*.

The amount of transfection efficiency in animal transfection models depends on the applied species.

Activated polyamidoamin dendrimers are alternative nonviral vectors. Like cationic liposomes, they build complexes with negatively charged DNA molecules. Dendrimers achieved transfection rates of 6–10% in corneal endothelial cells. *In vitro*, a maximum of transgene expression could be achieved at d 3 posttransfection *(30)*. Activated dendrimer vectors have relatively low intraocular toxicity.

In conclusion, liposomes and dendrimers are effective gene transfer vehicles to endothelial cells but have low transfection efficiency in comparison to viral vectors, and the transgene expression is predominantly transient.

2. Materials

2.1. Chemicals

1. Cell culture medium: 100.0 mL Dulbecco's modified Eagle's medium (DMEM), 2.5 mL HEPES, 1.0 mL L-glutamine, 1.0 mL penicillin/streptomycin, 0.1 mL gentamycin, 10.0 mL FCS (fetal calf serum) (Gibco).

2. HEPES buffer: 2.383 g in 10 mL H_2O; adjust to pH 7.2 with 5 N NaOH. Sterile filtration can be stored 1 yr at 4°C.

3. L-Glutamin (30 mg/mL): 1.5 g in 50 mL H_2O; sterilize filtrate and freeze in 1.2-mL aliquots (sterile Eppendorf tubes) at –20°C; can be stored 3–4 mo.

4. Trypsin (Gibco).

5. Polyvinylpyrrolidine (PVP)-iodine (Merck) (*see* **Note 1**).

6. $Na_2S_2O_3$ (Merck) (*see* **Note 1**).

7. Phosphate-buffered saline (PBS; Merck).

2.2. Liposomes

1. Dimethyldioctadecylammoniumbromide (DDAB) (Sigma).

2. Dioleoylphosphatidylethanolamine (DOPE) (Sigma).

3. Dicarbobenzoxy-spermine-carbamoyl-cholesterol (SP-Chol) (Sigma).

4. $CHCl_3$ (Merck).

2.3. Recombinant Adenovirus

1. Adenoviral construct pACCMV.

2. Adenoviral plasmid pJM17.

3. Ca-phosphate (Merck).

4. Sepharose CL6B (Pharmacia, Freiburg, Germany).

5. DMEM (Gibco).

6. FCS (Gibco).

7. Glycerol (Merck).

3. Methods

3.1. Preparation of Bovine Corneal Endothelial Cells

Bovine corneal endothelial cells (BCECs) can be isolated and grown as described by Gospardorowicz *(31)*. For this procedure, bovine eye globes must be disinfected in PVP-iodine; then, iodine has to be degraded by $Na_2S_2O_3$ (*see* **Note 1**), and globes have to be drained in PBS. Corneoscleral disks can be obtained by trephination; then, endothelial cells have to be harvested using a hockey knife (*see* **Note 2**). To disrupt remaining cell adhesions, cells should be trypsinized, then seeded in 12-well plates in DMEM (*see* **Note 3**) culture medium containing 5% FCS, HEPES, and penicillin/streptomycin. After growing to subconfluence at d 10, cells have to be transferred to 96-well plates (5000 cells per well). Cells must be grown to subconfluence (70% confluence) for 5 d. Liposomal transfection (*see* **Note 4** and **5**) should be performed in the culture medium with low FCS content (e.g., 2.5%). After transfection (*see* **Note 6** and **7**), cells should be cultivated in DMEM/10% FCS.

3.2. Preparation of Liposomes

Among the multitude of liposome-forming lipid compounds, the choice of the most suitable compound is one of the most important and difficult step of the procedure. The transfection efficacy depends on the cell type (*see* **Note 8**) and especially on the species from which the cells are isolated. In bovine endothelial cell cultures (as described in Materials), we found two lipid formulations with the highest transfection efficiency for marker gene transfer: DDAB/DOPE 30/70 and SP-Chol/DOPE 20/80 *(14)*. Other lipid formulations previously tested in bovine endothelial cell cultures are DAC/Chol, DOSGA, DMRIE, SP-SIT, DOAPβAla, DOTMA, Superfect, Unifectin, DOCSPER, and Lipofectin. The DDAB/DOPE 30/70 lipid formulation consists of DDAB and the neutral helper lipid DOPE in a molar ratio of 30:70; the second compound applied in this study consists of SP-Chol) and DOPE in a molar ratio of 20:80.

The preparation of the liposomes has to be performed as follows: Lipids must be dissolved in $CHCl_3$; then, the organic solvent has to be evaporated. The resulting lipid film must be vacuum dried. Lipids then have to be suspended in deionized water, and liposomes are formed by shaking for several hours. The final lipid concentration should be 1 mg/mL. Both described lipid formulations form cationic liposomes.

3.3. Preparation of Recombinant Adenoviruses

For adenovirus-based transduction experiments, first recombinant adenovirus encoding for the desired transgene (e.g., cytokine) has to be constructed. The cDNA for the transgene has to be subcloned into a suitable replication-deficient adenoviral construct such as pACCMV. This construct contains 1.3 map units of sequence from the left end of the adenovirus (Ad5) genome, the CMV early promoter, the pUC19 polylinker, SV40 poly(A) signal sequences, and map units 9 through 17 of the Ad5 genome. Then, the adenovirus plasmid has to be cotransfected into the 911 cell line *(32)* together with the large adenoviral plasmid pJM17 using a standard Ca-phosphate precipitation technique. Adenoviral genomes formed by homologous recombination between the pJM17 vector and the pACCMV vector contain the desired transgene's cDNA, are replication defective, and are efficiently packaged to form infectious virus. The adenovirus plaques have to be picked and propagated in 60-mm plates of 911 cells *(32)*. For the propagation and purification of recombinant adenovirus, 911 cells should be grown in DMEM/10%FCS, and 1% penicillin/streptomycin in 15-cm plates and infected at a multiplicity of 5–10. After 36–48 h when the cytopathic effect is complete, cells should be harvested; virus should be released by freeze-thawing five times and then purified over a CsCl gradient. Banded virus must be recovered and spin dialyzed over Sepharose CL6B and should be stored in aliquots at −80°C after the addition of 10% glycerol. To titre the final preparation, an aliquot of virus has to be serially diluted and assayed for its ability to form plaques on 911 cell monolayers. The capacity of forming plaques from different virus preparations is generally between 1×10^{10} and 1×10^{11} plaque-forming units (PFU) per millilitre.

3.4. Liposomal Transfection of BCECS

The lipids in different concentrations have to be incubated with the plasmid DNA for 20 min to form lipid/DNA complexes. Then, the liposomes must be transferred to the BCECs. The applied lipid concentrations should cover a certain range (e.g., 1.25 to 40 mg per well) because the optimal liposomal concentration can differ among different liposomes and cell types. In our hands in BCEC experiments, the optimal liposomal concentration is 10 mg per well; toxic effects were found for liposomal concentration of over 20 mg per well. The plasmid concentration should be 1.0 mg per well.<p>

As controls, cells in additional wells should be incubated with the plasmid only (without lipid), and others should be incubated with "empty" liposomes (without DNA). The cells have to be incubated for 4 h with the liposomes. Serum concentration should be low (e.g., 2.5%) during the transfection period.

After completing the transfection period, the cells must be washed with PBS, and fresh DMEM plus 10% FCS has to be added (0.2 mL per well). Then, cells have to be further cultivated depending on the time the transgene expression should be measured.

To quantify the transgene product (e.g., in case of cytokine), the growth medium must be separated from the cells, and the cytokine concentration in the supernatant can be determined (e.g., using specific enzyme-linked immunosorbent assay [ELISA]).

All transfection experiments should be done at least in triplicate.

3.5. Adenoviral Transduction of BCECS

For adenoviral transduction of BCECs, a modified protocol published in 1999 by Ritter et al. (5) is recommended. For transduction of BCECs (2×10^4 cells per well), adenoviral stocks should be serially diluted in DMEM with 2% FCS, and cells should be incubated with different adenoviral concentrations, with multiplicity of infection (MOI) from 1 to 1000, for 2 h at 37°C. After washing in DMEM with 10% FCS, cells should be cultured in DMEM for 2 d more. After the appropriate expression time (e.g., 2 d), supernatants have to be harvested and can be analysed for the transgene product (e.g., cytokine) content by specific ELISA.

4. Notes

1. PVP-iodine and $Na_2S_2O_3$ are toxic for endothelial cells. Therefore, globes have to be rinsed very thoroughly before excision of the corneoscleral buttons.

2. When the endothelial cells are harvested using a hockey knife, the knife has to be drawn flat and not at a 90° angle so that the descemet's membrane (basal membrane) is not scratched and destroyed. Otherwise, the harvested endothelial cells may be mixed with keratocytes (fibroblasts), which could overgrow the endothelial cells.

3. The cell culture medium (DMEM) should contain antibiotics. Otherwise, the risk of infection of the cell cultures is very high.

4. The liposomes have a limited storage time. After preparation, they should be used within several weeks.

5. Liposomes have to be stored at +4°C.

6. The cells have to be grown to subconfluence. If the cells are confluent, they decrease their proliferation activity by undergoing contact inhibition. Liposomal transfection efficiency in nonproliferating cells is very low. If the cells are confluent at the day of the scheduled transfection experiment, they need another passage before they are suitable for the transfection procedure.

7. The incubation time to form DNA/liposome complexes must not exceed the time given in the protocol. Otherwise, the lipids could aggregate to larger particles.

8. The optimal lipid concentration varies among different cell types, the same cell types from different species, cell densities, and the type of the applied lipid. Therefore, different lipid concentrations should be used in each experiment. Moreover, the optimal type of lipids varies among the same cell types of different species. So, the optimal lipids found in our experiments for BCECs (SP-Chol/DOPE 20/80 and DDAB/DOPE 30/70) may be different for endothelial cells from other species. In an immortalized human endothelial cell culture, Lipofectin was the most effective liposomal formulation *(23)*.

References

1. Larkin, D. F. P., Oral, H. B., Ring, C. J. A., Lemoine, N. R., George, A. J. T. (1996) Adenovirus-mediated gene delivery to the corneal endothelium. *Transplantation* 61, 363–370.

2. Fehervari, Z., Rayner, S. A., Oral, H. B., George, A. J. T., Larkin, D. F. P. (1997) Gene transfer to *ex vivo* stored corneas. *Cornea* 16, 459–464.

3. Oral, H. B., Larkin, D. F. P., Fehervari, Z., et al. (1997) *Ex vivo* adenovirus-mediated gene transfer and immunomodulatory protein production in human cornea. *Gene Ther* 4, 639–647.

4. Arancibla-Carcamo, C. V., Oral, H. B., Haskard, D. O., Larkin, D. F. P., George, A. J. T. (1998) Lipoadenofection-mediated gene delivery to the corneal endothelium. *Transplantation* 65, 62–67.

5. Ritter, T., Vogt, K., Rieck, P., et al. (1999) Adenovirus-mediated gene transfer of IL-4 to corneal endothelial cells and organ cultured corneas leads to high IL-4 expression. *Exp Eye Res* 69, 563–568.

6. Pleyer, U., Ritter, T. (2003) Gene therapy in immune-mediated diseases of the eye. *Prog Ret Eye Res* 22, 277–293.

7. Stechschulte, S. U., Joussen, A. M., von Recum, H. A., et al. (2001) Rapid ocular angiogenic control via naked DNA delivery to the cornea. *Invest Ophthalmol Vis Sci* 42, 1975–1979.

8. Moore, J. E., McMullen, T. C., Campbell, I. L., et al. (2002) The inflammatory milieu associated with conjunctivalized cornea and its alteration with IL-1RA gene therapy. *Invest Ophthalmol Vis Sci* 43, 2905–2915.

9. Williams, K. A., Jessup, C. F., Coster, D. J. (2004) Gene therapy approaches to prolonging corneal allograft survival. *Expert Opin Biol Ther* 4, 1059–1071.

10. Davies, J. B., Ciavatta, V. T., Boatright, J. H., Nickerson, J. M. (2003) Delivery of several forms of DNA, DNA-RNA hybrids, and dyes across human sclera by electrical fields. *Mol Vis* 9, 569–578.

11. Oshima, Y., Sakamoto, T., Hisatomi, T. (2002) Targeted gene transfer to corneal stroma *in vivo* by electric pulses. *Exp Eye Res* 74, 191–198.

12. Tanelian, D., Barry, M. A., Johnston, S. A., Le, T., Smith, G. (1997) Controlled gene gun delivery and expression of DNA within the cornea. *BioTechniques* 23, 484–488.

13. Klebe, S., Stirling, J. W., Williams, K. A. (2000) Corneal endothelial cell nuclei are damaged after DNA transfer using a gene gun. *Clin Exp Ophthalmol* 28, 58–59.

14. Pleyer, U., Groth, D., Hinz, B., et al. (2001) Efficiency and toxicity of liposome-mediated gene transfer to corneal endothelial cells. *Exp Eye Res* 73, 1–7.

15. Yang, Y., Nunes, F. A., Berencsi, K., Furth, E. E., Gonczol, E., Wilson, J. M. (1994) Cellular immunity to viral antigens limits E1-deleted adenoviruses for gene therapy. *Proc Natl Acad Sci U S A* 91, 4407–4411.

16. Somia, N., Verma, I. M. (2000) Gene therapy: trials and tribulations. *Nat Rev Genet* 1, 91–99.

17. Hudde, T., Rayner, S. A., De Alwis, M., et al. (2000) Adeno-associated and herpes simplex viruses for gene transfer to the corneal endothelium. *Cornea* 19, 369–373.

18. Tsai, M. L., Chen, S. L., Chou, P. I., Wen, L. Y., Tsai, R. J. F., Tsao, Y. P. (2002) Inducible adeno-associated virus vector delivered transgene expression in corneal endothelium. *Invest Ophthalmol Vis Sci* 43, 751–757.

19. Klebe, S., Sykes, P. J., Coster, D. J., Bloom, D. C., Williams, K. A. (2001) Gene transfer to ovine corneal endothelium. *Clin Exp Ophthalmol* 29, 316–322.

20. Wang X, Appukuttan B, Ott S. (2000) Efficient and sustained transgene expression in human corneal cells mediated by a lentiviral vector. *Gene Ther* 7, 196–200.

21. Bainbridge, J. W., Stephens, C., Parsley, K., et al.(2001) *In vivo* gene transfer to the mouse eye using an HIV-based lentiviral vector; efficient long-term transduction of corneal endothelium and retinal pigment epithelium. *Gene Ther* 8, 1665–1668.

22. Felgner, P. L., Gadek, T. R., Holm, M., et al. (1987) Lipofection: a highly efficient, lipid-mediated DNA-transfection procedure. *Proc Natl Acad Sci U S A* 84, 7413–7417.

23. Dannowski, H., Bednarz, J., Reszka, R., Engelmann, K., Pleyer, U. (2005) Lipid-mediated gene transfer of acidic fibroblast growth factor into human corneal endothelial cells. *Exp Eye Res* 80, 93–101.

24. Blumenthal, R., Ralston, E., Dragsten, P. (1982) Lipid vesicle-cells interactions: analysis of a model for transfer of contents from adsorbed vesicles to cells. *Membr Biochem* 4, 283–303.

25. New, R. R. G., Black, C. D. V., Parker, R. J. (1990) Liposomes in biological systems, in (New, R. R. C., ed.) *Liposomes, a Practical Approach.* Oxford University Press, New York.

26. Martin, A., Clynes, M. (1991) Acid phosphatase: endpoint for *in vitro* toxicity tests. *In vitro Cell Dev Biol* 27A, 183–184.

27. Groth, D., Keil, O., Schneider, M., Reszka, R. (1998) Transfection assay for dual determination of toxicity and gene expression. *Anal Biochem* 258, 141–143.

28. Bertelmann, E., Ritter, T., Vogt, K., Reszka, R., Hartmann, C., Pleyer, U. (2003) Efficiency of cytokine gene transfer in corneal endothelial cells and organ-cultured corneas mediated by liposomal vehicles and recombinant adenovirus. *Ophthal Res* 35, 117–124.

29. Masuda, I., Matsuo, T., Yasuda, T., Matsuo, N. (1996) Gene transfer with liposomes to the intraocular tissues by different routes of administration. *Invest Ophthalmol Vis Sci* 37, 1914–1920.

30. Hudde, T., Rayner, S. A., Comer, R. M., (1999) Activated polyamidoamine dendrimers, a non-viral vector for gene transfer to the corneal endothelium. *Gene Ther* 6, 939–943.

31. Gospodarowicz, D., Mescher, A. L., Bird-well, C. R. (1977) Stimulation of corneal endothelial cell proliferation *in vitro* by fibroblast and epidermal growth factors. *Exp Eye Res* 25, 75–89.

32. Fallaux, F. J., Kranenburg, O., Cramer, S. J., (1996) Characterization of 911: a new helper cell line for the titration and propagation of early region 1-deleted adenoviral vectors. *Hum Gene Ther* 20, 215–222.

Chapter 14

Creation of Human Skin Equivalents for the *In Vitro* Study of Angiogenesis in Wound Healing

Ira M. Herman and Alice Leung

Abstract

In our efforts aimed at studying the cellular responses to injury, including the angiogenesis of wound healing, we have developed a novel three-dimensional (3D) skin equivalent that is comprised of multiple cell types found in normal human skin or chronic wound beds. The *in vitro* model contains a microvascular component within the dermis-like extracellular matrix and possesses an intact epithelial covering comprised of skin-derived epithelial cells. Capillary endothelial cells can be labeled with fluorescent vital tracers prior to being embedded within a 3D matrix and overlaid with a monolayer of keratinocytes (normal or transformed). Once embedded in the matrix, the endothelial cells demonstrate capillary-like tube formation mimicking the microvasculature of true skin. Angiogenesis and the reepithelialization, which occur in response to injury and during wound healing, can be quantified using fluorescence-based and bright-field digital imaging microscopic, biochemical, or molecular approaches.

Key words: Biomatrix, imaging, injury, migration, neovascularization, repair, wounds.

1. Introduction

As is widely recognized, the human body's response to injury is complex, dependent on a panoply of signaling pathways expressed by several cell and tissue types over an extended period of time. Importantly, the wound "bed" undergoes significant remodeling as reepithelialization ensues concomitant with dermal angiogenesis. And, while wound healing typically occurs as a natural, uneventful process leaving the individual with neither noticeable scars nor wounds that chronically persist, this is not the case for 2–3% of the U.S. population. For these individuals, excessive scarring and chronic wounds are sustained as medical issues requiring specialized

S. Martin and C. Murray (eds.), *Methods in Molecular Biology, Angiogenesis Protocols, Second edition, Vol. 467*
© Humana Press, a part of Springer Science+ Business Media, LLC 2009
DOI: 10.1007/978-1-59745-241-0_14

treatment, individualized care, or in some cases, hospitalization. Thus, while acute wound healing may occur in a matter of days or weeks, chronic wounds can remain in an open state for months and even years.

1.1. Acute Healing: Cellular Responses to Injury

Under normal circumstances, the process of human wound healing can be broken down into three phases. An initial inflammatory phase, which is followed by robust tissue remodeling and proliferation (the proliferative phase), is ultimately succeeded by a "maturational phase" in which reepithelialization, dermal angiogenesis, and wound closure ensue *(1–3)*. The inflammatory phase is characterized by hemostasis, with a provisional matrix contributed by the blood itself creating the initial wound bed. As basement membrane and interstitial collagens are exposed during injury, blood platelets are stimulated to release multiple chemokines, including epidermal growth factor (EGF), fibronectin, fibrinogen, histamine, platelet-derived growth factor (PDGF), serotonin, and von Willebrand factor, to name several. These factors help to stabilize the wound through clot formation and control bleeding, therein limiting the extent of injury. Platelet degranulation also initiates the complement cascade, specifically via C5a, which is a potent chemoattractant for neutrophils.

The timeline for cell migration in a normal wound-healing process is also well ordered, with an inflammatory phase beckoning the migration of immune response cells. For example, neutrophils function to decontaminate the wound from foreign debris via phagocytosis with support from immigrating leukocytes and macrophages. In turn, macrophages release numerous enzymes and cytokines that locally function to debride the wound bed, while stimulating a robust proliferative response required for tissue morphogenesis and healing *(4–6)*. It is during the proliferative phase that reepithelialization and angiogenesis predominate *(7, 8)*. The entire process represents a dynamic continuum, with continued tissue remodeling until the wound site reaches maximal strength, perhaps as long as 1 year post-injury.

1.2. Injury-Induced Angiogenesis: A Critical Component of Wound Healing

Throughout life, the vasculature undergoes significant morphogenesis *(9)*. Two independent but related processes govern the formation of the adult vasculature: vasculogenesis and angiogenesis. Initially during vasculogenesis, immature vessels are formed *de novo* from endothelial cell precursors, the angioblasts, which proliferate and coalesce, creating a capillary plexus. Local differentiation of endothelial cells serves as an initiating event for the subsequent rounds of vascular "budding" or "sprouting," angiogenesis, which gives rise to the system of arteries, veins, arterioles, venules, and capillaries. Interestingly, in the adult, physiologic angiogenesis occurs during the female reproductive cycle, but otherwise the predominant form of physiologic angiogenesis during adult life occurs during wound healing *(7, 9)*.

Many positively and negatively acting factors influence the angiogenesis of wound healing, including the microenvironment in which vascular morphogenesis occurs. Soluble polypeptides, cell-cell and cell-matrix interactions, and hemodynamic and biomechanical forces all play strategic roles. More recently, we have learned that blood vessel sprouting during wound healing is likely to be critically dependent on a well-ordered signaling cascade responsible for regulating microvascular cytoskeletal function. In addition, clear-cut roles for the extracellular matrix and the repertoire of metalloproteinases controlling matrix remodeling also play modulatory roles in fostering wound-healing angiogenesis.

It has long been recognized that the association of pericytes with the nascent vessel is a marker of blood vessel maturation *(10–12)*, but the role that the pericyte plays in regulating wound-healing angiogenesis has yet to be established. Recent studies using tissue coculture models and transgenic animals have begun to elucidate the mechanisms that underlie the role of pericyte-endothelial interactions in vessel stabilization. Proliferating endothelial cells both *in vitro* and *in vivo* synthesize PDGF B, which acts to induce the proliferation of pericytes or their precursors and their migration toward the endothelium *(13–15)*. Culture studies suggest that when the pericytes, or their precursors, contact the endothelial cells, transforming growth factor-β (TGF-β) is activated, which in turn acts to induce pericyte differentiation *(16)*, inhibit endothelial cell and pericyte proliferation *(17)*, and stimulate the production of extracellular matrix, which in turn modulates pericyte phenotype *(18)*.

Finally, vessel stability appears to be influenced by the action of the Tie2-angiopoietin (ang) pathway. Observation of Tie2 and ang-1 knockout mice reveals defects in vessel remodeling *(19)*. Thus, the process of vessel formation and remodeling involves a variety of growth factors and intercellular interactions. As angiogenesis is likely to be stimulated or inhibited by multiple moieties and signaling cascades, this represents a critical rate-limiting step in the process of wound healing.

For these reasons, we have developed realistic three-dimensional (3D) tissue constructs that should help to fill important gaps in our understanding of the prevailing mechanisms regulating tissue repair and regeneration by enabling quantitative dissection and interrogation of the molecular and cellular mechanisms that control microvascular morphogenesis during wound healing.

2. Materials

2.1. Microvascular Endothelial Cells

Microvascular endothelial cells can be derived from a variety of tissue sources, including brain, retina, and dermis, using previously published procedures (20–22). For all isolations, tissue pieces are minced into 2 × 2 to 4 × 4 mm squares. For human dermal microvascular endothelial cell isolation, neonatal foreskins, which are discarded from clinical procedures, are placed aseptically in cold phosphate-buffered saline, pH 7.2, prior to cutting into 2 × 2 mm pieces. After removal of the epidermis by scraping the upper side of the dermis with a blunt scalpel blade, dermal tissue (including microvascular fragments) is digested for several hours at 37°C in 0.1% type II collagenase (Worthington, Freehold, NJ) prior to filtration through 100- and 30-μm Nitex nylon mesh (Sefar America, Kansas City, MO) (20). The cellular elements, largely capillary-derived vessel fragments, are retained by the 30-μm filter. These microvascular fragments are then plated (20). Following several days in tissue culture, human microvascular endothelial colonies are readily identified by their nonoverlapping, contiguous morphology and their ability to process di-acetylated (di-I-) low density lipoprotein (LDL) (20) as observed via fluorescence microscopy. Two- and three-cell colonies, reflective of capillary fragment-derived endothelial cells, are then isolated using cloning wells (see **Note 1**), which physically separate the capillary endothelial cells from the dermal pericytes present within these mixed-cell cultures. Dermal microvascular endothelial cells can then be propagated and subcultured using standard cell culture techniques. Bovine retinal endothelial cells can also be substituted (20).

2.2. Normal Human Epidermally Derived Karatinocytes

Normal human epidermally derived keratinocytes (NHEKs) can be purchased from Clonetics (Walkersville, MD) and cultured according to the manufacturer's instructions in complete keratinocyte growth medium (KGM; Clonetics) in 175-cm² tissue culture flasks as previously described (23). NHEKs are routinely subcultured in KGM at a ratio of 1:4 for cell propagation and at specified cell numbers for all experiments as described (23).

3. Methods

3.1. 3D Extracellular Matrices

Production of 3D matrices can be accomplished in several ways. For the creation of a dermal equivalent for injury repair studies assessing the angiogenesis of wound healing, we routinely create 3D matrices fabricated within eight-well cell culture chamber slides (Lab-Tek, EMS, Hatfield, PA, or BD Biosciences, San Jose, CA).

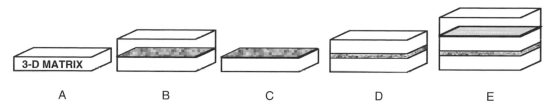

Fig. 14.1. Construction of *in vitro* skin analogs. Three-dimensional matrices are fabricated using defined extracellular matrix components taking advantage of a layer-on-layer design approach. Cells can be added between or within layers, as appropriate.

Growth factor-reduced Matrigel (BD Biosciences), and type I collagen (1.0 mg/mL) from rat-tail tendon (BD Biosciences) are used and solubilized according to manufacturers' recommendations prior to dilution in Dulbecco's modified Eagle's medium (DMEM) with 5% heat-inactivated serum and delivery to the eight-well chamber slide. Typically, growth factor-reduced Matrigel and collagen are mixed at a 2:1 v/v ratio prior to delivery into the culture wells. Each polymerized 3D matrix can then be inoculated with cells of interest (e.g., capillary endothelial cells) and then overlaid with other cellular components (e.g., NHEKs) as shown in **Figs. 14.1–Figs. 14.3**.

3.2. Embedding Capillary Endothelial Cells in 3D Matrix: Creation of Human Skin Equivalents for In Vitro Angiogenesis

Capillary cells are plated on the first layer of Matrigel/collagen (2:1) matrix and incubated at 37°C for 1 h. For eight-well chamber slides, endothelial cells are plated at 50–100,000 cells/well in 0.25 mL Dulbecco's minimum essential medium containing 5% calf serum supplemented with antibiotics *(20)*. Following 1-h incubation at 37°C, unattached cells are removed by tilting the chamber slide so that the unattached cells are suspended and excess culture media pools to one well corner (*see* **Note 2**). Thereafter, 225 µL Matrigel/collagen (2:1) is delivered to each well, thus overlaying the capillary endothelial cells (cf. **Fig. 14.1**). The overlaid 3D matrix with capillary endothelial cells embedded is then incubated for 1 h at 37°C prior to injury and refilling of the full-thickness injury site with Matrigel/collagen (2:1) for analysis of wound-healing angiogenesis (*see* **Note 3**). Typically, NHEKs (250,000) are plated within each well of the eight-well chamber slide and incubated at 37°C with serum-containing or defined media (**Figs. 14.2** and **Figs. 14.3**), which enables the analysis of reepithelialization as well as the angiogenesis of wound-healing responses.

4. Notes

1. Cloning wells must be placed to surround microvascular colonies on plates that are sparsely populated. Typically, 10- to 15-mm diameter Petri dishes are used so that the colonies can

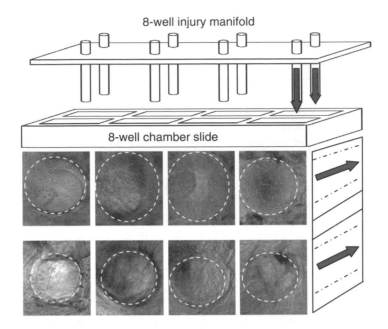

Fig. 14.2. Injury and repair analysis of three-dimensional (3D) tissue constructs comprised of epithelial-derived kerati-nocytes and capillary endothelial cells. 3D human skin constructs are created within the eight-well chamber slides comprised of defined extracellular matrices (Matrigel/collagen I). Human keratinocytes and capillary endothelial cells are wounded using vacuum-assisted punch manifold (arrows indicate the full-thickness nature and direction on wounding) Circular full-thickness wounds are created and then refilled with Matrigel/collagen I mixtures prior to analysis of the cellular responses to injury by monitoring cellular migration, cellular proliferation, and the angiogenesis of wound healing. Dashed lines in the chambers indicate the original wound circumference, which has been filled in by matrix and repopulated with cells (*en face* view, phase contrast optics).

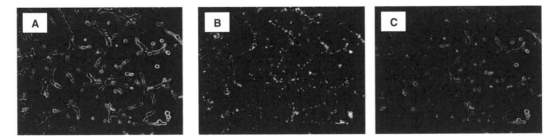

Fig. 14.3. *In vitro* angiogenesis. The angiogenesis of wound healing can be quantified using the *in vitro* fabricated layer-on-layered three-dimensional (3D) skin construct. Capillary endothelial cells are prelabeled by incubating monolayer cultures with 50–100 μg/mL tetramethylrhodamine-dextran sulfate (10,000 MW at 37°C, Molecular Probes). Prelabeled endothelial cells can then be trypsinized and plated within or on the 3D matrix for *in vitro* angiogenesis analyses. Shown are phase contrast (**A**), rhodamine-fluorescence (**B**), and merged, pseudocolored composite image (**C**; *green* phase contrast, *red* rhodamine dextran). (*See Color Plates*)

be readily identified using phase contrast optics. We typically use UV-sterilized high-vacuum grease to affix the cloning well (cylinder) to the plate, therein surrounding the colony or cell(s) of interest. It is essential that medium is quickly aspirated from the plate so that the cloning cylinder can be affixed prior to replacement of culture medium to the plate and cloning well simultaneously. Cells proliferating within cloning wells can then be allowed to become confluent prior to trypsinization and replating. The regimen followed is from a cloning well, to a 24-well plate, and then, as appropriate, to flasks or plates.

2. When removing unattached capillary endothelial cells from the Matrigel/collagen surface prior to the embedment of the attached cells by the addition of the next layer of Matrigel/collagen, it is critical to avoid touching the gel surface.

3. Creation of a full-thickness wound through the 3D matrix, together with refilling of the injury site with freshly prepared Matrigel/collagen is readily accomplished via a number of mechanisms. We have designed a manifold that can reproducibly create full-thickness injuries of defined diameters by applying slight negative pressure to the vacuum-assisted injury manifold (cf. **Fig. 14.2**).

Acknowledgments

Studies described in this chapter were supported in part by NIH EY15125, NIH EY 09033, and GM 55110 (I. M. H.) and a Williams Research Fellowship (A. L.).

References

1. Odland, G., Ross, R. (1968) Human wound repair. I. Epidermal regeneration. *J Cell Biol* 39, 135–151.

2. Abraham, J. A., Klagsbrun, M. (1996) Modulation of wound repair by members of the fibroblast growth factor family. Mol Cell Biol Wound Repair. New York: Plenum Press, pp. 195–248.

3. Cross, K. J., Mustoe, T. A. (2003) Growth factors in wound healing. *Surg Clin North Am* 83, 531–545.

4. Stadelmann, W. K., Digenis, A. G., Tobin, G. R. (1998) Physiology and healing dynamics of chronic cutaneous wounds. *Am J Surg* 176, 26S–38S.

5. Marikovsky, M., Breuing, K., Liu, P. Y., (1993) Appearance of heparin-binding EGF-like growth factor in wound fluid as a response to injury. *Proc Natl Acad Sci U S A* 90, 3889–3893.

6. Marikovsky, M., Vogt, P., Eriksson, E., (1996) Wound fluid-derived heparin-binding EGF-like growth factor (HB-EGF) is synergistic with insulin-like growth factor-I for Balb/MK keratinocyte proliferation. *J Invest Dermatol* 106, 616–621.

7. Papetti, M., Herman, I. M. (2002) Mechanisms of normal and tumor-derived angiogenesis. *Am J Physiol Cell Physiol* 282, C947–C970.

8. Galeano, M., Altavilla, D., Cucinotta, D., (2004) Recombinant human erythropoietin stimulates angiogenesis and wound healing in the genetically diabetic mouse. *Diabetes* 53, 2509–2517.

10. Crocker, D. J., Murad, T. M., Geer, J. C. (1970) Role of the pericyte in wound healing. An ultrastructural study. *Exp Mol Pathol.* 13, 51–65.

11. D'Amore, P. A., Herman, I. M. (2001) Molecular and cellular control of angiogenesis, in (Arias, I. M., ed.) *The Liver.* Plenum University Press, New York.

12. Kolyada, A., Riley, K., Herman, I. M. (2003) Rho GTPase regulates cell shape and contractile phenotype in an isoactin-specific manner. *Am J Physiol* 285, 1116–1121.

13. Hirschi, K. K., Rohovsky, S. A., Beck, L. H., Smith, S. R., D'Amore, P. A. (1999) Endothelial cells modulate the proliferation of mural cell precursors via platelet-derived growth factor-BB and heterotypic cell contact. *Circ Res* 84, 298–305.

14. Hirschi, K. K., Rohovsky, S. A., D'Amore, P. A. (1998) PDGF-BB, TGFβ, and heterotypic cell-cell interactions mediate endothelial cell-induced recruitment of 10T1/2 cells and their differentiation to a smooth muscle fate. *J Cell Biol* 141, 805–814.

15. Hellstrom, M., Kalen, M., Lindahl, P., Abramsson, A., Betsholtz, C. (1999) Role of PDGF-B and PDGFR-beta in recruitment of vascular smooth muscle cells and pericytes during embryonic blood vessel formation in the mouse. *Development* 126, 3047–3055.

16. Sieczkiewicz, G. J., Herman, I. M. (2003) TGFβ1 signaling controls retinal pericyte contractile protein expression. *Microvasc Res* 66, 190–196.

17. Antonelli-Orlidge, A., Saunders, K. B., Smith, S. R., D'Amore, P. A. (1989) An activated form of transforming growth factor β is produced by cocultures of endothelial cells and pericytes. *Proc Natl Acad Sci U S A* 86, 4544–4548.

18. Newcomb, P. M., Herman, I. M. (1993) Pericyte growth and contractile phenotype: modulation by endothelial-synthesized matrix and comparison with aortic smooth muscle. *J Cell Physiol* 155, 385–393.

19. Suri, C., Jones, P. F., Patan, S., (1996) Requisite role of angiopoietin-1, a ligand for the TIE2 receptor, during embryonic angiogenesis. *Cell* 87, 1171–1180.

20. Herman, I. M., D'Amore, P. A. (1985) Microvascular pericytes contain muscle and nonmuscle actins. *J Cell Biol* 101, 43–52.

21. Herman, I. M., Jacobson, S. (1988) *In situ* analysis of microvascular pericytes in hypertensive rat brains. *Tissue Cell* 20, 1–12.

22. Helmbold, P., Nayak, R. C., Marsch, W., Herman, I. M. (2001) Characterization of clonal human dermal microvascular pericytes. *Microvasc Res* 61, 160–166.

23. Riley, K. N., Herman, I. M. (2005) Collagenase promotes the response wound healing in vivo. *J Burns Wounds* 4, 141–59.

Section V

In Vivo Techniques

Chapter 15

Measurement of Angiogenic Phenotype by Use of Two-Dimensional Mesenteric Angiogenesis Assay

Andrew V. Benest and David O. Bates

Abstract

Successful therapeutic angiogenesis requires an understanding of how the milieu of growth factors available combine to form a mature vascular bed. This requires a model in which multiple physiological and cell biological parameters can be identified. The adenoviral-mediated mesenteric angiogenesis assay as described here is ideal for that purpose. Adenoviruses expressing growth factors (vascular endothelial growth factor [VEGF] and angiopoietin 1 [Ang-1]) were injected into the mesenteric fat pad of adult male Wistar rats. The clear, thin, and relatively avascular mesenteric panel was used to measure increased vessel perfusion by intravital microscopy. In addition, high-powered microvessel analysis was carried out by immunostaining of features essential for the study of angiogenesis (endothelium, pericyte, smooth muscle cell area, and proliferation), allowing functional data to be obtained in conjunction with high-power microvessel ultrastructural analysis. A combination of individual growth factors resulted in a distinct vascular phenotype from either factor alone, with all treatments increasing the functional vessel area. VEGF produced shorter, narrow, highly branched, and sprouting vessels with normal pericyte coverage. Ang-1 induced broader, longer neovessels with no apparent increase in branching or sprouting. However, Ang-1-induced blood vessels displayed a significantly higher pericyte ensheathment. Combined treatment resulted in higher perfusion, larger and less-branched vessels, with normal pericyte coverage, suggesting them to be more mature. This model can be used to show that Ang-1 and VEGF use different physiological mechanisms to enhance vascularisation of relatively avascular tissue.

Key words: Angiogenesis, angiopoietin 1, pericyte, VEGF

1. Introduction

The proangiogenic effects of endothelial growth factor overexpression have been characterised in a wide range of models. For instance, for vascular endothelial growth factor (VEGF), the rat

S. Martin and C. Murray (eds.), *Methods in Molecular Biology, Angiogenesis Protocols, Second edition, Vol. 467*
© Humana Press, a part of Springer Science+ Business Media, LLC 2009
DOI: 10.1007/978-1-59745-241-0_15

mesentery *(1, 2)*, skeletal muscle *(3, 4)*, cardiac muscle *(5)*, trachea *(6)*, skin *(7–9)*, and corneal eye pocket *(10, 11)* have all been used. With the exception of the trachea and mesenteric models, a reductionist approach has relied on quantifying the increase in vessel density as the angiogenic response. Although the primary end point of angiogenesis should be increased vessel density, this does not reveal any details regarding the manner in which angiogenesis proceeds, such as whether growth is by a sprouting or nonsprouting mechanism. Moreover, in many of these systems, it has proven difficult, if not impossible, to quantify the increase in parameters such as vessel branch point density, proliferating endothelial cells (PECs), and so on. Furthermore, the physiological characterisation of vessels undergoing angiogenesis within the microvessel network is normally not possible. The mesentery is currently used for physiological recordings of microvessel permeability *(12)*, compliance *(13)*, vasoreactivity *(14)*, and conducted vasodilation *(15)*, so an angiogenesis assay in the same tissue allows the determination of functionality of the neovessels formed, measuring parameters such as vascular reactivity, hydraulic conductivity, vessel compliance, and more.

To illustrate this, we describe an angiogenesis assay that compares two different growth factors: angiopoietin 1 (Ang-1) and VEGF. Details of the work here have been published *(16)*, but the methodology is described here in detail *(1)*. For a full description of the angiogenic phenotype to be considered (e.g., type of vessel growth, time course, mechanisms), the use of a two-dimensional microvascular network such as the mesentery offers significant advantages. Although models using the tracheal microvessel network are able to rely on a well-studied system, the method of perfusion fixation and the three-dimensional networks mean that *(1)* only perfused vessels are analysed, *(2)* measurements of perfusion are indirect, and *(3)* the tracheal vasculature cannot be visualised *in vivo*.

For therapeutic purposes, the most successful end result of angiogenesis would be a direct increase in tissue perfusion. The rat mesentery surpasses the limitations of the tracheal system and offers the ability to quantify the vessel phenotype and how this might be manifested anatomically. We can image the perfused mesenteric microvasculature on d 1 and d 7 *in vivo* and if required perform physiological measurements such as those for permeability *(12)*, perfusion *(1)*, and vessel reactivity (Dietrich, 1989 #667; Schreihofer, 2005 # 581), then stain and image the mesentery by confocal fluorescence or electron microscopy. This enables quantitative assessments of vessel density, branching, sprouting, length and diameter, and the degree and constituents of mural support, and this can be linked to the degree of tissue perfusion.

2. Methods

2.1. Angiogenesis and Arteriogenesis Preparation

All surgical procedures need to be performed using sterile equipment. Male Wistar rats (300–350 g) are anaesthetised by 5% halothane inhalation, and anaesthesia is maintained with 3% halothane. A rectal temperature probe is inserted and attached to a thermostatically controlled heated blanket (Harvard CCM79 animal blanket control unit, Kent, UK) to maintain core body temperature at 37°C. A small portion of the ventral surface is shaved and sterilised with Betadine (7.5% povidine-iodine, Napp, Lodi, NJ, USA) and 70% ethanol. A laparotomy is performed, creating an incision approximately 1.5–2 cm long.

A small region of the small intestine is gently teased out and draped over a quartz pillar (*see* **Fig. 15.1**). A mesenteric panel with a flowing microvessel bed is visualised using a Leica DC350F (Leica, Bucks, UK) connected to an intravital microscope (Leica DMIL) and a personal computer. Then, 25 µL of adenovirus are injected into the surrounding fat pad using a 30-gauge needle and Hamilton syringe (VWR, Leicester, UK). Surrounding panels are tattooed with 0.6% w/v Monastral blue in mammalian Ringer (millimolar concentrations: 132.0 NaCl, 4.6 KCl, 1.27 $MgSO_4$, 2.0 $CaCl_2$, 25.0 $NaHCO_3$, 5.5 D-glucose, 3.07 HEPES acid, and 2.37 HEPES sodium salt, pH corrected to 7.45 ± 0.02 with $0.115M$ NaOH). This enables the same panel to be located on d 7 or 14. The following viruses have been used for the angiogenesis assay adenovirus-cytomegalovirus (Ad-CMV)–VEGF$_{165}$, titre $8*10^8$ plaque-forming units (PFU)/mL; Ad-CMV-Ang-1,

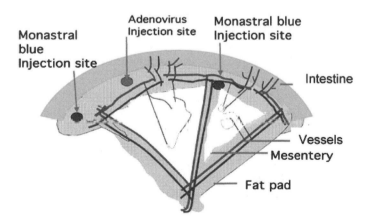

Fig. 15.1. A schematic representation of rat mesenteric angiogenesis assay demonstrating the intestine (grey), the fat pad (grey), the transparent mesentery (white). Within the mesenteric panel are the microvessels, showing approximately equal distributions of pre-, post-, and true capillary order vessels. The adenovirus is injected into the fat pad (marked checked circle), and the panel is located by placing a tattoo on either side of the injected panel.

titre $8*10^8$ PFU/mL; Ad-CMV-eGFP (enhanced green fluorescent protein), titre $5*10^8$ PFU/mL. In experiments requiring multiple viruses to be used, 25 μL of each virus are used but injected as a mixture.

Throughout the procedure (which should last approximately 2–5 min), the mesentery and surrounding intestine are superfused with mammalian Ringer, kept at a constant 37°C ± 1°C. Temperature is maintained by connecting the superfusate to a heat exchange coil fed by a thermostatically controlled water bath. The abdominal cavity and skin are sutured and the animal allowed to recover on 100% O_2 before being transferred to a warmed, clean animal cage. Analgesia is provided by a single intramuscular injection of buprenorphine (Temgesic®, Schering-Plough, Harefield, UK). All postoperative animals are caged individually and checked hourly and, after 24 h postprocedure, daily.

2.2. Immunofluorescence on Whole-Mount Mesentery

For angiogenesis assays, 6/13 d later (d 7/14) the same animal is anaesthetised, a laparotomy is performed, and the same mesenteric panel is located (by the Monastral blue tattoo). The panel is imaged as before and fixed *in vivo* with 4% paraformaldehyde (in 1*phosphate-buffered saline [PBS], pH 7.40) for 5 min. The animal is then killed by cervical dislocation, and the mesenteric panel is excised. The fat pads are removed and snap frozen in liquid nitrogen for later examination of protein expression and stored at –20°C. The panel is washed in mammalian Ringer and refixed for a maximum of 1 h at room temperature in the same fixative.

The panels are washed with 0.5% Triton X-100 in PBS (0.5% PBX), for 1 h, changing solutions every 10 min at room temperature. The panels are blocked in 1% bovine serum albumen (BSA)-0.5% PBX, or 1.5% normal goat serum for 1 h. The mesentery is incubated overnight at 4°C with 10 μg/mL biotinylated *Griffonia simplicifolia* isolectin IB4 (GSI-IB4, Molecular Probes, Cambridge, UK) and mouse monoclonal antibodies to either Ki-67 (Novocastra Lab, Newcastle upon Tyne, UK, NCL-L-Ki67-MM1) for dividing cells, NG2 (Chemicon, Temecula, CA, USA, MAB5384, 5 μg/mL) for pericytes, or α-smooth muscle actin (DAKO, Glostrup, Denmark, M 0851, 1.4 μg/mL) for smooth muscle.

Panels are washed as before and (TRITC) Tetramethyl Rhodamine ISO-Tiocynate-labelled streptavidin (1 μg/mL, S-870, Molecular Probes) and Alexa Fluor 488, 350 goat antimouse immunoglobulin G (IgG; 2 μg/mL, Molecular Probes) are used as secondary detection antibodies to lectin, Ki67, and VEGF and smooth muscle actin, respectively. Primary antibodies and lectin are incubated overnight on a rocker at 4°C. The panels are washed for 1 h, changing the PBX solution six times and incubating with secondary antibodies for 2 h at room temperature on a rocker. The panels are washed a further six times, and if appropriate Hoechst 33324 (1 μ*M*, Molecular Probes) is added to stain mesenteric nuclei.

The panels are carefully manipulated using fine forceps and flattened on a glass slide before being mounted in Vectashield (Vector Lab, Peterborough, UK), and a coverslip is carefully placed on the tissue. The tissue is imaged using a Leica confocal microscope (Leica confocal TCS-NT DMIRBE). Five random sections of each panel are imaged and analysed offline at ×40 magnification with oil immersion.

2.3. Microvessel Analysis

Data obtained from intravital microscopy are analysed to provide information regarding the maturity of the neovessels formed. The red blood cells flowing through the microvessel bed contrast with the clear mesothelial layer surrounding the vessels. Consequently, fractional vessel area (FVA) is measured as the area of flowing blood vessels per area of mesenteric tissue. All microvessel analysis is carried out using Openlab 3.1 (Improvison, Coventry, UK). The vessel area is selected using the Wand tool, with a pixel threshold of 32. The FVA is a measure of the percentage of vessel area per mesenteric panel. The angiogenesis index (AI) is expressed as the following equation:

$$AI(\%) = [(FVA\ Day\ 7 - FVA\ Day\ 1)/FVA\ Day\ 1]*100$$

2.4. Microvessel Measurement

The images obtained from the confocal imaging are also analysed using Openlab software. The scale is calibrated using the scale obtained from the confocal microscope. Measurements are made for vessel diameter, vessel length (distance of vessel not broken by a branch or sprout point), vessel number, sprout point number, branch point number, and PEC number (*see* **Fig. 15.2**). The mean value from four or five panels is taken, and the density (mm ^{2}) can then be calculated. The fractional pericyte area (FPA) is calculated as the percentage of the vessel covered by pericyte (NG2 positive). Staining for α-smooth muscle actin is used to confirm the presence of vascular smooth muscle cells (vSMCs); therefore, the density of vSMC-positive vessels can also calculated as well as the fractional smooth muscle area (FSMA). See **Fig. 15.2** for an illustration of the analyses. To calculate the frequency histogram of vessel diameters, all vessel diameters are pooled, and a contingency table is calculated as either absolute values or as a percentage of maximum frequency.

2.5. Adenovirus Amplification

Although all adenoviruses we have used have been previously used and characterised, it is necessary to ensure that they are still biologically active. For this, plaque assays are carried out to minimise the amplification of defective adenovirus.

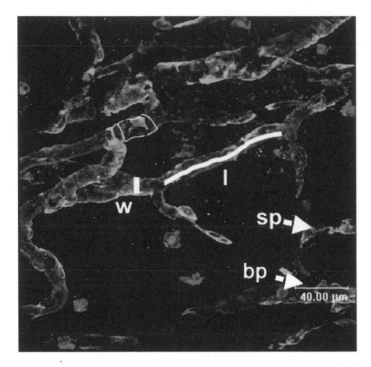

Fig. 15.2. Measurement of microvascular parameters on stained mesenteries. *Red* isolectin-stained vessels, *green* NG2-stained pericytes. Parameters of width (measured line by w), length (measured line by l), sprouts (sp), branches (bp), and pericyte coverage (delineated green divided by delineated red areas by pericyte). Note that the pericytes can be seen right up to the tips of the sprouts (grey arrowhead). (*See Color Plates*)

2.6. Plaque Purification

Under aseptic conditions, HEK-293 cells are grown in Dulbecco's modified Eagle's medium (DMEM) supplemented (per 500-mL bottle) with 50 mL foetal bovine serum (FBS), 10 mL L-glutamine, 10 mL HEPES buffer, and 10 mL penicillin/streptomycin. For normal passage and amplification, cells are grown in 75-cm^2 cell culture plates. Cells are split at 80% confluency. Cell splitting is carried out by removing the growth media and washing in 3*PBS before the addition of 5 mL trypsin-EDTA (ethylenediaminetetraacetic acid). The flasks are incubated at 37°C for 5 min. The cell suspension is then added to 10 mL DMEM and centrifuged for 5 min. The medium is removed and the cell pellet resuspended and added to new flasks.

HEK-293 cells are plated and grown on 60-mm Petri dishes in DMEM. When cells reach 80% confluency, the medium is removed. Then, 10 μL adenovirus are diluted in 990 μL PBS^{2+} (0.1 mg/mL CaCl$_2$, 0.1 mg/mL MgCl$_2$ in PBS); this is then serially diluted an additional five times. To each place, carefully add 200 μL of the viral solution. The plate is carefully agitated every 5 min for 1 h.

A sterile 1% w/v agarose solution (Bioline, London) is melted and an equal volume added to 25 mL of 0.2 µm filter-sterilised 2*DMEM overlay medium (5 mL 10*DMEM), 5 mL heat-inactivated FCS, 2.5 mL 7.5% w/v sodium bicarbonate, 12.5 mL ddH$_2$O [double-distilled water]). The 10 mL agarose overlay is gently added to each dish when it cools to 37°C. The plates are left at room temperature until the overlay sets, and then they are placed at 37°C in 5% CO$_2$. Cytolytic plaques can be observed by eye and microscope by 8 d postinfection.

2.7. Plaque Isolation

Well-isolated plaques that appear 8 d after transfection are chosen for screening. A plug of agarose around each viral plaque is removed and placed in 1 mL sterile PBS. The plaque is then lysed in liquid nitrogen and thawed before vortexing the tube. Low-passage Chinese hamster ovary (CHO) cells are previously grown to 80% confluency in 25-cm^2 culture flasks with 3 mL F-12 medium (supplemented with 5 mL penicillin/streptomycin and 50 mL FBS; Gibco). The medium is removed, and 500 µL of PBS^{2+}/plaque mix are spread across the cell monolayer. The flask is then incubated at 37°C for 4 d. The remainder of the plaque solution is stored at –80°C until further use.

2.8. Large-Scale Amplification

Following verification of gene product expression, 500 µL of the PBS/agarose mix are added to a 75-cm^2 (Nunc, Rothewas UK) flasks of 80% confluent HEK-293 cells. Low-passage HEK-293 cells are grown to 80% confluency in 10*175 cm^2. When the first flasks reaches full cytopathic effect (CPE) and begins detaching from the culture plate, the media and cell suspension are removed and added to the 10*175 cm^2. Approximately 3 d postinfection, the cells reach CPE but have not detached. The cells are removed and centrifuged at 2000 rpm (approx. 600g) for 10 min at 22°C. The cell pellets are pooled and resuspended in 5 mL of 200 mM Tris-HCl (pH 7.5) solution and stored at –20°C overnight.

2.9. Adenovirus Purification

Viruses are liberated from HEK-293 by a single freeze/thaw cycle followed by sonication on ice for four 30-s pulses. The suspension is then centrifuged at 2000 rpm for 10 min to remove all cell debris. The supernatant (now containing released viral particles) is added to 0.6 volumes of CsCl-saturated 100 mM Tris-HCl (pH 7.5). This is transferred to a 11.2-mL OptiSeal polyallomer ultracentrifuge tube (Beckman, Fullerton, CA, USA), and the volume is made up to 11.2 mL with a solution containing 0.6 volumes CsCl-saturated 100 mM Tris-HCl at pH 7.5 to one volume of 100 mM Tris-HCl. The tube is centrifuged at 65,000 rpm (NVTi 65.1 ultracentrifuge rotor, Beckman) for at least 8 h at 25°C. The resultant white virus particle band (approximately 200 µL) is removed from the tube with an 18-gauge needle and 5-mL syringe. To ease removal of the band, three 27-gauge needles are placed in the top of the tube. The viral isolate is then added to approximately 11 mL of CsCl-Tris-HCl solution as before and centrifuged at 65,000 rpm for at least 8 h at 25°C. The resultant viral band is removed as before, and the volume made up to 2.5 mL with 100 mL Tris-HCl (pH 7.5) in a sterile bijou.

Salt residues are removed from the viral isolate with a Sephadex PD10 column (GE Healthcare Amersham, UK) is equilibrated with 25 mL of storage buffer (10 mM Tris-HCl, 1 mM MgCl$_2$, pH 8.0) and the viral isolate added to the column. The purified virus is eluted from the column by addition of a further 3.5 mL storage buffer and collected in a sterile bijou. The desalted viral suspension is then filtered using a 0.2-μm sterile filter (Sartorius, Edgewood, NY, USA), aliquoted into sterile Eppendorf tubes, snap-frozen in liquid nitrogen, and stored at –80°C.

2.10. Recombinant Adenovirus Titration by Tissue Culture Infectious Dose 50

To determine the viral titre, the tissue culture infectious dose 50 (TCID$_{50}$) method is used. This has the advantage of having greater reliability, ease, and speed of results and more consistency between individuals.

The following procedures are carried out in duplicate. A 75-cm^2 flask of HEK-293 cells is taken, suspended in 20 mL of 2% FBS-DMEM, and counted using a haemocytometer. Add 100 μL of cell suspension per well (approximately 10^4 cells) of a 96-well plate. The cells are allowed to attach overnight.

Add 990 μL of 2% FBS-DMEM to 0.1 μL viral stock and agitate, forming a 10^{-2} dilution. Eight 15-mL Falcon tubes with 1.8 mL of 2% FBS-DMEM are taken, and 200 μL of the 10^{-2} viral solution are added to the first tube. This is then serially diluted from 10^{-3} to 10^{-10}. A new filtered pipette tip (Starlab, Milton Keynes) is used each time. two columns are used as controls, with only cells and 2% FBS-DMEM in each well.

Add 100 μL of each viral solution to each well and incubate at 37°C for 10 d, with each row representing increased dilution. At 10 d postinfection, the wells that underwent CPE are counted, and a fraction of cells with CPE per row is measured as a ratio. The following equation is used to calculate titre:

$$\text{Titre} = 10^{1 + d(S - 0.5)}$$

where S is the sum of the ratios starting at $10^{(-1)}$ dilution. If all cells died, then the ratio is 1; d is the log dilution factor. The average of both plates is taken as the viral titre. Conversion to PFU per millilitre is carried out by dividing the TCID$_{50}$ by 0.7.

3. Results

3.1. Ang-1 and VEGF Increase Functional Vessel Area

To quantify functional changes brought about by growth factor expression (as confirmed by enzyme-linked immunosorbent assay [ELISA]; *(16)*, intravital microscopy is used to image the microvessel bed before and after administration of the growth factor. On d 1, clear, functional vessels (FVA) are seen (**Fig. 15.3a,c,e**). The area covered by these vessels (FVA) is measured using Open-Lab. After 6 d of infection with Ad-GFP, a similar functional vessel area Is seen (**Fig. 15.3b,g**), but following injection of angiogenic growth factors there is an increase in the perfused area (**Fig. 15.3d,f,g,h**). Application of eGFP does not result in a significant increase in FVA (**Fig. 15.3g**), and consequently there is no angiogenic response (AI%, **Fig. 3h**). VEGF induces a significant increase in FVA (**Fig. 15.3g**) and significantly increases AI compared with control (**Fig. 15.3h**). Ang-1 also increases FVA (**Fig. 15.3g**) and increases the AI compared with control (**Fig. 15.3h**), but this increase is no different from VEGF.

In summary, both growth factors are capable of increasing vessel perfusion (as determined by an increased area of visible blood vessel), but to determine if this is due to a haemodynamic effect (VEGF is a known vasodilating agent *(18)* or if there is an angiogenic response (e.g., increase in vessel density, sprouting, proliferation, etc.), a detailed examination of the microvessel bed is necessary, and this can be done by immunofluorescence and confocal microscopy. Images obtained from immunofluorescent confocal microscopy (**Fig. 15.4**) using antibodies to Ki67, with lectin and Hoechst 33324 counterstaining, are used to examine neovessel phenotype. Overlaid images of the three image stacks are used to identify the density of PECs.

The analysis of the stained mesenteries can produce a lot of information about the nature of the angiogenesis. This is summarised in **Fig. 15.5**. To begin, it is clear that the increase in vessel perfusion brought about by growth factors must be due to vessel proliferation, not just vasodilation. The assay shows that both VEGF and Ang-1 increase PEC density compared with control as the Ki67-positive endothelial nuclei are increased. However, there is no difference between the growth factors. In the relatively avascular mesentery, generation of new blood vessels will increase the vessel density (vessels per area of tissue) as the new vessels are more likely to invade into previously unvascularised mesothelial tissue. VEGF and Ang-1 both increase vessel density compared with eGFP, and VEGF also increased the vessel density compared with Ang-1. However, this method allows a subtler interpretation. The manner in which the microvessels are formed, the branch point density, and the sprout point density can be measured. Consistent with VEGF inducing sprout formation, there

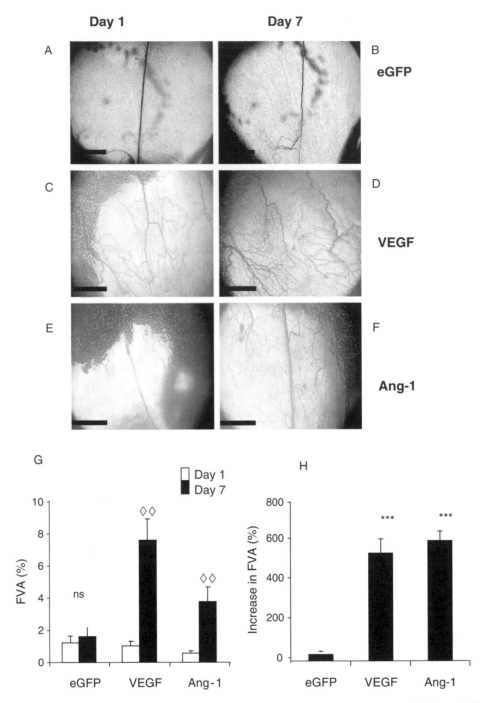

Fig. 15.3. Intravital images of mesenteric panels before and after growth factor overexpression (**A–F**). Intravital images obtained by ×4 objective lens of patent vessels. Vessels on d 1 (**A, C, E**) and the same vessels imaged 6 d later (**B, D, F**). Vessels are visible due to the contrast of the blood flow. An increase in functional vessel area (FVA) is determined by an increase in patent vessel area (**G**), the magnitude of this increase is measured in **H**. ***$p < 0.001$ versus enhanced green fluorescent protein (eGFP), analysis of variance (ANOVA), ..$p < 0.01$ d 1 versus d 7, t-test. Scale bar 1 mm, n = 5 all groups, mean ± standard error of the mean (SEM). *Ang-1* angiopoietin 1, *VEGF* vascular endothelial growth factor

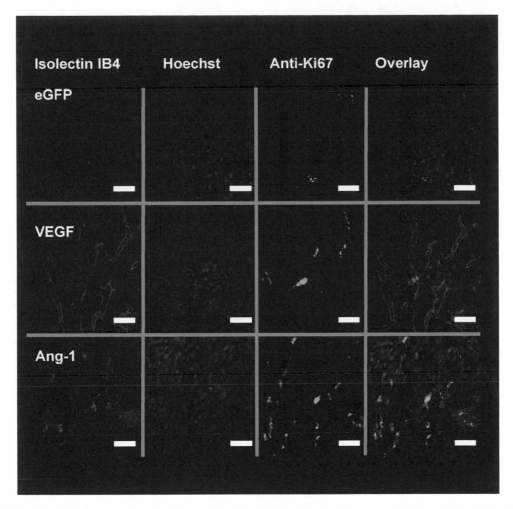

Fig. 15.4. Growth factor expression stimulates endothelial cell (EC) proliferation. Fluorescent staining of mesenteries after injection of adenovirus-enhanced green fluorescent protein (Ad-eGFP) (control) or Ad-growth factors. Isolectin IB4-TRITC (red) stains endothelial cells ECs, Hoechst 33324 (blue) stains all mesenteric nuclei and antibodies to Ki67-AF488 (green) to detect proliferating cells. Overlaying of the stack images Is used to calculate the number of proliferating endothelial cells (PECs). Images are triple stained with TRITC-streptavidin and biotinylated GSL (Griffonia Simplicifolia Lectin, Ec, red) lectin IB4 (EC, red). Alexa Fluor 488-labeled goat antimouse immunoglobulin G and mouse monoclonal anti-Ki67 antibody (proliferating cells, green) and overlay. The PECs can be distinguished from other cells by their position within the vessel wall. White arrow demonstrates a PEC. *VEGF* vascular endothelial growth factor. Scale bar 40 µm (*See Color Plates*)

is a significant increase in sprout point density following VEGF but not Ang-1 application compared with control. Moreover, the branch point density follows an identical pattern. Taken together, this assay can demonstrate that Ang-1 does not induce sprout formation or increase branching, whereas VEGF does, although both VEGF and Ang-1 increased the FVA to approximately the same level. As blood flow is largely dependent on the vessel diameter, this increase in flow could be brought about by an increase

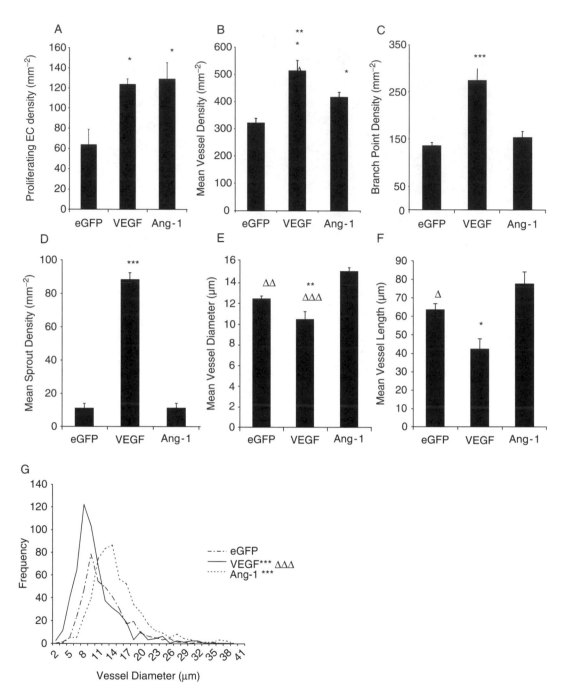

Fig. 15.5. Vascular endothelial growth factor (VEGF) and angiopoietin 1 (Ang-1) are proangiogenic, yet result in different vessel phenotypes. Analysis of blood vessel parameters, from confocal stack data of mesenteries stained for lectin, Hoechst, and Ki67. Analysis of proliferation data (**A**), microvessel density (**B**), branch point (**C**), and sprout point (**D**) revealed that VEGF induced a hyperplasic response, but Ang-1 increased proliferation, but vessel density only increased by a moderate degree. Additional analysis of microvessels from confocal data. VEGF reduced vessel diameter and vessel length, reflective of increases in branching and sprouting (**A, B**). Consistent with increased branching and sprouting, Ang-1 increased mean vessel diameter (**E**) but did not alter vessel length (**F**). VEGF produced shorter vessels, and Ang-1 produced longer ones. Frequency histogram of mean vessel diameters of each vessel measured demonstrates a significant difference in the distribution of each treatment group (**G**). *$p < 0.05$, ***$p < 0.001$ versus enhanced green fluorescent protein (EGFP), Δ $p < 0.05$, .. $p < 0.01$, ... $p < 0.001$ versus Ang-1. All values shown are mean ± standard error of the mean (SEM), n = 5. *EC* endothelial cell

in diameter of the vessels. This assay also allows measurement of microvessel diameter. Consistent with the notion that VEGF induces sprout formation and then remodels them to form neo-vessels, VEGF produces narrower microvessels, which were significantly narrower than with eGFP and Ang-1. Of note, Ang-1 significantly increases the vessel diameter. Ang-1 produces significantly longer vessels than VEGF, which may be representative of the reduced sprouting and branching behaviour rather than independent effect. In addition to the increased mean vessel diameter, frequency histograms of microvessel diameters can be generated, and these demonstrate that VEGF produces a significantly different distribution compared with eGFP and Ang-1 (analysed by χ^2 analysis).

It has been shown that VEGF selectively affects microvessels according to their size in the mesentery *(1)*, and in the tracheal system it has been demonstrated that Ang-1 affects venules specifically *(19–21)*. This assay can be used to analyse if there is a difference between exchange (<16-μm diameter) and larger, conduit capillaries (16–35 μm). This is based on the antique works of Schleier, who analysed the diameters of all vessels present in the dog mesentery, leading to the classification of vessels smaller than 16 μm to be exchange (terminal capillaries) and 16- to 35-μm vessels to be conduit capillaries *(22)*. For instance, Ang-1 produces more conduit vessels compared with eGFP and VEGF, whereas VEGF produces more exchange capillaries than eGFP or Ang-1 (**Fig. 15.6**). It can also be determined that Ang-1 induces a greater degree of proliferation in the larger microvessels than eGFP and VEGF. Taken together, these results suggest that the assay can be used to show that Ang-1 is a more potent inducer of microvascular hypertrophy, and VEGF is more potent at neovascular angiogenesis.

3.2. Pericyte Support

The role of periendothelial support in angiogenesis is not clear, with the majority of data coming from tumor studies. But, transgenic studies indicated that the Ang-1/Tie2 system has a role in regulating the degree of periendothelial support *(23–25)*. This assay enables analysis of mural cell association with the micro vessels. **Figure 15.7** demonstrates pericytes wrapping around the vessel, with what appears to be a close association between the endo- and periendothelium. The absence of any αSMA-containing cells in the presence of NG2-labelled pericytes demonstrates that the periendothelium was entirely pericyte derived. VEGF-stimulated vessels appear to have pericytes not only along the length of the vessel but also at branch and sprout points. Ang-1 induces vessels that have a considerably higher degree of association, with the majority of the vessel covered in pericyte.

The pericyte coverage (FPA) can be calculated as the percentage of endothelium covered by pericyte. Ang-1 increases the

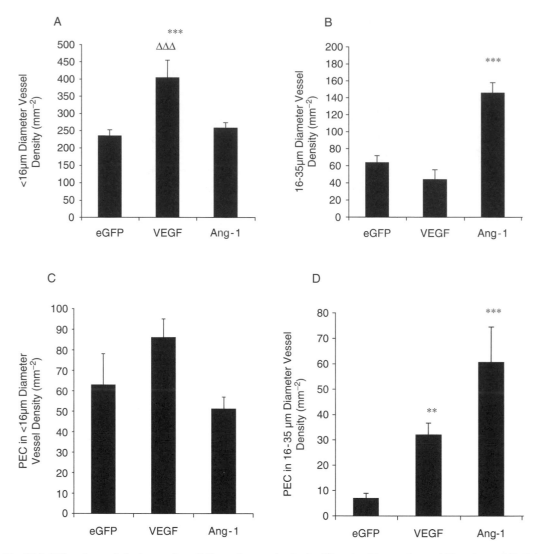

Fig. 15.6. Different growth factors preferentially produce and activate different subtypes of vessel. Vascular endothelial growth factor (VEGF) stimulated the generation of sub-16-μm diameter vessels compared with angiopoietin 1 (Ang-1) or green fluorescent protein (GFP) (**A**). Ang-1 transfections stimulated the generation of 16- to 35-μm diameter vessels to a greater extent than enhanced GFP (eGFP) or VEGF (**B**).In larger vessels, VEGF and Ang-1 stimulated endothelial proliferation, Ang-1 to a greater extent (**C**), whereas in smaller vessels endothelial cell proliferation was not increased (**D**). All values shown are mean ± standard error of the mean (SEM), n = 5. *$p < 0.05$, **$p < 0.01$, ***$p < 0.001$ versus EGFP, ... $p < 0.001$ versus Ang-1. *PEC* proliferating endothelial cells

FPA (almost double) compared with eGFP and VEGF, whereas VEGF does not change the coverage compared with control. By plotting the vessel area (mm²) against the pericyte area (mm²) for the different treatment groups, the relationship between pericyte and endothelial cells can be investigated. This is increased by Ang-1 (**Fig. 15.8**).

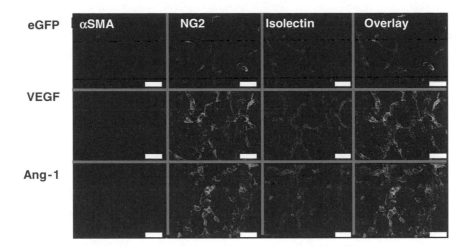

Fig. 15.7. Growth factor expression stimulates endothelial cell (EC) proliferation. Fluorescent staining of mesenteries after injection of adenovirus-enhanced green fluorescent protein (Ad-EGFP) (control) or Ad-growth factors. Isolectin IB4-TRITC (red) stains ECs, Hoechst 33324 (blue) stains all mesenteric nuclei and antibodies to Ki67-AF488 (green) to detect proliferating cells. Overlaying of the stack images is used to calculate the number of proliferating endothelial cells (PECs). Images are triple stained with TRITC-streptavidin and biotinylated GSL lectin IB4 (ECs, red). Alexa Fluor 488-labeled goat antimouse immunoglobulin G and mouse monoclonal anti-Ki67 antibody (proliferating cells, green) and overlay. The PECs can be distinguished from other cells by their position within the vessel wall. White arrow demonstrates a PEC. Scale bar 40 μm. *SMA* smooth muscle area (*See Color Plates*)

4. Discussion

Growth factor-induced angiogenesis can be characterised in many models. However, the advantage of the rat mesenteric adenovirus assay *(1)* is that it allows detailed and subtle phenotypic differences between growth factors to be analysed as shown here. The key advantage of this assay is to marry such detailed and subtle phenotypic measurements with functional assessment, including flow measurements, permeability, compliance, and ultrastructural investigations. For instance, we provide evidence that VEGF does increase the FVA; in addition to this, we can also record an increase in vessel density. Likewise, although the therapeutic potential of Ang-1 has been tested in rabbit hindlimb ischaemia models, and Ang-1 was seen to enhance tissue perfusion via naked plasmid transfection *(26, 27)*, the potential impact of the previously induced ischaemia might obscure any measurable role of Ang-1.

4.1. Increased Vascularity

The use of angiography as a measure of "angiogenesis" relies on the caveat that there was a vasodilatating effect of the growth factors used. VEGF is a potent vasodilator *(18)*, but so far there have been no reports of such an effect by Ang-1. Using this assay, we can show that both VEGF and Ang-1 significantly increase

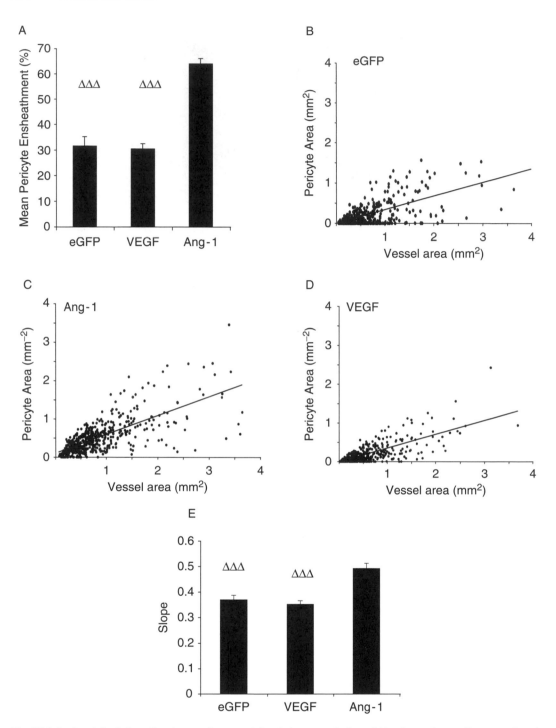

Fig. 15.8. Angiopoietin 1 (Ang-1) enhances the association between pericyte and blood vessel area. The area of each blood vessel covered by pericyte (fractional pericyte area, **A**) is increased by Ang-1 but is not changed between enhanced green fluorescent protein (eGFP) or vascular endothelial growth factor (VEGF). Correlation analysis between vessel area and pericyte area eGFP (**B**), Ang-1 (**C**), VEGF (**D**). Comparison of the slope by analysis of variance (ANOVA) reveals Ang-1 has a greater association (**E**), that is, larger vessels have a better pericyte coverage. ... $p < 0.001$ versus Ang-1. All values shown are mean ± standard error of the mean (SEM), n = 5

the vessel density over eGFP. Moreover, it should be noted that the effect of Ang-1 was only modest in comparison to VEGF. VEGF is currently the most proangiogenic molecule identified, and it is known to lead to an increase in vessel density in a number of models (1, 6, 7, 28). The increase in vessel density seen here is modest compared with that seen with transgenic overexpression in the skin (25), adenoviral-mediated Ang-1 in rabbit carotid adventitia (29), or COMP-Ang-1 (N-terminal portion of Ang1 with the short Coiled-Coil domain of cartilage oligomeric matrix protein) pellets in the cornea (30). The work by Suri et al. used whole-mount, perfusion-fixed, and stained skin tissue, which only allowed analysis of perfused vessels, and the efficacy of staining would be limited by haemodynamic factors and the risk of antibodies/lectins diffusing into the mesenchyme (25). The work by Bhardwaj et al. used a heavily inflamed tissue, which would prevent interpretation of the role of Ang-1 in the absence of other inflammatory cytokines. Cho et al. used the rabbit cornea, which is normally avascular and is not a physiological model of angiogenesis as the tissue does not normally support blood vessel growth due to endogenous inhibition of angiogenesis by soluble flt-1 (31).

4.2. Vessel Morphology

VEGF-treated dermal tissue produced irregular-formed, pericyte-devoid, hypertrophic, and heterogenic vessels (32). Our whole-mount analyses revealed that the vessels formed were generally of quite small calibre (0.10 μm), and we demonstrated a highly significant degree of sprout and branch formation. It can be interpreted that the sprouts are re-formed into neovessels (leading to increased vessel density and branch point formation). The narrow vessel diameter may be a factor of the narrow sprout being remodelled, and the vessel reaches optimum diameter as perfusion is achieved (33). The sprouting phenotype observed in this assay is consistent with other findings examining the phenotypic effects of VEGF (1, 2, 6, 21, 34–36). In contrast, this assay shows that Ang-1 increases vessel diameter (15 μm), with no increase in sprout point or branch point formation. Ang-1 has been described to induce in vitro sprouting activity and associated protease release (26, 37). However, Ang-1-mediated sprouting is difficult to assess in vivo, and a lack of vessel sprouting has been previously described using the mouse tracheal microcirculation (21) and mouse skin (25, 38).

4.3. Endothelial Cell Proliferation In Vivo

The use of the assay to show that both Ang-1 and VEGF induced a high degree of endothelial cell proliferation demonstrated the utility of the assay for proving angiogenesis. Ang-1 was previously not thought to be proliferative in vitro (26, 39), although it has been reported in vivo that Ang-1 treatment led to increased expression of proliferation markers (21, 30).

4.4. Periendothelium

The investigation of pericyte function and form is of increasing interest. This assay clearly demonstrates that Ang-1 is able to recruit pericytes to the growing endothelium. Despite tumor studies that have demonstrated that altering Ang-1 overexpression is capable of altering pericyte recruitment (e.g., *40, 41)* there is no agreement regarding whether this is conducive for angiogenesis. Knockout studies have confirmed a role for Ang-1 in endothelial-pericyte support *(24)*, but the effects of Ang-1 on the endothelium are probably not direct. However, evidence for a role in Erb signalling exists as Ang-1 could upregulate Erb, (also known as EGF-receptor or HER1) which in turn is known to attract mural cells and induce mural cell proliferation through heparin binding EGF-line growth factor (HB-EGF) *(42–45)*. Other candidates for recruitment of periendothelial cells to the vessel were 5-HT, by which Ang-1 can induce production and release from the endothelial cell *(43, 46, 47)*. Serotonergic receptors are found on smooth muscle cells from human and rodent origin *(47)* and are believed to be essential for the migration and proliferative pathways. It will be interesting to determine whether there are hydroxytryptamine (5-HT) receptors in this preparation.

4.5. Pericyte Coverage as a Regulator of Angiogenesis

In this assay, pericyte coverage does not appear to be an essential regulator of angiogenesis. Pericyte coverage does not change when VEGF is overexpressed; that is, there are still pericytes associated with the growing vessel, but the microanatomical architecture is altered. The physiological role of pericytes is still not clear. It appears as though they might regulate vessel morphology and endothelial cell number *(48)* and release prosurvival molecules such as Ang-1 to the endothelium *(49, 50)*. This may result in the prevention of remodelling, so-called termination of the plasticity window for vessel remodelling *(51, 53)*. Consequently, it is been suggested that angiogenesis requires pericytes to be removed from the vessel wall, and Ang-2 has been proposed to be the critical mediator of this event *(5, 54)*. Although this is clearly the case when considering VEGF-induced angiogenesis, which is a sprouting phenotype, this assay has been used to demonstrate that Ang-1-induced angiogenesis does not need this to occur. The plasticity argument would suggest that pericytes are recruited to end angiogenesis, whereas we and others provided evidence that pericytes can be attracted to sprouts *(55, 56)*. Either way, vessels with enhanced pericyte coverage are considered capable of perfusing a neovascular bed more effectively *(57)*.

In conclusion, the mesenteric angiogenesis assay is a highly sensitive, flexible, and subtle assay for investigation of blood vessel growth *in vivo* in a physiological vascular bed. The experiments are not onerous or technically difficult and do not require special equipment. They do require substantial analysis time, but that results in a very useful analysis of growth characteristics and can be linked to physiological data.

References

1. Woolard, J., Wang, W. Y., Bevan, H. S., Qiu, Y., Morbidelli, L., Pritchard-Jones, R. O., Cui, T. G., Sugiono, M., Waine, E., Perrin, R., Foster, R., Digby-Bell, J., Shields, J. D., Whittles, C. E., Mushens, R. E., Gillatt, D. A., Ziche, M., Harper, S. J., and Bates, D. O. (2004) *Cancer Res* 64, 7822–35.

2. Wang, W. Y., Whittles, C. E., Harper, S. J., and Bates, D. O. (2004) *Microcirculation* 11, 361–75.

3. Rissanen, T. T., Markkanen, J. E., Gruchala, M., Heikura, T., Puranen, A., Kettunen, M. I., Kholova, I., Kauppinen, R. A., Achen, M. G., Stacker, S. A., Alitalo, K., and Yla-Herttuala, S. (2003) *Circ Res* 92, 1098–106.

4. Vajanto, I., Rissanen, T. T., Rutanen, J., Hiltunen, M. O., Tuomisto, T. T., Arve, K., Narvanen, O., Manninen, H., Rasanen, H., Hippelainen, M., Alhava, E., and Yla-Herttuala, S. (2002) *J Gene Med* 4, 371–80.

5. Visconti, R. P., Richardson, C. D., and Sato, T. N. (2002) *Proc Natl Acad Sci U S A* 99, 8219–24.

6. Baluk, P., Lee, C. G., Link, H., Ator, E., Haskell, A., Elias, J. A., and McDonald, D. M. (2004) *Am J Pathol* 165, 1071–85.

7. Detmar, M., Brown, L. F., Schon, M. P., Elicker, B. M., Velasco, P., Richard, L., Fukumura, D., Monsky, W., Claffey, K. P., and Jain, R. K. (1998) *J Invest Dermatol* 111, 1–6.

8. Sundberg, C., Nagy, J. A., Brown, L. F., Feng, D., Eckelhoefer, I. A., Manseau, E. J., Dvorak, A. M., and Dvorak, H. F. (2001) *Am J Pathol* 158, 1145–60.

9. Thurston, G., Rudge, J. S., Ioffe, E., Zhou, H., Ross, L., Croll, S. D., Glazer, N., Holash, J., McDonald, D. M., and Yancopoulos, G. D. (2000) *Nat Med* 6, 460–3.

10. Cursiefen, C., Chen, L., Borges, L. P., Jackson, D., Cao, J., Radziejewski, C., D'Amore, P. A., Dana, M. R., Wiegand, S. J., and Streilein, J. W. (2004) *J Clin Invest* 113, 1040–50.

11. Ziche, M., Morbidelli, L., Choudhuri, R., Zhang, H. T., Donnini, S., Granger, H. J., and Bicknell, R. (1997) *J Clin Invest* 99, 2625–34.

12. Glass, C. A., Harper, S. J., and Bates, D. O. (2006) *J Physiol* 572, 243–57.

13. Bates, D. O. (1998) *J Physiol* 513 (Pt 1), 225–33.

14. Dietrich, H. H. (1989) *Microvasc Res* 38, 125–35.

15. Takano, H., Dora, K. A., Spitaler, M. M., and Garland, C. J. (2004) *J Physiol* 556, 887–903.

16. Benest, A. V., Salmon, A. H., Wang, W., Glover, C. P., Uney, J., Harper, S. J., and Bates, D. O. (2006) *Microcirculation* 13, 423–37.

17. Schreihofer, A. M., Hair, C. D., and Stepp, D. W. (2005) *Am J Physiol Regul Integr Comp Physiol* 288, R253–61.

18. Horowitz, J. R., Rivard, A., van der Zee, R., Hariawala, M., Sheriff, D. D., Esakof, D. D., Chaudhry, G. M., Symes, J. F., and Isner, J. M. (1997) *Arterioscler Thromb Vasc Biol* 17, 2793–800.

19. Baffert, F., Le, T., Thurston, G., and McDonald, D. M. (2006) *Am J Physiol Heart Circ Physiol* 290, H107–18.

20. Thurston, G., Suri, C., Smith, K., McClain, J., Sato, T. N., Yancopoulos, G. D., and McDonald, D. M. (1999) *Science* 286, 2511–4.

21. Thurston, G., Wang, Q., Baffert, F., Rudge, J., Papadopoulos, N., Jean-Guillaume, D., Wiegand, S., Yancopoulos, G. D., and McDonald, D. M. (2005) *Development* 132, 3317–26.

22. Schleier, J. (1918) *Archiv fur die Gesamte Physiologie* 173, 172–204.

23. Sato, T. N., Tozawa, Y., Deutsch, U., Wolburg-Buchholz, K., Fujiwara, Y., Gendron-Maguire, M., Gridley, T., Wolburg, H., Risau, W., and Qin, Y. (1995) *Nature* 376, 70–4.

24. Suri, C., Jones, P. F., Patan, S., Bartunkova, S., Maisonpierre, P. C., Davis, S., Sato, T. N., and Yancopoulos, G. D. (1996) *Cell* 87, 1171–80.

25. Suri, C., McClain, J., Thurston, G., McDonald, D. M., Zhou, H., Oldmixon, E. H., Sato, T. N., and Yancopoulos, G. D. (1998) *Science* 282, 468–71.

26. Chae, J. K., Kim, I., Lim, S. T., Chung, M. J., Kim, W. H., Kim, H. G., Ko, J. K., and Koh, G. Y. (2000) *Arterioscler Thromb Vasc Biol* 20, 2573–8.

27. Shyu, K. G., Manor, O., Magner, M., Yancopoulos, G. D., and Isner, J. M. (1998) *Circulation* 98, 2081–7.

28. Cao, R., Eriksson, A., Kubo, H., Alitalo, K., Cao, Y., and Thyberg, J. (2004) *Circ Res* 94, 664–70.

29. Bhardwaj, S., Roy, H., Karpanen, T., Hi, Y., Jauhiainen, S., Hedman, M., Alitalo, K., and Yla-Herttuala, S. (2005) *Gene Ther* 12, 388–94.

30. Cho, C. H., Kim, K. E., Byun, J., Jang, H. S., Kim, D. K., Baluk, P., Baffert, F., Lee, G. M., Mochizuki, N., Kim, J., Jeon, B. H., McDonald, D. M., and Koh, G. Y. (2005) *Circ Res* 97, 86–94.

31. Ambati, B. K., Nozaki, M., Singh, N., Takeda, A., Jani, P. D., Suthar, T., Albuquerque, R. J., Richter, E., Sakurai, E., Newcomb, M. T., Kleinman, M. E., Caldwell, R. B., Lin, Q., Ogura, Y., Orecchia, A., Samuelson, D. A., Agnew, D. W., St Leger, J., Green, W. R., Mahasreshti, P. J., Curiel, D. T., Kwan, D., Marsh, H., Ikeda, S., Leiper, L. J., Collinson, J. M., Bogdanovich, S., Khurana, T. S., Shibuya, M., Baldwin, M. E., Ferrara, N., Gerber, H. P., De Falco, S., Witta, J., Baffi, J. Z., Raisler, B. J., and Ambati, J. (2006) *Nature* 443, 993–7.

32. Pettersson, A., Nagy, J. A., Brown, L. F., Sundberg, C., Morgan, E., Jungles, S., Carter, R., Krieger, J. E., Manseau, E. J., Harvey, V. S., Eckelhoefer, I. A., Feng, D., Dvorak, A. M., Mulligan, R. C., and Dvorak, H. F. (2000) *Lab Invest* 80, 99–115.

33. Djonov, V., Baum, O., and Burri, P. H. (2003) *Cell Tissue Res* 314, 107–17.

34. Baffert, F., Thurston, G., Rochon-Duck, M., Le, T., Brekken, R., and McDonald, D. M. (2004) *Circ Res* 94, 984–92.

35. Gerhardt, H., and Betsholtz, C. (2003) *Cell Tissue Res* 314, 15–23.

36. Ruhrberg, C., Gerhardt, H., Golding, M., Watson, R., Ioannidou, S., Fujisawa, H., Betsholtz, C., and Shima, D. T. (2002) *Genes Dev* 16, 2684–98.

37. Kim, I., Kim, H. G., Moon, S. O., Chae, S. W., So, J. N., Koh, K. N., Ahn, B. C., and Koh, G. Y. (2000) *Circ Res* 86, 952–9.

38. Baffert, F., Le, T., Sennino, B., Thurston, G., Kuo, C. J., Hu-Lowe, D., and McDonald, D. M. (2006) *Am J Physiol Heart Circ Physiol* 290, H547–59.

39. Davis, S., Aldrich, T. H., Jones, P. F., Acheson, A., Compton, D. L., Jain, V., Ryan, T. E., Bruno, J., Radziejewski, C., Maisonpierre, P. C., and Yancopoulos, G. D. (1996) *Cell* 87, 1161–9.

40. Stoeltzing, O., Ahmad, S. A., Liu, W., McCarty, M. F., Parikh, A. A., Fan, F., Reinmuth, N., Bucana, C. D., and Ellis, L. M. (2002) *Br J Cancer* 87, 1182–7.

41. Stoeltzing, O., Ahmad, S. A., Liu, W., McCarty, M. F., Wey, J. S., Parikh, A. A., Fan, F., Reinmuth, N., Kawaguchi, M., Bucana, C. D., and Ellis, L. M. (2003) *Cancer Res* 63, 3370–7.

42. Iivanainen, E., Nelimarkka, L., Elenius, V., Heikkinen, S. M., Junttila, T. T., Sihombing, L., Sundvall, M., Maatta, J. A., Laine, V. J., Yla-Herttuala, S., Higashiyama, S., Alitalo, K., and Elenius, K. (2003) *Faseb J* 17, 1609–21.

43. Iwamoto, R., Yamazaki, S., Asakura, M., Takashima, S., Hasuwa, H., Miyado, K., Adachi, S., Kitakaze, M., Hashimoto, K., Raab, G., Nanba, D., Higashiyama, S., Hori, M., Klagsbrun, M., and Mekada, E. (2003) *Proc Natl Acad Sci U S A* 100, 3221–6.

44. Kobayashi, H., DeBusk, L. M., Babichev, Y. O., Dumont, D. J., and Lin, P. C. (2006) *Blood* 108, 1260–6.

45. Nykanen, A. I., Pajusola, K., Krebs, R., Keranen, M. A., Raisky, O., Koskinen, P. K., Alitalo, K., and Lemstrom, K. B. (2006) *Circ Res* 98, 1373–80.

46. Eddahibi, S., Guignabert, C., Barlier-Mur, A. M., Dewachter, L., Fadel, E., Dartevelle, P., Humbert, M., Simonneau, G., Hanoun, N., Saurini, F., Hamon, M., and Adnot, S. (2006) *Circulation* 113, 1857–64.

47. Sullivan, C. C., Du, L., Chu, D., Cho, A. J., Kido, M., Wolf, P. L., Jamieson, S. W., and Thistlethwaite, P. A. (2003) *Proc Natl Acad Sci U S A* 100, 12331–6.

48. Hellstrom, M., Gerhardt, H., Kalen, M., Li, X., Eriksson, U., Wolburg, H., and Betsholtz, C. (2001) *J Cell Biol* 153, 543–53.

49. Erber, R., Thurnher, A., Katsen, A. D., Groth, G., Kerger, H., Hammes, H. P., Menger, M. D., Ullrich, A., and Vajkoczy, P. (2004) *Faseb J* 18, 338–40.

50. Reinmuth, N., Liu, W., Jung, Y. D., Ahmad, S. A., Shaheen, R. M., Fan, F., Bucana, C. D., McMahon, G., Gallick, G. E., and Ellis, L. M. (2001) *Faseb J* 15, 1239–41.

51. Benjamin, L. E., Golijanin, D., Itin, A., Pode, D., and Keshet, E. (1999) *J Clin Invest* 103, 159–65.

52. Benjamin, L. E., Hemo, I., and Keshet, E. (1998) *Development* 125, 1591–8.

53. Holash, J., Wiegand, S. J., and Yancopoulos, G. D. (1999) *Oncogene* 18, 5356–62.

54. Maisonpierre, P. C., Suri, C., Jones, P. F., Bartunkova, S., Wiegand, S. J., Radziejewski, C., Compton, D., McClain, J., Aldrich, T. H., Papadopoulos, N., Daly, T. J., Davis, S., Sato, T. N., and Yancopoulos, G. D. (1997) *Science* 277, 55–60.

55. Lindblom, P., Gerhardt, H., Liebner, S., Abramsson, A., Enge, M., Hellstrom, M., Backstrom, G., Fredriksson, S., Landegren, U., Nystrom, H. C., Bergstrom, G., Dejana, E., Ostman, A., Lindahl, P., and Betsholtz, C. (2003) *Genes Dev* 17, 1835–40.

56. Gerhardt, H., Golding, M., Fruttiger, M., Ruhrberg, C., Lundkvist, A., Abramsson, A., Jeltsch, M., Mitchell, C., Alitalo, K., Shima, D., and Betsholtz, C. (2003) *J Cell Biol* 161, 1163–77.

57. Furuhashi, M., Sjoblom, T., Abramsson, A., Ellingsen, J., Micke, P., Li, H., Bergsten-Folestad, E., Eriksson, U., Heuchel, R., Betsholtz, C., Heldin, C. H., and Ostman, A. (2004) *Cancer Res* 64, 2725–33.

Chapter 16

Quantitative Estimation of Tissue Blood Flow Rate

Gillian M. Tozer, Vivien E. Prise, and Vincent J. Cunningham

Abstract

Tissue blood flow rate ($\mathbf{F}$) is a critical parameter for assessing functional efficiency of a blood vessel network following angiogenesis. This chapter aims to provide the principles behind estimation of $\mathbf{F}$ and a practical approach to its determination in laboratory animals using small, readily diffusible, and metabolically inert radiotracers. The methods described require relatively nonspecialized equipment. However, the analytical descriptions apply equally to complementary techniques involving sophisticated noninvasive imaging. Two techniques are described for the quantitative estimation of $\mathbf{F}$ using the tissue uptake following intravenous administration of radioactive iodoantipyrine (or other suitable radiotracer). The tissue equilibration technique is the classical approach, and the indicator fractionation technique, which is simpler to perform, is a practical alternative in many cases. The experimental procedures and analytical methods for both techniques are given, as well as guidelines for choosing the most appropriate method.

Key words: Backflux, blood flow rate, cannulation, distribution volume, indicator fractionation, iodoantipyrine, partition coefficient, radiotracer, tissue equilibration.

1. Introduction

The maturation phase of angiogenesis results in a functional blood vessel network. Blood flow rate through the network is a measure of its functional efficiency, knowledge of which is central to understanding the angiogenic process. Blood flow rate determines the efficiency of delivery of oxygen, nutrients, and drugs to the tissue. In cancer therapy, these parameters are critical for determining treatment outcome, so that pharmacological modification of tumor blood flow rate has a potential therapeutic benefit. In terms of determining the efficiency of vascular-disrupting approaches in cancer therapy, blood flow rate is the

S. Martin and C. Murray (eds.), *Methods in Molecular Biology, Angiogenesis Protocols, Second edition, Vol. 467*
© Humana Press, a part of Springer Science+ Business Media, LLC 2009
DOI: 10.1007/978-1-59745-241-0_16

most sensitive and relevant pharmacodynamic end point. This chapter aims to provide the principles behind, and a practical approach to, the quantitative estimation of blood flow rate in experimental mice and rats.

Blood flow rate is the rate of delivery of arterial blood to the capillary beds within a particular mass of tissue. It is typically measured in units of millilitres of blood per gram of tissue per minute ($mL \cdot g^{-1} \cdot min^{-1}$), or, alternatively, per unit volume of tissue ($mL \cdot mL^{-1} \cdot min^{-1}$).

Several experimental techniques can be used to estimate the blood volume of tissue, and intravital microscopy enables measurement of red cell velocity ($\mu m \cdot s^{-1}$) in individual capillaries. However, neither of these parameters is a direct reflection of blood flow rate as defined here. The average time taken for blood to pass through a particular capillary bed (capillary mean transit time, τ, is the parameter that relates tissue blood flow rate (F in $mL \cdot g^{-1} \cdot min^{-1}$) to fractional blood volume of the tissue (V in $mL \cdot g^{-1}$). This classical relationship is known as the *central volume principle(1)*:

$$\tau = V/F \tag{1}$$

For different tissues, **F** can vary widely; for example, it is approximately 0.1 $mL \cdot mL^{-1} \cdot min^{-1}$ in rat skin and 4.0 $mL \cdot mL^{-1} \cdot min^{-1}$ in rat kidney. From **Eq. 1** and using a value for **V** of 0.03 $mL \cdot g^{-1}$ for skin and 0.06 $mL \cdot g^{-1}$ for kidney, τ is approximately 18 s and 0.9 s, respectively.

To estimate blood flow rate **F**, the most accurate approach is to measure the rate of delivery of an agent carried to the tissue by the blood. A contrast agent is injected into the bloodstream; its concentration time course in arterial blood (input function) together with the kinetics of its uptake into tissue (tissue response function) are measured. **F** is then estimated from a mathematical model relating the tissue response function to the input function (*see* **Subheading 3.3** below). The contrast agent can be radioactive, with tissue concentrations measured by gamma or scintillation counting or by an external imaging system (e.g., a positron emitter for positron emission tomography [PET]). Alternatively, a contrast agent that is suitable for external magnetic resonance imaging, computed tomography, or ultrasound imaging can be used. Radioactive agents have the advantage that they can be administered at true tracer concentrations, therefore not interfering with physiological processes, and they do not necessarily need sophisticated imaging technology.

Some common methods for determining blood perfusion parameters are given, not all of which provide fully quantitative estimates of blood flow rate:

1. Laser Doppler flowmetry (LDF) provides a means of estimating relative changes in red cell velocity (e.g., following treatment) via surface or tissue-inserted probes. This measures a frequency shift in light reflected from moving red cells, which is a measure of average red cell velocity *(2)*. However, changes in red cell velocity may not accurately reflect the discussed changes in **F**.

2. The fluorescent DNA-binding dye Hoechst 33342 and certain carbocyanine dyes are examples of rapidly binding agents that have been used to determine a "perfused vascular volume" (as a fraction of the total tissue volume) rather than blood flow rate per se. This method has been used especially in tumor studies *(3,4)*. In this case, tissues are excised after several circulation times, following intravenous injection of the dye, and functional vessels appear in tissue sections as fluorescent halos. Conventional Chalkley point counting *(5,6)* or image analysis provides the fractional tissue volume occupied by fluorescence. This is a useful measure of vascular function in many circumstances but is insensitive because it cannot discriminate between perfused vessels with different flow rates.

3. For contrast agents that are confined to the bloodstream, methods based on **Eq. 1** can be used to calculate **F***(7)*. However, this is difficult in practice because τ is only a few seconds, requiring highly sensitive techniques for its measurement. Radioactive or coloured "microspheres" of diameters around 15–25 µm are a special class of contrast agents that are confined to the bloodstream because they should be trapped on first pass through tissue. Therefore, following injection directly into a major artery, they distribute to tissues in direct proportion to the fraction of the cardiac output received by the tissues, enabling calculation of blood flow rate *(8)*. With this technique, care needs to be taken to ensure adequate mixing of the microspheres in the arterial blood (which is challenging in mice, for instance) and enough microspheres are lodged in the tissue regions of interest to obtain statistical validity. In the case of tumors, care needs to be taken to determine and correct for microspheres that are recirculated due to lack of trapping in large-diameter vessels *(9,10)*.

4. The principles used to calculate blood flow rate using microspheres can sometimes be used even when the indicator crosses the vascular wall into the tissue and recirculates after the first pass through the tissue. If the tissue concentration of the indicator reaches a constant level that is maintained for the first minute or so after injection, this indicates that

the extraction fraction by the tissue is equal to that of the whole body *(11)*. The fractional uptake of the indicator into the tissue must therefore equate to the blood flow fraction of the cardiac output received. In the original description of the technique, potassium and rubidium chloride behaved as "pseudomicrospheres" in most normal tissues, with the notable exception of brain *(11)*.

5. Small, lipid-soluble, metabolically inert molecules, which rapidly cross the vascular wall and diffuse through the extravascular space, are also useful as blood flow markers. In this case, the fraction of marker crossing the capillary vascular wall from the blood in a single pass through the tissue (extraction fraction, E) is close to 1.0, and for fully perfused tissue the accessible volume fraction α of the tissue is also close to 1.0. The inert radioactive gas ^{133}xenon or hydrogen can be administered by inhalation *(12,13)*. However, safety issues with ^{133}xenon and the necessity for tissue insertion of polarographic electrodes for hydrogen have limited their use. A practical approach, which has utility for accessing the spatial heterogeneity of tissue blood flow rate, is the intravenous administration of a small, lipid-soluble, inert molecule dissolved in saline. In this case, net uptake rate into tissue over a short time (seconds) after intravenous injection is determined primarily by blood flow rate. Methods for quantitative estimation of tissue blood flow rate and related parameters using these agents are described (see **Note 1**).

2. Materials

2.1. Radioactive Tracer Preparation

1. Any small, lipid-soluble molecule that can be suitably labelled and is not metabolized in tissue over the short time of the experiment can be used. Suitable radioisotopes include ^{125}I and ^{14}C, for which tissue and blood counts can be obtained using standard techniques, and autoradiography/phosphorimaging can be applied if spatial variation in blood flow rate across a tissue of interest is required (see **Note 2**).

2. One that has been used commonly for both normal tissue studies (primarily brain *(14)*) and tumor studies *(15)* is iodoantipyrine (IAP) (**Fig. 16.1**). ^{125}I-IAP is commercially available, for example, from MP Biomedicals) and ^{14}C-IAP (4-iodo[**N**-methyl-14C]antipyrine) from GE Healthcare. Alternatively, a technique for labelling IAP with ^{125}I was described by Trivedi *(16)* (see **Note 3**).

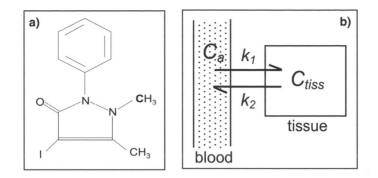

Fig. 16.1. Chemical structure of 4-iodo[**N**-methyl-^{14}C]antipyrine (**a**) and the compartmental model used for the quantitative estimation of **F** (**b**). When the extraction fraction **E** of a blood-borne tracer is 1.0, the rate constant **k1** represents **F**. **k2** represents the backflux, and **Ca** and **Ctiss** represent the arterial blood and tissue concentrations of the tracer, respectively. In this model, the tissue is a single well-mixed compartment.

2.2. Animal Preparation

Large-vessel cannulation is required for intravenous administration and arterial blood sampling. Materials required are as follows:

1. General anaesthetic.

2. Heparinised saline for cannulae (add 0.3 mL of 1000 U · mL^{-1} heparin to 10 mL saline).

3. General surgical equipment plus fine angled forceps, small spring scissors, microvascular clip.

4. Polythene tubing cut to suitable lengths, size appropriate to that of the vessel being cannulated (e.g., for rat 0.58-mm internal diameter [id], 0.96-mm outside diameter [od]).

5. Dissecting microscope.

6. Cold light source.

7. Thermostatically controlled heating blanket with rectal thermometer.

2.3. Blood Flow Assay

1. ^{125}I-labelled IAP (^{125}I-IAP) and suitable protective equipment.

2. Minimal dead-space glass syringe.

3. General dissecting instruments for excising tissue.

4. Stopclock.

5. Lidded container containing saline-soaked gauze.

6. Gamma counter and suitable vials for tissue and blood samples.

7. Analytical balance.

8. Anaesthetic (see **Note 4**).

9. Fraction collector set to collect at 1-s intervals.

10. Syringe pump for infusion and withdrawal.

11. 1000 U/mL heparin (use neat for rats and diluted 1 in 10 with saline for mice).

12. Injection saline.

13. High-concentration solution of sodium pentobarbitone (e.g., Euthatal™).

3. Methods

3.1. Animal Preparation

In the rat, a tail artery and vein are most suitable for cannulation. In the mouse, either the carotid artery and jugular vein or femoral artery and vein can be used.

1. Prepare 30-cm lengths of cannulae. For rat, use 0.96-mm od; 0.58-mm id. The cannula wall may be shaved down at the tip, and the end may be slightly bevelled to aid insertion. Use a microscope to ensure that there are no sharp edges. For mouse, use a short length of 0.61-mm od, 0.28-mm id cannula, stretched to a smaller diameter at the tip and connected to a longer length of 0.96-mm od, 0.58-mm id cannula to reduce resistance to flow. Attach each length to a 1-mL syringe filled with heparinised saline.

2. Anaesthetize the animal, insert rectal thermocouple, and place on heating blanket. An overhead lamp is a useful additional heat source.

3. Illuminate surgical area with a cold light source.

4. Expose the relevant vessel. For the rat tail, this involves making two 2-cm incisions through the skin, each side of the vessel, approximately 5 mm apart and approximately 2 cm from the base of the tail. Use artery forceps to clear the skin from the underlying connective tissue and cut the skin at the distal end to create a flap (*see* **Note 5**).

5. Keep exposed vessels moist at all times using warmed saline.

6. Free the vessel from surrounding connective tissue using fine blunt-end forceps.

7. Place two lengths of suture under the vessel, tying off the most distal to the heart, which can be used to apply slight tension to the vessel.

8. Occlude the vessels as far proximal as possible using a microvascular clip.

9. Using spring scissors, make a V-shaped cut in the vessel close to the distal knot and insert the cannula. Advance the cannula into the vessel approximately 2 cm or more by removing clip (see **Note 6**).

10. Aspirate gently to ensure that blood is free flowing. It may not be possible to aspirate the vein, but a small volume of saline can be injected to check for patency.

11. Tie both sutures securely around the cannula. Use tape or tissue-compatible glue to secure the cannula to the skin distal to the distal suture. Close the wound.

3.2. Blood Flow Assay

Two alternatives are described for the blood flow assay: the classic tissue equilibration method for rats and the indicator fractionation method for rats or mice. The main advantages and disadvantages of these techniques are given in **Table 16.1.**

3.2.1. Tissue Equilibration Technique

Cannulation of two tail veins and one tail artery are required as described in **Subheading 3.1.**

1. Remove ^{125}I-IAP from the freezer and bring slowly to room temperature. Using suitable containment and a low dead-space glass syringe, carefully remove required volume of ^{125}I-IAP (0.2–0.3 MBq per rat) and dispense into a vial.

2. Evaporate the methanol using a very gentle stream of nitrogen and slowly add injection saline to the ^{125}I-IAP (0.8 mL per rat plus extra to account for syringe dead spaces, etc.). Gently mix.

3. Load a syringe of size suitable for infusion with the ^{125}I-IAP solution (needle must be suitable size for the venous cannula).

4. For anaesthetized animals, keep warm as described above in **Subheading 3.1** step 2. Check arterial blood pressure and heart rate by connecting the arterial cannula to a pressure transducer and recording device. Then, clamp off the cannula and connect it to the fraction collector loaded with preweighed glass tubes for subsequent blood collection.

Table 16.1
Comparison of radiotracer methods for estimation of tissue blood flow rate F

Tissue Equilibration Technique	Indicator Fractionation Technique
Requires blood vessel cannulation	Requires blood vessel cannulation
Time-consuming	Relatively easy to perform
Calculations require curve fitting algorithms	Calculations require only a simple formula
Backflux is taken into account in the model	Very sensitive to errors associated with backflux
Relatively sensitive to timing errors and delay and dispersion effects	Reasonably tolerant of timing errors and delay and dispersion effects
Reasonably tolerant of imprecision in λ	Tolerant of imprecision in λ
Method of choice if λ is well-defined	Method of choice if λ is ill-defined, especially if $\mathbf{F}$ is low and α large

5. Inject and flush in 0.1 mL neat heparin (= 100 IU) via one of the venous cannulae to ensure the blood flows freely from the arterial cannula.

6. Cut one of the venous cannulae to approximately 3 cm long and connect a syringe containing approximately 0.5 mL Euthatal. A small "T-connector" may also be used to allow drug administration via this cannula.

7. Set syringe pump speed to 1.6 mL/min (see **Note 7**). Carefully place the ^{125}I-IAP-containing syringe in the pump and connect it to the second venous cannula.

8. Start the stopclock and unclamp the artery, checking that blood is free flowing. At 5 s, start syringe pump and fraction collector (see **Note 8**). At 35 s, inject Euthatal and stop pump; rapidly excise tissues of interest and stop fraction collector (see **Note 9**). Place tissues in the lidded container to prevent drying. Weigh the blood tubes and cap them. Weigh the tissues and place them in gamma counting tubes.

9. Count the blood and tissue samples on the gamma counter.

3.2.2. Indicator Fractionation Technique

Cannulations of one artery and two veins are required. Alternatively, cannulae attached to shafts of hypodermic needles can be inserted into tail veins percutaneously instead of cannulating veins.

1. Follow **Subheading 3.2.1, steps 1–3**, preparing 0.07 MBq ^{125}I-IAP in 0.05 mL saline per mouse.

2. For anaesthetized animals, keep warm as described above in Subheading 3.1 Step 2. Check arterial blood pressure and heart rate by connecting the arterial cannula to a pressure transducer and recording device.

3. Set syringe pump speed to 150 µL · min^{-1} (mouse) or 1.6 mL · min^{-1} (rat).

4. Load a 250-µl syringe (for mouse) or 2-mL syringe (for rat) with approximately 100 µL saline, attach a 23-gauge needle, and position in the pump.

5. Cut the venous cannula as short as possible and inject 0.05 mL of diluted heparin (≡ 5 IU) for mouse or 0.1 mL neat heparin (≡ 100 IU) for rat. Disconnect heparin syringe and attach ^{125}I-IAP syringe and a syringe containing injection saline via a T-piece. Connect syringe containing Euthatal to second venous cannula.

6. Clamp the artery cannula, disconnect it from the pressure transducer, and connect it to the syringe pump. Ensure that the pump is set to withdraw and allow it to withdraw very briefly to ensure that the cannula is patent and the syringe is positioned correctly.

7. Start the stop clock and pump simultaneously. Check that the blood is flowing freely. At 3 s, inject 0.05 mL of ^{125}I-IAP

for mouse or 0.2 mL for rat as a rapid bolus via the venous cannula, followed immediately by 0.05 mL saline from the second syringe. At 13–18 s (see **Note 10**), inject Euthatal and immediately pull out full length of arterial cannula and excise the tissue of interest. All the blood should be retained in the cannula. Place the tissue in a preweighed gamma counting tube.

8. Attach a saline-filled syringe to the arterial cannula and eject all the blood into a gamma counting tube together with 1 mL of saline.

9. Count the blood and tissue samples on the gamma counter.

3.3. Blood Flow Analysis

3.3.1. Tissue Equilibration Technique

1. Analysis is based on a model that assumes a vascular compartment from which the input function derives and a single (extravascular) well-mixed tissue compartment (Fig. 16.1). A small, highly soluble, and inert tracer, such as IAP, is assumed to rapidly equilibrate between all blood components and the tissue compartment. In this case, the model, based on Kety *(17)* describes the relationship between the tissue concentration of the tracer at time **t**, $\mathbf{C}_{tiss}(\mathbf{t})$, and the arterial blood concentration of the tracer at time **t**, $\mathbf{C}_a(\mathbf{t})$, by the equation

$$\mathbf{C}_{tiss}(\mathbf{t}) = \mathbf{k}_1 \mathbf{C}_a(\mathbf{t}) \otimes \exp(-\mathbf{k}_2 \mathbf{t}) \qquad (2)$$

where $\mathbf{k}_1$ is tissue blood flow rate **F**, and $\mathbf{k}_2$ is $\mathbf{k}_1/\alpha\lambda$; α is the accessible volume fraction of tissue (i.e., the effectively perfused fraction), and λ is the equilibrium partition coefficient of the tracer between tissue and blood; $\otimes$ denotes the convolution integral; $\alpha\lambda$ is equivalent to the apparent volume of distribution **VDapp** of the tracer in the tissue *(18)*. $\mathbf{C}_{tiss}(\mathbf{t})$ and $\mathbf{C}_a(\mathbf{t})$ are expressed in radioactivity counts per gram tissue and per millilitre blood, respectively, using 1.05 for the density of blood.

2. In this method, $\mathbf{C}_{tiss}(\mathbf{t})$ is measured at only one time point (i.e., after tissue excision). Hence, only one parameter, $\mathbf{k}_1(\mathbf{F})$, can be estimated from the data (see **Note 11**). λ is approximated from literature values or estimated from a separate experiment *(19)*, and α is taken as 1.0 (see **Note 12**). Studies have shown that the method is relatively insensitive to small changes in λ because of the short timescale of the experiment *(20)*. Also see **Table 16.1**

3. Solving **Eq. 2**: Data can be fitted to **Eq. 2** using a simple table lookup method. In this method, since the input function is known, then the expected tissue activity at the time of excision $\mathbf{C}_{tiss}(\mathbf{T})$ can be calculated for each of a range of realistic values of *F* using **Eq. 2**. Direct comparison of the observed $\mathbf{C}_{tiss}(\mathbf{T})$ against the table then gives the required estimate of **F** (**Fig. 16.2**). Evaluation of the integral

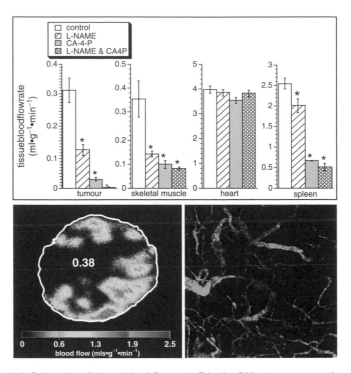

Fig. 16.2. Estimation of tissue blood flow rate **F** in the P22 rat sarcoma and several normal rat tissues using the tissue equilibration method with ^{125}I-IAP (**a**) and ^{14}C-IAP (**b**). Results in panel (**a**) were obtained by calculating **C**$_{tiss}$ from gamma counts of large tissue samples. Data, from left to right, shows the effects of saline, the vascular disrupting agent combretastatin A4-P, the nitric oxide synthase inhibitor, nitro-L-arginine (L-NNA) and tumor the combination of the two. Asterices represent a significant difference from the control group. The image in panel (**b**) was obtained from an untreated P22 tumor by calculating multiple values for **C**$_{tiss}$ from quantitative autoradiography of tumor sections. The mean **F** is 0.8 mL·g^{-1}·min^{-1}. The image in panel (**c**) illustrates the vascular networks in the P22 tumor obtained by multiphoton fluorescence microscopy.(*See Color Plates*)

in **Eq. 2** requires a numerical integration routine, which are commonly available in statistical analysis packages or can be programmed using computer applications such as MAT-LAB® (The Mathworks, USA©). A further issue that needs to be taken into account when assessing the accuracy of this type of technique is the possibility that the input function time course may not be accurately measured because of a time delay between the radioactivity reaching the tissue and reaching the blood collection tubes and because of smearing or dispersion effects occurring on the arterial cannula before blood collection. These delay and dispersion effects can be corrected for (see **Note 13**) but do add further complication to the analysis.

3.3.2. Indicator Fractio-nation Technique

1. This method was first used by Goldman and Sapirstein *(21)*, with later modifications *(22)*. It simplifies the model used in **Eq. 2** by assuming that the backflux of the tracer from tissue

into the blood is negligible compared with its influx into the tissue for a short period of time after injection of the tracer (see **Note 14**). Under these conditions, **Eq. 2** reduces to

$$\mathbf{k_1 = C_{tiss}(T)} / \int_0^T \mathbf{C_a(t)dt} \qquad (3)$$

where $\mathbf{k_1}$, $\mathbf{t}$, and $\mathbf{Ca(t)}$ are as defined, and $\mathbf{C_{tiss}(T)}$ is the concentration of tracer in the tissue at the end of the experiment (at time $\mathbf{t = T}$).

2. $\mathbf{T}$ is typically set at 10–15 s, during which time collection of sufficient blood samples, as described for the tissue equilibration technique, is difficult. Instead, the constant-withdrawal technique can be used. Here, blood is withdrawn from an artery at a constant rate using a pump from $\mathbf{t = -T'}$ to $\mathbf{t = T}$, where $\mathbf{-T'}$ is the time when the pump is started. A bolus injection of tracer is given at $\mathbf{t = 0}$. Under these conditions,

$$\int_0^T \mathbf{C_a(t)dt = C_c(T+T) = C_c V_b / r = X / r} \qquad (4)$$

where $\mathbf{C_c}$ is the mean concentration of tracer in the blood sample; $\mathbf{V_b}$ is the volume of blood collected; $\mathbf{X}$ is the total radioactive counts in the collected blood sample; and $\mathbf{r}$ is the rate of withdrawal of blood.

3. $\mathbf{X/r}$ can be substituted into **Eq. 3** and blood flow rate $\mathbf{k_1}$ can be calculated from a knowledge of the counts $\mathbf{X}$, pump rate $\mathbf{r}$, and concentration of the tracer in the tissue at the end of the experiment $\mathbf{C_{tiss}(T)}$. As for the tissue equilibration technique, there are inaccuracies in the measurement of the input function using this technique associated with delay and dispersion along the plastic cannula. However, the constant-withdrawal method means that definition of a concentration-time curve is not required, and only the last part of the actual arterial time course is lost by the blood-sampling method. In addition, the experimental setup of the indicator fractionation method means that the arterial cannula can be kept short, minimizing the delay involved.

3.3.3. Comparison of the Two Blood Flow Methods

The advantages and disadvantages of the classic tissue equilibration method and the indicator fractionation method are summarized in **Table 16.1**. Patlak et al. *(20)* carried out an evaluation of errors involved in the two techniques, which can be summarized as follows:

1. Errors in the tissue equilibration method are minimized if an optimal infusion schedule is used (a ramped schedule is best, but a constant infusion is reasonable), timing is measured precisely, and corrections are made for delay and dispersion. Under these conditions, 10–15% inaccuracy in

the value used for λ is well tolerated in the calculation of F. If precise measurement of λ can be made, this is the method of choice.

2. If λ cannot be measured reasonably accurately, the indicator fractionation technique maybe the better option for estimating F. Errors associated with backflux are minimized by a short experimental time. Errors associated with imprecise timing are minimized by bolus administration of the tracer (so that arterial concentration is low at tissue excision). The constant withdrawal method is an added advantage for its simplicity and accuracy. However, backflux cannot be completely prevented (especially with bolus administration) and may introduce significant errors where F is high or α is low. Delay and dispersion effects are reasonably well tolerated.

3. Both techniques require accurate measurement of tracer concentration in the tissue (C_{tiss}).

4. Notes

1. The basic experimental principles and analytical methods described here also apply to various external imaging techniques that are now available for use in small animals, such as PET *(23)*. These techniques allow repeated evaluation of blood flow rate (and other pharmacodynamic end points) in the same animal as long as the biological and radioactive half-lives of the tracer are compatible with the timescale of the experiment. In addition, they allow definition of more than one vascular parameter (see **Note 11**).

2. Instead of obtaining a single value for the blood flow rate within a tissue (usually by calculating C_{tiss} from gamma counts of tissue activity), the variation of blood flow rate within a tissue can be obtained at high spatial resolution ($\sim$50 μm) by using a radiotracer that is suitable for autoradiography or phosphorimaging (Fig. 16.2). In these cases, **Ctiss** is obtained in raster fashion across tissue sections for calculation of corresponding $k_1(F)$ values *(24)*.

3. Local radiation safety procedures need to be followed for all the techniques described to avoid contamination of personnel and equipment.

4. General anaesthesia seriously affects mean arterial blood pressure in rodents, especially mice, and its effects on tissue blood flow rates need to be considered in planning experiments. Animals can be allowed to recover from general anaesthesia induced by inhalational anaesthetics following

cannulation, but in this case, procedures for preventing cannula disturbance and minimising pain and distress to the animals need to be implemented *(25)*.

5. Tail cannulations: The tail artery lies relatively deep within a cleft in the cartilage and requires an incision to be made through the overlying connective tissue for it to be accessible. Once freed, it is robust for cannulation; the vein is much more superficial and easily located, although more fragile than the artery and liable to constriction and tearing.

6. A topical vasodilator such as procaine can be used to aid cannula insertion.

7. Volumes of saline solutions of IAP for intravenous administration are chosen to compensate for rate of blood loss during the course of the experiments.

8. A constant infusion schedule for delivery of the radiotracer is described as it is simple to achieve in practice. However, an infusion schedule that increases with time (is ramped) could be employed because this reduces the influence of an incorrect value for λ on the calculated value of $\mathbf{F}$ *(20)*. The movement of the fraction collector and the infusion/withdrawal rates of the pump need to be carefully calibrated prior to experiments.

9. Timing errors can be significant if blood flow to tissues of interest is not stopped at the instant that the pump is stopped (**also see Table 16.1**).

10. A short duration increases timing errors, but a longer one increases errors associated with backflux of the tracer from the tissue into blood (**also see Table 16.1**).

11. A disadvantage of this particular technique is that it only allows the tissue concentration of the tracer to be assayed at one time point. Hence, as noted, only one parameter ($\mathbf{F}$) can be estimated, whereas values for α and λ have to be assumed. Other, more sophisticated techniques involving noninvasive imaging, such as PET (see **Note 1**) allow a full-time course of the tissue to be assayed, allowing estimation of $\alpha\lambda$ (**VDapp**), for example, as well as F. This is of particular interest in the case of tumor blood flow, where the perfused fraction α, which is immediately accessible to the tracer, is often less than 1.0 because of large intercapillary distances or ischaemic regions *(18)*. However, the spatial resolution of noninvasive imaging cannot compete with the high spatial resolution achievable with the invasive techniques described here (see **Note 2**). It should also be noted that, in the invasive technique, α is assumed to be 1.0. If it is actually less than 1.0 for a sampled tissue region, the measured tissue concentration of the tracer $\mathbf{C}_{tiss}$ will be low because it is averaged over the whole region, including the inaccessible part.

This gives rise to a low value for **F** for the region, so that even when the assumption that α is 1.0 is incorrect, the calculated value of *F* will be an accurate reflection of the average blood flow rate to that region.

12. Beyond the short timescale of these experiments, IAP redistributes in tissue in a space-dominated rather than a flow-dominated pattern. At equilibrium, IAP might be expected to distribute in proportion to the tissue water content, such that λ would be similar in all tissues and close to 1.0. However, experimental evidence indicates that, although λ\ for IAP is close to 1.0 in many tissues, it is somewhat variable *(19)*. This may relate to its high lipid solubility, or it may be a reflection of problems in accessing true values of λ because of loss of the radioactive label from the molecule at long times (hours) after injection.

13. A simple model that can be used to describe dispersion effects is given by the following equation:

$$C_m(t) = C_a(t) \otimes k_d \exp(-k_d t) \qquad (5)$$

where $C_m(t)$ is the measured tracer concentration at the cannula outflow, $C_a(t)$ is the inflow concentration, and k_d (min^{-1}) is a dispersion constant that is dependent on flow rate, length, and internal diameter of the cannula and the interaction with blood on its internal surface. k_d for a particular cannula and flow rate of blood can be calculated from **Eq. 5** if experiments are undertaken in vitro, with the blood pumped through a cannula at a particular rate, switched rapidly between labelled blood and unlabelled blood and the dispersion effect measured in the outflow. Results from such an experiment are presented in **Table 16.2**.

Table 16.2
Example of expressions used to calculate the dispersion constant k_d (min^{-1}) from the linear speed (v in centimetres per minute) of blood flowing down various types of cannulae

Type of Cannula	200 mm length	300 mm length
0.50-mm id	$6.72 + 0.048v$	$3.84 + 0.036v$
0.58-mm id	$9.54 + 0.040v$	$-0.6 + 0.042v$

These expressions were obtained by pumping blood at several known flow rates through each type of cannula and rapidly switching between labelled and unlabelled blood. The time-activity curves for the blood flowing out of the cannulae were compared with the known inflow time-activity curves, using **Eq. 5**, to estimate k_d for each condition. The relationship between k_d and v was described by the equation for a straight line, as shown. The cannulae used were prepared from Portex™ low-density polyethylene tubing.

Delay *td* can be estimated directly from the known volume of the cannula and the blood flow rate down the cannula. These values for k_d and t_d can be incorporated in **Eq. 2** to give a working form of the equation:

$$C_{tiss}(t)=(k_1/k_d)C_m(t+t_d)+(1k_2/k_d)k_1C_m(t+t_d)\otimes exp(-k_2t) \qquad (6)$$

If dispersion effects are small (i.e., large k_d), then this equation reduces to **Eq. 2**. If dispersion effects are marked (i.e., small k_d), then the equation illustrates that failing to take dispersion effects into account has a marked effect on estimates of **F**. These issues were first quantitatively described by Meyer *(26)* (but note that a dispersion time constant equivalent to $1/k_d$ was used in this case).

14. In addition to using a short experimental time **T**, errors associated with backflux in this technique are minimized if blood flow rate is low and there is a big accessible space in the tissue for the tracer (high α). Note that the short **T** gives the potential for large timing errors, and great care must be taken to time the experiment accurately. As for the tissue equilibration technique, this means that blood flow to tissues needs to be stopped precisely at **T**. Interestingly, timing errors appear to be less of an issue with this technique than with the tissue equilibration technique for measurement of cerebral blood flow rate *(20)*.

References

1. Stewart, G. N. (1894) Researches on the circulation time in organs and on the influenced which affect it. *J Physiol (London)* 15, parts I–III.

2. Stern, M. D. (1975) *In vivo* evaluation of microcirculation by coherent light scattering. *Nature* 254, 56–58.

3. Smith, K. A., Hill, S. A., Begg, A. C., Denekamp, J. (1988) Validation of the fluorescent dye Hoechst 33342 as a vascular space marker in tumours. *Br J Cancer* 57, 247–253.

4. Hill, S. A., Tozer, G. M., Chaplin, D. J. (2002) Preclinical evaluation of the antitumour activity of the novel vascular targeting agent Oxi 4503. *Anticancer Res* 22, 1453–1458.

5. Chalkley, H. W. (1943) Method for quantitative morphologic analysis of tissues. *J Natl Cancer Inst* 4, 47–53.

6. Vermeulen, P. B., Gasparini, G., Fox, S. B., (2002) Second international consensus on the methodology and criteria of evaluation of angiogenesis quantification in solid human tumours. *Eur J Cancer* 38, 1564–1579.

7. Weiskoff, R. M. (1993) Pitfalls in MR measurement of tissue blood flow with intravascular tracers: which mean transit time? *Magn Reson Med* 29, 553–559.

8. Messmer, K. (1979) Radioactive microspheres for regional blood flow measurements. Actual state and perspectives. *Bibl Anat* 18, 194–197.

9. Jirtle, R. L. (1980) Blood flow to lymphatic metastases in conscious rats. *Eur J Cancer* 17, 53–60.

10. Jirtle, R. L., Hinshaw, W. M. (1981) Estimation of malignant tissue blood flow with radioactively labelled microspheres. *Eur J Cancer Clin Oncol* 17, 1353–1355.

11. Sapirstein, L. A. (1958) Regional blood flow by fractional distribution of indicators. *Am J. Physiol* 193, 161–168.

12. Obrist, W. D., Thompson, H. K., King, C. H., Wang, H. S. (1967) Determination of regional cerebral blood flow by inhalation of 133-xenon. *Circ Res* 20, 124–135.

13. Young, W. (1980) H2 clearance measurement of blood flow: a review of technique

and polarographic principles. *Stroke* 11, 552–564.

14. Sakurada, O., Kennedy, C., Lehle, J., Brown, J. D., Carbin, J. L., Sokoloff, L. (1978) Measurement of local cerebral blood flow with iodo [14C] antipyrine. *Am J Physiol* 234, H59–H66.

15. Tozer, G. M., Shaffi, K. M. (1993) Modification of tumour blood flow using the hypertensive agent, angiotensin II. *Br J Cancer* 67, 981–988.

16. Trivedi, M. A. (1996) A rapid method for the synthesis of 4-iodoantipyrine. *J Labelled Comp Radiopharm* 38, 489–496.

17. Kety, S. S. (1960) Theory of blood tissue exchange and its application to measurements of blood flow. *Methods Med Res* 8, 223–227.

18. Tozer, G. M., Shaffi, K. M., Prise, V. E., Cunningham, V. J. (1994) Characterisation of tumour blood flow using a "tissue-isolated" preparation. *Br J Cancer* 70, 1040–1046.

19. Tozer, G. M., Morris, C. (1990) Blood flow and blood volume in a transplanted rat fibrosarcoma: comparison with various normal tissues. *Radiother Oncol* 17, 153–166.

20. Patlak, C. S., Blasberg, R. G., Fenstermacher, J. D. (1984) An evaluation of errors in the determination of blood flow by the indicator fractionation and tissue equilibration (Kety) methods. *J Cerebr Blood Flow Metab* 4, 47–60.

21. Goldman, H., Sapirstein, L. A. (1973) Brain blood flow in the conscious and anaesthetized rat. *Am J Physiol* 224, 122–126.

22. Gjedde, S. B., Gjedde, A. (1980) Organ blood flow rates and cardiac output of the Balb/c mouse. *Comp Biochem Physiol* 67A, 671–674.

23. Herrero, P., Kim, J., Sharp, T. L., (2006) Assessment of myocardial blood flow using 15O-water and 1–11C-acetate in rats with small-animal PET. *J Nucl Med* 47, 477–485.

24. Tozer, G. M., Prise, V. E., Wilson, J., (1999) Combretastatin A-4 phosphate as a tumor vascular-targeting agent: early effects in tumors and normal tissues. *Cancer Res* 59, 1626–1634.

25. Richardson, C. A., Flecknell, P. A. (2005) Anaesthesia and post-operative analgesia following experimental surgery in laboratory rodents: are we making progress? *Altern Lab Anim* 33, 119–127.

26. Meyer, E. (1989) Simultaneous correction for tracer arrival delay and dispersion in CBF measurements by the H215O autoradiographic method and dynamic PET. *J Nucl Med* 30, 1069–1078.

Chapter 17

In Vivo Matrigel Migration and Angiogenesis Assay

Katherine M. Malinda

Abstract

The search for rapid and reproducible *in vivo* angiogenesis and antiangiogenesis assays is an area of intense interest. These types of assays are extremely useful in testing putative drugs and biological agents and for the comparison and enhancement of *in vitro* tests. The Matrigel plug assay is one such assay and has proved to be a relatively quick and easy method to evaluate both angiogenic and antiangiogenic compounds *in vivo*. Initial indications of the levels of activity of strong angiogenic or antiangiogenic compounds can be visually assessed even as the plugs come out of the mouse because there are color differences in the plugs compared to the controls. Further quantitation is then needed to determine levels of angiogenic/antiangiogenic activity, and this can be performed using a variety of methods. This chapter presents an overview of the basic methods used to set up both angiogenic and antiangiogenic assays, discusses factors influencing variability, and discusses the methods for quantitating the plugs obtained. The Matrigel plug assay provides another useful tool in angiogenesis research.

Key words: Angiogenesis, antiangiogenesis, in vivo assay, plug assay, Matrigel

1. Introduction

Angiogenesis, the process of new blood vessels forming from pre-existing vessels, is an important feature in developmental processes, wound healing, and pathologic conditions such as cancer and vasculitis diseases. Due to the importance of angiogenesis, a relatively simple and rapid *in vivo* method to determine the angiogenic potential of compounds is desirable to augment *in vitro* findings.

One such quantitative method is the murine Matrigel plug assay, which can measure both angiogenesis and antiangiogenesis. Matrigel, an extract of the Engelbreth-Holm-Swarm tumor composed of basement membrane components, is liquid at 4°C and

S. Martin and C. Murray (eds.), *Methods in Molecular Biology, Angiogenesis Protocols, Second edition, Vol. 467*
© Humana Press, a part of Springer Science+ Business Media, LLC 2009
DOI: 10.1007/978-1-59745-241-0_17

forms a gel when warmed to 37°C *(1)*. When plated on Matrigel, human umbilical vein endothelial cells (HUVECs) undergo differentiation into capillary-like tube structures *in vitro (2, 3)*. *In vivo*, Matrigel is injected either alone or mixed with potential angiogenic compounds. When injected subcutaneously into the ventral region of mice, it solidifies, forming a "Matrigel plug." When known angiogenic factors, such as basic fibroblast growth factor (bFGF), are mixed with the Matrigel and injected, mouse endothelial cells migrate into the plug-forming vessels. The level of angiogenesis is typically viewed by embedding and sectioning the plugs in paraffin and staining using Masson's Trichrome, which stains the Matrigel blue and the endothelial cells/vessels red (**Fig. 17.1**). These vessels contain erythrocytes, indicating that they form functional capillaries. In addition, these capillaries

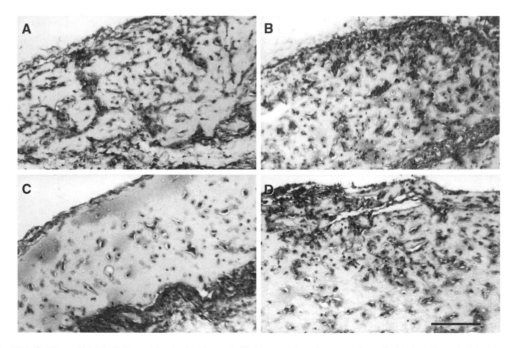

Fig. 17.1. Sections of Matrigel plugs stained with Masson's Trichrome. All sections are oriented with the side underlying the skin at the top of the figure. (**A**) and (**B**) Representative fields of plugs containing 5 μg/mL thymosin β_4. (**C**) Field showing Matrigel alone. (**D**) Representative field of a plug with 10 ng/mL endothelial cell growth supplement (ECGS) (positive control). Sections with thymosin β_4 contain many more cells than Matrigel alone and have cells with a similar morphology to those in the ECGS control. Bar 100 μm. (With permission from *ref. 11*, Malinda, K. M., Goldstein, A. L., Kleinman, H. K., 1997, Thymosin β_4 stimulates directional migration of human umbilical vein endothelial cells, *FASEB J 11*, 474–481.

stain positive for factor VIII *(4, 5)*. In unsupplemented Matrigel, few cells invade the plug. Strong angiogenic compounds result in yellowish plugs so that initial indications of activity can be made at the time when plugs are removed from the mice.

The assay also can be utilized when putative antiangiogenic compounds are being tested. In this assay, Matrigel is premixed with bFGF (angiogenic compound), and the test substances are then added. Thus, test antiangiogenic substances inhibit the formation of vessels induced by bFGF in the Matrigel plug. In this case, when removed from the mouse the plugs are relatively colorless, and when viewed by Masson's Trichrome staining should contain few endothelial cells. Because of the strong vascular response to bFGF, it is also possible to measure hemoglobin levels with the Drabkin assay *(4)*. An additional method for the quantitation of plasma volume in the plugs involves using fluorescein isothiocyanate dextran (145,000–200,000 MW) injected intravenously into the tail vein *(6–9)*. Alternatively, putative vessels can be visualized and quantitated with a fluorescent microscope and imaging software *(10)*.

One important consideration in using either of the above assays is that some variability will be observed. Differences in the mice and in the basement membrane preparations will affect the background levels of blood vessel formation one observes. Age and gender of the mice selected also can result in differences in the results between experiments. Vessel formation in young mice (6 mo old) is reduced compared to older mice (12–24 mo old) *(4)*. Also, variability results if the Matrigel is injected into different sites in the mouse. Lower angiogenic response is observed if the material is injected into the dorsal surface of the animal, while one of the best areas in terms of angiogenic response is the ventral surface of the mouse in the groin area close to the dorsal midline (**Fig. 17.2**). Regardless of these potential problems, this assay is one of the best for the rapid screening of potential angiogenic and antiangiogenic compounds.

2. Materials

2.1. General

1. C57Bl/6N female mice. Three or four mice are needed for each test group.
2. Additional cages and supplies, one for each test group.
3. Matrigel (BD Biosciences, Bedford, MA) stored at –20°C. Enough will be needed to inject 0.5 or 1 mL per mouse.
4. 3-cc syringes (Monoject).
5. 25-gauge needles (Monoject).
6. 14-mL Falcon round polypropylene bottom tubes.

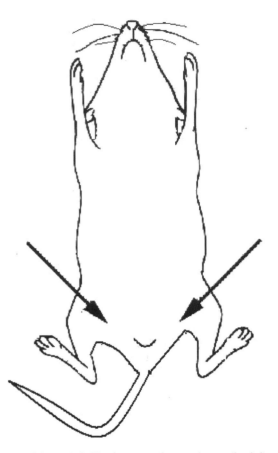

Fig. 17.2. Diagram of the ventral side of a mouse. Arrows show optimal sites for Matrigel injection.

7. 10X phosphate-buffered saline (PBS), pH 7.2 pH, diluted to 1X in sterile distilled water.

8. Endothelial cell growth supplement (ECGS; bovine brain extract containing aFGF and bFG) (no. 354006, BD Biosciences). Prepare 1 mg/mL solution just before use in sterile distilled water and store unused portion in aliquots at –20°C. It loses activity after 1 mo according to the manufacturer. Also, one can use bFGF (no. 356037, BD Biosciences). Prepare a 1 mg/mL solution just before use in sterile distilled water and store unused portion in aliquots at –20°C. Final concentration used in experiment is either 10 and 100 ng/mL ECGS or 150 ng/mL bFGF.

9. Tabletop vortex

10. Sterile pipette tips

11. Sterile 5-mL disposable pipettes.

12. Scalpel or dissecting scissors.

2.2. Embedding and Quantitation

1. Scintillation or specimen vials.
2. 10% formalin in PBS.
3. Histology service for paraffin embedding, sectioning, and staining with Masson's Trichrome.
4. Microscope equipped with a video system linked to a computer with NIH Image (free to the public) or similar image acquisition and analysis software.
5. Spreadsheet software.

3. Methods

3.1. Screening Putative Angiogenic Compounds

1. Thaw Matrigel on ice or overnight at 4°C. At no time should Matrigel be warmed to room temperature during handling.
2. Mix Matrigel on vortex and place on ice (*see* Notes 1 and 2).
3. Pipette equal amounts of Matrigel for each test condition in separate 14-mL tubes on ice. Each mouse injection requires 0.5 to 1 mL, with three or four mice for each condition (*see* Notes 2 and 3).
4. Tubes should be included for the positive and negative controls. The negative control is Matrigel alone; positive controls contain 10 ng/mL and 100 ng/mL ECGS or 150 ng/mL bFGF added to Matrigel.
5. Equalize the volumes using cold PBS so that all tubes, after test substances have been added, contain Matrigel at equal dilutions (*see* Note 4). Vortex.
6. Load syringe with each solution and inject 0.5 or 1 mL subcutaneously into the ventral area of each mouse (*see* Fig. 17.2) (*see* Notes 5 and 6).
7. Place animals of each test group in separate labeled cages.
8. After 7–10 d, sacrifice the animals and remove the plugs. Plugs appear as bumps on the ventral side of the animal and are removed using a sharp pair of scissors or a scalpel. Remove the plug with some surrounding tissue for orientation when histology is performed.

3.2. Screening Putative Antiangiogenic Compounds

1. Thaw Matrigel on ice overnight at 4°C.
2. Mix Matrigel on vortex and place on ice (*see* Note 1).
3. Pipette equal amounts of Matrigel for each test condition in separate 14-mL tubes on ice. Each mouse injection requires 0.5 to 1 mL, with three or four mice for each condition (*see* Notes 2 and 3).

4. Add 150 ng/mL bFGF to all tubes except the one used for the negative control, Matrigel alone.

5. Equalize the volumes using PBS so that all tubes, after test substances have been added, contain Matrigel at equal dilutions (*see* **Note 4**). Vortex.

6. Load syringe with each solution and inject 0.5 or 1 mL subcutaneously into the ventral area of each mouse (*see* Fig. 17.2) (*see* **Notes 5** and **6**).

7. Place animals of each test group in separate labeled cages.

8. After 7–10 d, sacrifice the animals and remove the plugs. Plugs appear as bumps on the ventral side of the animal and are removed using a sharp pair of scissors or a scalpel. Remove the plug with some surrounding tissue for orientation when histology is performed.

3.3. Fixation and Quantitation by Microscopy

1. Place plugs in labeled scintillation or specimen vials filled with 10% formalin. Allow plugs to fix overnight before embedding in paraffin.

2. Section and stain with Masson's Trichrome stain, which stains the endothelial cells/vessels dark red and the Matrigel blue (*see* **Note 7**).

3. Visually inspect the sections to determine if the test substance is stimulatory and if further quantitation is necessary (Fig. 17.1).

4. Capture images of the sections at ×10 in the area underlying the skin (*see* **Note 8**).

5. Adjust the image to selectively highlight the gray level corresponding to the endothelial cells/vessels using the density slice function of the software and the light level of the microscope.

6. Use the measure option to measure the area occupied by all cells selected by the density slice function (*see* **Notes 9** and **10**).

7. Count at least three random fields of the area underlying the skin on each slide.

8. Load the numbers into a spreadsheet and calculate the total area occupied by endothelial cells in the three fields and then average these areas among the three or four replicate mouse plugs.

9. Compare the resulting numbers for statistical significance.

4. Notes

1. Bubbles form when Matrigel is vortexed. Minimize vortexing to keep bubbles to a minimum, or injection problems will occur. Also, allowing the tubes to stand on ice for 10 min will reduce the number of bubbles.

2. All reagents should be kept as sterile as possible to prevent animal infection.

3. Always use gloves when working with reagents.

4. Keep dilution of Matrigel to a minimum. If too dilute, it will not gel when injected.

5. Use caution when handling animals. Only approved personnel should handle animals. Contact your animal safety officer for the appropriate safety regulations you must follow. Wear gloves, masks, gowns, hair covers, and booties when appropriate.

6. Inject solutions slowly; a bump should form under the skin. If the skin is punctured during injection, the material will flow out. Allow the Matrigel to gel about 10 s before removing the needle after finishing injection; otherwise, injected material will leak out.

7. Fixation according to these methods can inactivate epitopes so that immunohistochemical analysis may not work. An alternative is to make frozen sections of the tissue and eliminate the fixation and paraffin steps.

8. An alternative method of quantitating the vessel ingrowth is to measure the hemoglobin content using a Drabkin reagent (D5941) (Sigma-Aldrich, St. Louis, MO). In addition, the plasma volume in the plugs can be quantitated by using fluorescein isothiocyanate dextran (145,000–200,000 MW) (Sigma-Aldrich) injected intravenously into the tail vein, or the fluorescent putative vessels can be visualized and quantitated with a fluorescent microscope and imaging software.

9. Selecting a subset of the picture maybe required to avoid counting areas that may contain bubbles, cracks, or other artifacts that refract the light and are picked up by the density slice feature. If this is necessary, use the drawing tool to draw a box, note the size in the "info" window, and draw the same size box on all sections for the entire experiment.

10. Since animals will have different responses to the injected material, it is not unusual for there to be variability in the responses of animals to different test substances.

References

1. Kleinman, H. K., McGarvy, M. L., Liotta, L. A., Gehron-Robbey, P., Tryggvasson, K., Martin, G. R. (1982) Isolation and characterization of type IV procollagen, laminin, and heparan sulfate proteoglycan from the EHS sarcoma. *Biochemistry* 24, 6188–6193.

2. Kubota, Y., Kleinman, H. K., Martin, G. R., Lawley, T. J. (1988) Role of laminin and basement membrane in the morphological differentiation of human endothelial cells into capillary-like structures. *J Cell Biol* 107, 1589–1598.

3. Grant, D. S., Kinsella, J. L., Fridman, R., (1992) Interaction of endothelial cells with a laminin A chain peptide (SIKVAV) *in vitro* and induction of angiogenic behavior *in vivo*. *J Cell Physiol* 153, 614–625.

4. Passaniti, A., Taylor, R. M., Pili, R., (1992) A simple, quantitative method for assessing angiogenesis and antiangiogenic agents using reconstituted basement membrane, heparin, and fibroblast growth factor. *Lab Invest* 67, 519–528.

5. Kibbey, M. C., Corcoran, M. L., Wahl, L. M., Kleinman, H. K. (1994) Laminin SIKVAV peptide-induced angiogenesis *in vivo* is potentiated by neutrophils. *J Cell Phys* 160, 185–193.

6. Auerbach, R., Lewis, R., Shinners, B., Kubai, L., Akhtar N. (2003) Angiogenesis assays: a critical overview. *Clin Chem* 49, 32–40.

7. Johns, A., Freay, A. D., Fraser, W., Korach, K. S., Rubanyi, G. M. (1996) Disruption of estrogen receptor gene prevents 17 β estradiol-induced angiogenesis in transgenic mice. *Endocrinology* 137, 4511–4513.

8. Prewett, M., Huber, J., Li, Y., (1999) Antivascular endothelial growth factor receptor (fetal liver kinase 1) monoclonal antibody inhibits tumor angiogenesis and growth of several mouse and human tumors. *Cancer Res* 59, 5208–5218.

9. Babaei, S., Teichert-Kuliszewska, K., Zhang, Q., Jones, N., Dumont, D. J., Stewart, D. J. (2003) Angiogenic actions of angiopoietin-1 require endothelium-derived nitric oxide. *Am J Pathol* 1927–1936.

10. Akhtar, N., Dickerson, E. B., Auerbach, R. (2002) The sponge/Matrigel angiogenesis assay. *Angiogenesis* 5, 75–80.

11. Malinda, K. M., Goldstein, A. L., Kleinman, H. K. (1997) Thymosin β4 stimulates directional migration of human umbilical vein endothelial cells. *FASEB J* 11, 474–481.

Chapter 18

The Sponge Implant Model of Angiogenesis

Silvia Passos Andrade and Mônica Alves Neves Diniz Ferreira

Abstract

The host response observed after the application of an appropriate stimulus, such as mechanical injury or injection of neoplastic or normal tissue implants, has allowed the cataloguing of a number of molecules and cells involved in the vascularization of normal repair or neoplastic tissue. Implantation of sponge matrices has been adopted as a model for the accurate quantification of angiogenic and fibrogenic responses as they may occur during wound healing *in vivo*. Such implants are particularly useful because they offer scope for modulating the environment within which angiogenesis occurs. A sponge implantation model has been optimised and adapted to characterise essential components and their roles in blood vessel formation in a variety of physiological and pathological conditions. As a direct consequence of advances in genetic manipulation, mouse models (i.e., knockouts, severe combined immunodeficient [SCID], nude) have provided resources to delineate the mechanisms regulating the healing associated with implants. Here, we outline the usefulness of the cannulated sponge implant model of angiogenesis and provide a detailed description of the methodology.

Keywords: *N*-Acetylglucosaminidase, angiogenesis, blood flow, cannulated sponge model, chemokine, cytokine, myeloperoxidase.

1. Introduction

The realisation that advances in angiogenesis research depended on making the assays more quantitative and reproducible *in vitro* and *in vivo* led to the development of new techniques and improvement of old or current models to comply with such requirements. The implantation technique for assessment of the inflammatory process is an old surgical procedure. Grinlay and Waugh *(1)*, Woessner and Boucek *(2)*, and Edwards *(3)* were the first to use polyvinyl sponge implants in dogs, rats, and rabbits as a framework for the ingrowth of vascularised connective

S. Martin and C. Murray (eds.), *Methods in Molecular Biology, Angiogenesis Protocols, Second edition, Vol. 467*
© Humana Press, a part of Springer Science + Business Media, LLC 2009
DOI: 10.1007/978-1-59745-241-0_18

tissue and measurement of enzyme activities in the newly formed fibrovascular tissue.

This technique has further been developed to determine other biochemical variables of the fibrovascular tissue, including collagen metabolism *(4)*, fibronectin deposition *(5)*, and proteoglycan turnover *(6)*. In addition, the technique was employed to characterise the sequence of histological changes in granulation tissue formation *(7)* and to monitor the kinetics of cellular proliferation *(8)*. The extent of neutrophil and macrophage accumulation in the sponge compartment has also been possible by assaying the inflammatory enzymes myeloperoxidase (MPO) and N-acetylglucosaminidase (NAG) *(9–11)*. The effects of various anti-inflammatory agents on leukocyte migration and production of prostaglandin-like activity were evaluated in sponge implants *(12)*. The model of acute inflammation was particularly useful, allowing the collection and examination of both cellular and fluid phases of the exudate formed within the sponge *(12)*. The model was also modified to study more chronic inflammatory response and the evolution of granulation tissue *(12)*.

Sponge implantation has also been used as a framework to host rodent cell lines *(13–15)*. The advantage of the implantation technique for investigating tumor-induced angiogenesis is that the assessment of the relative contributions of the tumor cells to early changes in the implant blood flow can be detected even before visible growth of the tumor mass is evident. Using a ^{133}Xe clearancc technique or fluorescein diffusion method, the development of Colon 26, melanoma B16, or Ehrlich tumor has been evaluated regarding solid tumor haemodynamic features.

2. Materials

2.1. Sponge matrix

A number of different sponge matrices have been used for inducing fibrovascular growth and for hosting tumor cells. The synthetic materials are mainly polyvinyl alcohol, cellulose acetate, polyester, polyether, and polyurethane alone or in combination. In our laboratory, we use cannulated sponge disks made of polyether polyurethane. This type of material possesses the following characteristics: uniform pore size and intercommunicating pore structure, ability to resist chemical treatment, and biocompatibility.

2.2. Cannula

1. Polythene tubing for cannula: 1.4-mm internal diameter, 1.2 cm long.
2. Polythene tubing for plug: 1.2-mm internal diameter, 0.6 cm long.

3. 5-0 silk sutures for attachment of the cannula to the centre of the sponge disks, for holding the sponge disk in place following implantation, and for closing the surgical incision.

2.3. Anaesthesia

Rats are anaesthetized with intramuscular injection of 0.5 mL/kg of 0.315 mg/kg fentanyl citrate and 10 mg/mL fluanisone.

Mice are anaesthetized using the combination of fentanyl citrate and fluanisone acetate plus 5 mg/mL of midazolam hydrochloride, each at a dose of 0.5mL/kg. Ether inhalation can also be used.

3. Methods

3.1. Preparation of cannulated sponge implants

Circular sponge disks are cut from a sheet of sponge using a cork borer. Usually, the diameter and thickness of the disk depend on the animal used. For mice and rats, the recommended dimensions range from 8 × 4mm to 12 × 6 mm, respectively. The segment of polythene tubing (1.4-mm internal diameter) is secured to the interior of each sponge disk (i.e., midway through its thickness) by three 5-0 silk sutures in such a way that the tube is perpendicular to the disk face, and its open end is sealed with a removable plastic plug made of a smaller polythene tubing (1.2-mm internal diameter). The cannula allows accurate injection of tracers and test substances and withdrawal of fluid into and from the interior of the implant. The cannulated sponge disks are sterilised by boiling in distilled water for 10 min, then placed in sterile glass Petri dishes and are irradiated overnight under ultraviolet light in a laminar flow hood.

3.2. Surgical procedure for sponge disk Implantation

1. Implantation of sponge disks is performed with aseptic techniques following induction of anaesthesia.

2. The hair on the dorsal side is shaved and the skin wiped with 70% v/v ethanol in distilled water.

3. A 1-cm midline incision is made, and through it one subcutaneous pocket is prepared by blunt dissection using a pair of curved scissors.

4. A sterilized sponge implant is then inserted into the pocket; its cannula is pushed through a small incision that previously had been made on the cervical side of the pocket.

5. The base of the cannula is sutured to the animal skin to immobilize the sponge implant. Finally, the cannula is plugged with smaller sealed polythene tubing to prevent infection.

6. The midline incision is closed by three interrupted 5-0 silk stitches, and the animals are kept singly with free access to food and water after recovery from anaesthesia.

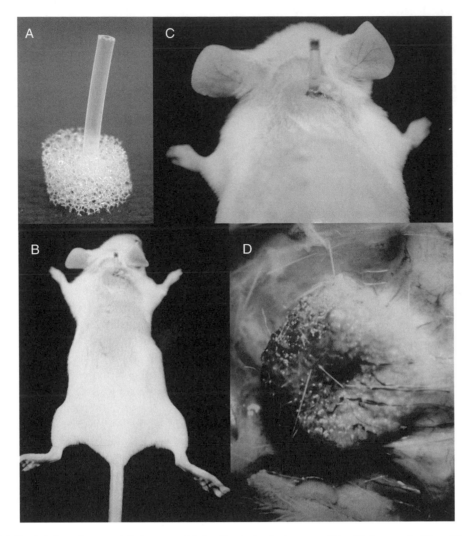

Fig. 18.1. Frontal view of a cannulated sponge disk *in situ*. An implant sponge disk with its cannula (**A**) and the subcutaneous arrangement of the implant in a mouse (**B** and **C**). Note that the cannula is exteriorised from and immobilised perpendicularly to dorsal skin by stitches. Substances to be tested are injected directly into the sponge disk via the cannula. (D) A vascularized sponge 14 d postimplantation is shown. The dorsal skin was removed to expose the implant. Blood vessels of various sizes can be seen infiltrating the sponge (×6.5).

The cannulated sponge disks can be left *in situ* for periods ranging from days to weeks. Schematic representation of the cannulated sponge disk and its arrangement after subcutaneous implantation is shown in **Fig. 18.1**.

3.3. Procedure for Estimating Blood Flow Development in the Sponge Implant by the ¹³³Xe Washout Technique

The sequential development of blood flow in the implanted sponges, originally acellular and avascular, can be determined by measuring the washout rate of ^{133}Xe injected into the implants, a technique developed to measure blood flow and thus monitor the vascular changes indirectly. This is based on the principle that the amount of a locally deposited radioactive tracer decreases at a rate proportional to the blood flow at the site of the injection. The decrease of radioactivity is exponential, and $t_{1/2}$ (time taken for the radioactivity to fall to 50% of its original value) for the washout is inversely related to the local blood flow *(16)*.

1. Anaesthetize the animals as before.

2. Inject ^{133}Xe (10 µL containing 1X 10^6 cps into the implant via the cannula).

3. Immediately after injection, plug the cannula to prevent evaporation of the tracer.

4. Monitor the washout of radioactivity from the implant using a collimated gamma scintillation detector for 6 min. The detector, a sodium iodide thallium-activated crystal (1 × 1 in.) is positioned directly above the site of the injection.

5. The radioactivity is recorded for 40-s epochs and printed on a scaler rate meter. The radioactivity-versus-time data are fitted to an exponential decline curve to derive $t_{1/2}$ (half time in minutes) after deduction of background radioactivity.

A particular advantage of the ^{133}Xe-$t_{1/2}$ assay is that it allows nondestructive, and thus repeated, measurements of blood flow in the same animal over the period of neovascularization of the sponge. This combination of techniques requires fewer animals and allows an estimate of variability in individual animals.

3.4. Procedure for Estimating Blood Flow Development by Efflux of Sodium Fluorescein Applied Intraimplant

Low molecular fluorochrome-complexed tracers or fluoregenic dyes have provided additional methods for detecting new blood vessels. Compared with radioactive isotope compounds, the advantages of fluorescent dyes are evident. Fluorescence is relatively atoxic, nonradioactive, and inexpensive *(17)*. The measurement of fluorochrome-generated emission in the bloodstream following its application in the sponge implant compartment at various intervals postimplantation reflects the degree of local blood flow development and the interaction of the angiogenic site with the systemic circulation *(18)*. This approach can be used to study sponge-induced angiogenesis quantitatively and to investigate the pharmacological reactivity of the neovasculature.

Measurements of the extent of vascularization of sponge implants are made by estimating $t_{1/2}$ (min) of the fluorescence peak in the systemic circulation following intraimplant injection of sodium fluorescein (50 µL/kg of a sterile solution of 10% sodium fluorescein) at fixed time intervals (e.g., d *1, 4, 7, 10,* and *14)* postimplantation.

1. Anaesthetize the animal.

2. Determine blood background fluorescence by piercing the extremity of the tail and collecting 5 µL of blood with a heparinized yellow tip. Transfer the blood sample to a centrifuge tube containing 1 mL of isotonic saline (0.9%).

3. At time 0, administer sodium fluorescein (50 µL of a sterile solution of 10% sodium fluorescein per kilogram weight) to anesthetized animals.

4. After 1 min, collect the first blood sample following dye injection as in **step 2**.

5. At 3 min, collect a second blood sample. Repeat this procedure every 2–3 min for 25–30 min.

6. Centrifuge the blood samples for 10 min at $1400g$ (2000 rpm). Keep the supernatant for fluorescence determination (excitation 485 nm/emission 519 nm).

7. From the fluorescence values, estimate the time for the fluorescence to peak in the bloodstream (absorption) and the time required for the elimination of the dye from the systemic circulation (elimination). These parameters are expressed in terms of half-time ($t_{1/2}$; time taken for the fluorescence to reach or to decay 50% of the peak value in the systemic circulation).

3.5. Procedure for Estimating Biochemical Parameters in the Implants

Quantitation of various biochemical parameters further supports the functional characterization of the fibrovascular tissue that infiltrates the implants and have been used to corroborate assessment of angiogenesis (12, 19–23).

1. Remove the implants at any time postimplantation as required.

2. Immediately on removal, weigh, homogenize in 2 mL of Drabkin reagent, and centrifuge the tissue in isotonic physiological solution (0.9% saline or phosphate-buffered saline [PBS]).

3. Store the supernatant of the homogenate at –20°C for later analysis of hemoglobin (vascular index), cytokines, and chemokines (angiogenic factors).

4. Store the pellet of the homogenate at –20°C for determination of the proinflammatory enzymes MPO (neutrophil influx) and NAG (macrophage recruitment).

3.5.1. Haemoglobin Determination

The vascularization of the implant can be assessed by measuring the amount of hemoglobin contained in the tissue using the Drabkin method.

1. After centrifuging the homogenate (implant plus 2 mL Drabkin reagent) at $12,000g$ for 20 min, filter it in a 0.22-µm Millipore filter.

2. Determine hemoglobin concentration by measuring absorbance at 540 nm using an enzyme-linked immunosorbent assay (ELISA) plate reader and compare against a standard curve of hemoglobin.

3.5.2. Cytokine Determination

1. Take 50 µL of the supernatant previously homogenized in Drabkin reagent (to remove hemoglobin) and centrifuged ($12,000g$, 20 min at 4°C) and add 500 µL of PBS at pH 7.4 containing 0.05% Tween-20 (Difco) and centrifuge at $12,000g$ at 4°C for 30 min.

2. The amount of the cytokines and the chemokines in each sample is determined using immunoassay kits (R & D Systems, USA) and following the manufacturer's protocol.

3. Dilutions of cell-free supernatants are added in duplicate to ELISA plates coated with a specific murine polyclonal antibody against the cytokine/chemokine, followed by the addition of a second polyclonal antibody against the cytokine/chemokine. After washing to remove any unbound antibody-enzyme reagent, a substrate solution (a 1:1 solution of hydrogen peroxide and tetramethylbenzidine) is added to the wells. The reaction is terminated with 50 µL/well of $1M$ H_2SO_4. Plates were read at 492 nm in a spectrophotometer. Standards of recombinant murine chemokines were diluted in a range from 7.5 to 1000 pg/mL. Express the results as picograms/milligram wet tissue.

3.5.3. Measurement of MPO and NAG Activities

The extent of neutrophil accumulation in the implants is measured by assaying MPO activity in whole tissue.

1. After processing the supernatant of the tissue for hemoglobin determination (*see* **Subheading 3.5.1.**), a part of the corresponding pellet is weighed, homogenized in pH 4.7 buffer ($0.1M$ NaCl, $0.02M$ Na_3PO_4, $0.015M$ Na_2 ethylenediaminetetraacetic acid [EDTA]), and centrifuged at $12,000g$ for 20 min at 4°C.

2. The pellets are then resuspended in $0.05M$ sodium phosphate buffer (pH 5.4) containing 0.5% hexa-1,6-bis-decyltrimethylammonium bromide (HTAB; Sigma). The suspensions are freeze-thawed three times using liquid nitrogen and finally centrifuged at $10,000g$ for 20 min at 4°C.

3. MPO activity in the resulting supernatant is assayed by mixing 25 µL of 3,3′-5,5′-tetramethylbenzidine [TMB]; Sigma), prepared in dimethylsulphoxide (DMSO; Merck) in a final concentration of 1.6 mM; add 100 µL H_2O_2 in a final concentration of 0.003% v/v, dissolved in sodium phosphate buffer (pH 5.4) and 25 µL of the supernatant from the tissue sample.

4. The assay is carried out in a 96-well microplate and is started by adding the supernatant sample to the H_2O_2 and TMB solution and incubated for 5 min at 37°C.

5. The reaction is terminated by adding 100 µL $4M$ H_2SO_4 at 4°C and is quantified colorimetrically at 450 nm in a spectrophotometer. Results can be expressed as change in optical density per gram of wet tissue.

Numbers of monocytes/macrophages are quantitated by measuring the levels of the lysosomal enzyme NAG, present in high levels in activated macrophages (9, 24).

1. Part of the pellet remaining after the hemoglobin measurement is kept for this assay. These pellets are weighed, homogenized in NaCl solution (0.9% w/v) containing 0.1% v/v Triton X-100 (Promega), and centrifuged (3000g for 10 min at 4°C).

2. Samples of the resulting supernatant (100 μL) are incubated for 10 min with 100 μL p-nitrophenyl-N-acetyl-D-glucosaminide (Sigma) prepared in 0.1M citrate/sodium phosphate buffer (pH 4.5) in a final concentration of 2.24 mM.

3. The reaction is terminated by the addition of 100 μL 0.2M glycine buffer, pH 10.6. Hydrolysis of the substrate is determined by measuring the colour absorption at 405 nm. NAG activity is finally expressed as optical density per gram wet tissue.

3.6. Histological Analysis of the Implants

To further establish the sequential development of granulation tissue and blood vessels in the implants, several histologic techniques have been employed.

1. Kill the animals bearing the implants.

2. Dissect the implants free of adherent tissue, fix in formalin (10% w/v in isotonic saline), and embed them in paraffin.

3. Cut the sections (5–8 μm) from halfway through the sponge's thickness.

4. Stain and process for light microscopy studies.

4. Notes

1. The attachment of the cannula to the centre of the sponge disk is facilitated by making in one end of the polythene tubing "teeth-like" (usually four) structures in such a way that the thread is secured in one of them and then sutured to the sponge.

2. To avoid leakage following injection of test substances via the cannula, it is important to hold the base of the cannula with a forceps. Sometimes, it is necessary to push the substances injected by blowing into the cannula.

3. To avoid infection, it is recommended to change the plugs every time they are removed from the cannula.

4. It is advisable that administration of drugs be performed 2 to 3 d postimplantation to allow the sponges to be encapsulated.

5. To eliminate acute effects of the vasoactive substances tested (vasodilation or vasoconstriction), they should be given 6–8 h prior to blood flow measurement.

6. Depending on the material, the inflammatory response can cause excessive matrix deposition and unwanted fibrosis. Because of the variety of materials used (size, structure, composition, porosity), the pattern of the response varies widely.

7. To avoid possible "contamination" of blood spilled during and after the surgical procedure or with surrounding preexisting vessels, removal of the implants must be done 30 min after death of the animals.

Acknowledgement

This work was supported by grant from CNPq - Brazil.

References

1. Grindlay, J. H.,Waugh, J. M. (1951) Plastic sponge which acts as a framework for living tissue; experimental studies and preliminary report of use to reinforce abdominal aneurysms experimental studies and preliminary report of use to reinforce abdominal aneurysms. *AMA Arch Surg* 63, 288–297.

2. Woessner, J. F., Jr., Boucek, R. J. (1959) Enzyme activities of rat connective tissue obtained from subcutaneously implanted polyvinyl sponge. *J Biol Chem* 234, 3296–3300.

3. Edwards, R. H., Sarmenta, S. S., Hass, G. M. (1960) Stimulation of granulation tissue growth by tissue extracts; study by intramuscular wounds in rabbits. *Arch Pathol* 69, 286–302.

4. Paulini, K., Korner, B., Beneke, G., Endres, R. (1974) A quantitative study of the growth of connective tissue: investigations on polyester-polyurethane sponges. *Connect Tissue Res* 2, 257–264.

5. Holund, B., Clemmensen, I., Junker, P., Lyon, H. (1982) Fibronectin in experimental granulation tissue. *Acta Pathol Microbiol Immunol Scand [A]* 90, 159–165.

6. Bollet, A. J., Goodwin, J. F., Simpson, W. F., Andreson, D. V. (1958) Mucopolysaccharide, protein and DNA concentration of granulation tissue induced by polyvinyl sponges. *Proc Soc Exp Biol Med* 99, 418–421.

7. Holund, B., Junker, P., Garbarsch, C., Christiffersen, P., Lorenzen, I. (1979) Formation of granulation tissue in subcutaneously implanted sponges in rats. *Acta Pathol Microbiol Scand [A]* 87, 367–374.

8. Davidson, J. M., Klagsbrun, M., Hill, K. E., (1985) Accelerated wound repair, cell proliferation and collagen accumulation are produced by a cartilage-derived growth factor. *J. Cell Biol* 100, 1219–1227.

9. Bailey, P. J. (1988) Sponge implants as models. *Meth Enzymol* 162, 327–334.

10. Belo, A. V., Barcelos, L. S., Ferreira, M. A. N. D., Teixeira, M. M., Andrade, S. P. (2004) Inhibition of inflammatory angiogenesis by distant subcutaneous tumor in mice. *Life Sci* 74, 2827–2837.

11. Ferreira, M. A. N. D., Barcelo, L. S., Campos, P. P., Vasconcelos, A. C., Teixeira, M. M., Andrade, S. P. (2004) Sponge-induced angiogenesis and inflammation in PAF receptor-deficient mice (PAFR-KO). *Br J Pharmacol* 141, 1185–1192.

12. Ford-Hutchinson, A. W., Walker, J. A., Smith, J. A. (1977) Assessment of anti-inflammatory activity by sponge implantation techniques. *J Pharmacol Meth* 1, 3–7.

13. Mahadevan, V., Hart, I. R., Lewis, G. P. (1989) Factors influencing blood supply in

wound granuloma quantitated by a new *in vivo* technique. *Cancer Res* 49, 415–419.

14. Andrade, S. P., Bakhle, Y. S., Hart, I., Piper, P. J. (1992) Effects of tumour cells and vasoconstrictor responses in sponge implants in mice. *Br J Cancer* 66, 821–826.

15. Lage , A. P., Andrade, S. P. (2000) Assessment of angiogenesis and tumor growth in conscious mice by a fluorimetric method. *Microvasc Res* 59, 278–285.

16. Kety, S. S. (1949) Measurement of regional circulation by local clearance of radioactive sodium. *Am Heart J* 38, 321–331.

17. McGrath, J. C., Arribas, S., Daly, C. J. (1996) Fluorescent ligands for the study of receptors. *Trends Pharmacol Sci* 17, 393–399.

18. Andrade, S. P., Machado, R. D., Teixeira, A. S., Belo, A. V., Tarso, A. M., Beraldo, W. T. (1997) Sponge-induced angiogenesis in mice and the pharmacological reactivity of the neovasculature quantitated by a fluorimetric method. *Microvasc Res* 54, 253–261.

19. Andrade, S. P., Vieira, L. B. G. B., Bakhle, Y., Piper, P. J. (1992) Effects of platelet activating factor (PAF) and other vasoconstrictors

on a model of angiogenesis in the mouse. *Int J Exp Pathol* 73, 503–513.

20. Andrade, S. P., Cardoso, C. C., Machado, R. D. P., Beraldo, W. T. (1996). Angitensin-II-induced angiogenesis in sponge implants in mice. *Int J Microcirc Clin Exp* 16, 302–307.

21. Hu, D.-E., Hiley, C. R., Smither, R. L., Gresham, G. A., Fan, T.-P. D. (1995) Correlation of 133Xe clearance, blood flow and histology in rat sponge model for angiogenesis. *Lab Invest* 72, 601–610.

22. Buckley, A., Davidison, J. M., Kamerath, C. D., Wolt, T. B., Woodward, S. C. (1985) Sustained release of epidermal growth factor accelerates wound repair. *Proc Natl Acad Sci U S A* 82, 7340–7344.

23. Plunkett, M. L., Halley, J. A. (1990) An in vivo quantitative angiogenesis model using tumor cells entrapped in alginate. *Lab Invest* 62, 510–517.

24. Belo, A. V., Barcelos, L. S., Teixeira, M. M., Ferreira, M. A., Andrade, S .P. (2004) Differential effects of antiangiogenic compounds in neovascularizationzleukocyte recruitment, VEGF production, and tumor growth in mice. *Cancer Invest* 22, 723–729.

Chapter 19

The Dorsal Skinfold Chamber: Studying Angiogenesis by Intravital Microscopy

Axel Sckell and Michael Leunig

Abstract

Intravital microscopy represents an internationally accepted and sophisticated experimental method to study angiogenesis, microcirculation, and many other parameters in a wide variety of neoplastic and nonneoplastic tissues. Since 1924, when the first transparent chamber model in animals was introduced, many other chamber models have been described in the literature for studying angiogenesis and microcirculation. Because angiogenesis is an active and dynamic process, one of the major strengths of chamber models is the possibility of monitoring angiogenesis *in vivo* continuously for up to several weeks with high spatial and temporal resolution. In addition, after the termination of experiments, tissue samples can be excised easily and further examined by various *in vitro* methods, such as histology, immunohistochemistry, and molecular biology. This chapter describes the protocol for the surgical preparation of a dorsal skinfold chamber in mice as well as the method to implant tumors in this chamber for further investigations of angiogenesis and other microcirculatory parameters. However, the application of the dorsal skinfold chamber model is not limited to the investigation of neoplastic tissues. To this end, the investigation of angiogenesis and other microcirculatory parameters of nonneoplastic tissues such as tendons, osteochondral grafts, or pancreatic islets have been objects of interest.

Key words: Angiogenesis, intravital microscopy, microcirculation, skinfold chamber.

1. Introduction

Since 1924, when the first transparent chamber model in animals was introduced by Sandison (1), many other chamber models have been described in the literature for studying angiogenesis and microcirculation in a wide variety of neoplastic and nonneoplastic tissues by means of intravital microscopy (for reviews, see refs. 2–4). Because angiogenesis is an active and dynamic process, one of the

S. Martin and C. Murray (eds.), *Methods in Molecular Biology, Angiogenesis Protocols, Second edition, Vol. 467*
© Humana Press, a part of Springer Science+ Business Media, LLC 2009
DOI: 10.1007/978-1-59745-241-0_19

major strengths of chamber models is the possibility of monitoring angiogenesis *in vivo* continuously for up to several weeks with high spatial and temporal resolution. In addition, after the termination of experiments, tissue samples can be excised easily and further examined by various *in vitro* methods, such as histology, immunohistochemistry, and molecular biology.

The advantages of using mice as experimental animals are, for instance, the availability of a large number of different well-defined mouse strains, including transgenic or knockout mice, and the wide variety of commercially generated agents suitable for mice, such as monoclonal antibodies, nanoparticles, and single-gene products.

This chapter describes the protocol for the surgical preparation of the dorsal skinfold chamber in mice as well as the method to implant tumors in this chamber for further investigations of angiogenesis and other microcirculatory parameters. The model *(5, 6)* presented here is the development of a similar model in hamsters *(7)*. In brief, take a fold of the depilated dorsal skin of an anesthetized mouse and cut out surgically a circular area of one skin layer (consisting of epidermis, dermis, subcutis, cutaneous muscle, and subcutaneous fatty tissue) completely. Then, fix the skinfold like a sandwich between the two titanium frames of the chamber and close the operation field with a sterile coverslip to avoid drying, infection, or mechanical damage of the inner layer (i.e., the cutaneous muscle) of the unprotected side of the opposite skin. For tissue implantation or other local treatments, the chamber can easily be opened again by removing the coverslip and be reclosed with a new sterile coverslip. The cutaneous muscle serves as site for implantation of tissues such as little chunks of solid tumors (*see* **Subheading 3.2**). From now on, intravital microscopy can be performed for monitoring angiogenesis and other parameters such as tumor growth, microvascular perfusion index, microcirculation, and leukocyte endothelium interaction. However, the application of the dorsal skinfold chamber model is not limited to the investigation of neoplastic tissues. To this end, the investigation of angiogenesis and other microcirculatory parameters of nonneoplastic tissues such as tendons *(8)*, osteochondral grafts *(6, 9)*, or pancreatic islets *(10)* has been object of interest. In principle, the implantation of these tissues follows the same rules as the implantation of tumors as described in this chapter (*see* **Subheading 3.2**).

2. Materials

Except for commonly used devices, all materials necessary for the preparation of the dorsal skinfold chamber in mice are listed with a detailed manufacturer's record. All materials given are only suggestions and may be modified for personal preferences.

2.1. Dorsal Skinfold Chamber Preparation

2.1.1. Facilities and Apparatus

1. Laminar flow hood (HPH 12, Merck Eurolab GmbH, Bruchsal, Germany).

2. Dry sterilizer (model IS-350, Inotech, Intergra Biosciences, Wallisellen, Switzerland).

3. Dissecting microscope (Leica MZ75, ×6.3–50; Leica Microsystems AG, Heerbrugg, Switzerland).

4. Two flat custom-made thermal pads (silicon, Therm TSW 3, Isopad GmbH, Heidelberg, Germany).

5. Halogen lamp with two flexible swan-neck light transmission tubes (Intralux, 150H, Volpi AG, Urdorf-Zürich, Switzerland).

6. Custom-made skin-spreading device. This consists of a heavy metal base and two flexible swan-neck tubes for applying tension to spread out the mouse skinfold prior to the fixation of the back titanium frame of the dorsal skinfold chamber to the skinfold of the mouse (Workshop, Department of Experimental Surgery, University of Heidelberg, Germany).

2.1.2. Drugs

1. Isotonic sodium chloride: 0.9% NaCl solution injectable.

2. Anesthesia: Mixture consisting of isotonic sodium chloride, ketamine hydrochloride (Ketalar®; Parke-Davis, Morris Plains, NJ), and xylazine (XylaJect®; Phoenix Pharmaceutical, St. Joseph, MO).

3. Depilatory cream (pilca med creme, Asid Bonz GmbH, Boeblingen, Germany).

2.1.3. Dorsal Skinfold Chambers

1. Custom-made dorsal skinfold chambers (Workshop, Department of Experimental Surgery, University of Heidelberg; **Fig. 19.1**) consisting of two titanium frames, three screws (M2 × 6), six nuts (size 4), and one tension ring to keep the sterile coverslip in position after closing of the chamber preparation.

2. Sterile coverslips (0.13- to 0.16-mm thick, 11.75-mm diameter, circular, Assistent, Sonthcim, Germany).

3. Special pair of pliers (Garant®, Germany) to bring the tension ring in position to keep the chamber closed and to remove it again.

4. Wrench (CHR-VAN, size 4, SKG, Germany).

2.1.4. Surgical Instruments

1. Electric hair clipper (Electra® II, GH 204 or 201, Aesculap®, Aesculap AG, Tuttlingen, Germany) equipped with a 1/20-mm cutting head (GH 700, Aesculap).

2. Two delicate hemostatic forceps (baby mosquito, BH 115, Aesculap).

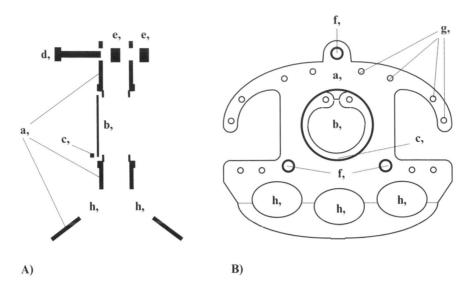

Fig. 19.1. Construction plan of the dorsal skinfold chamber: (**A**) Cross section; (**B**) lateral view: *a* titanium frame, *b* coverslip, *c* tension ring, *d* screw, *e* nut, *f* screw hole, *g* bore holes for holding sutures, *h* holes for weight reduction of the chamber.

3. One needle holder (Castroviejo, BM 2, Aesculap).

4. One delicate dissecting forceps (Micro-Adson, BD 220, Aesculap®).

5. Two microforceps (BD 331, Aesculap).

6. One pair of dissecting scissors, fine patterns (Cottle-Masing, sharp, OK 365, Aesculap).

7. One pair of microscissors (spring type) with round handles (FD 103, Aesculap).

8. One pair of microscissors (spring type) with flat handles and cross-serration (Vannas, FD 15, Aesculap).

9. Sterile scalpel blades (no. 15, BB 515, Aesculap).

10. Sutures: Polypropylene (Prolene monofil, 4-0, C1, EH7476H, Ethicon GmbH, Norderstedt, Germany).

2.1.5. Other Materials

1. Mouse (25–30 g body weight, 6–12 wk); depending on the research goal, inbreed, outbreed, immune-competent, immunedeficient, and so on.

2. One cage per animal (*see* **Note 1**).

3. Syringes (1 mL).

4. 26-gauge needles (26G3/8, 0.45 × 10).

5. Sterile nonwoven swabs (5 × 5 cm).

6. Sterile Q-tips (cotton pads on wooden sticks).

7. Fine black waterproof permanent pen.

8. Surgical masks.

9. Rubber gloves.

10. 70% alcohol to disinfect skin of the mouse, surgical instruments, and rubber gloves.

2.1.6. Additional Equipment Necessary for Tissue Implantation

1. Custom-made device consisting of a slitted polycarbon tube (24-mm internal diameter, 120 mm long) and a special mounting stage to fix the tube with the animal in it (Workshop, Department of Experimental Surgery, University of Heidelberg).

2. Adhesive tape (Transpore™ 3M, hypoallergenic, ≈1.2cm × 9.1 m, 3M Medical-Surgical Division, St. Paul, MN, USA).

3. Hank's balanced salt solution (H-9269, 100 mL, Sigma-Aldrich, Irvine, CA, UK) stored at 6°C.

4. Sterile Petri dishes (diameter ≈100 × 20 mm).

3. Methods

3.1. Surgical Preparation of the Dorsal Skinfold Chamber

A scheme of the setup for the surgical preparation of the dorsal skinfold chamber is shown in **Fig. 19.2**. All surgical procedures should be performed under aseptic conditions (*see* **Note 2**).

1. Anesthetize the mouse by an injection of a mixture of ketamine (100 mg/kg body weight) and xylazine (10 mg/kg body weight) intramuscularly into the limb (*see* **Note 3**).

2. For depilation of the entire dorsum of the mouse, carefully shave the anesthetized mouse with the electric hair clipper. Put the mouse on a thermal pad outside the hood and apply a thick layer (1–2 mm) of depilatory cream on the shaved skin area. After the cream is allowed to take effect for 5–10 min (*see* **Note 4**), it can be easily removed using a nonwoven swab soaked in hand-warmed, germ-free water, wiping in caudal-to-cranial direction (*see* **Notes 5** and **6**).

3. Dry the wet depilated skin with a dry sterile swab and disinfect the skin with a 70% alcohol-soaked swab.

4. Place the mouse between the surgeon and the light transmission tube (tube 2) of the halogen lamp (**Fig. 19.2**); the mouse should be in a prone position on an opened sterile nonwoven swab lying on the thermal pad inside the hood. The longitudinal axis of the mouse (with its head lined up to the left-hand side of the surgeon) should be parallel to the frontal plane of the surgeon. Illuminate the mouse from top with the light transmission tube (tube 1) of the halogen

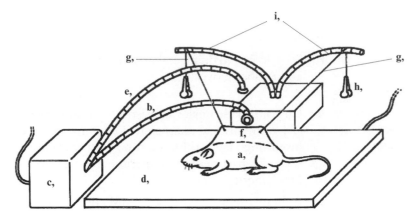

Fig. 19.2. Setup for the surgical preparation inside the hood: *a* mouse, *b* swan-neck light transmission tube 2 (for transillumination from behind), *c* halogen lamp, *d* thermal pad, *e* swan-neck light transmission tube 1 (for epi-illumination from above), *f* skinfold, *g* holding thread, *h* baby mosquito, *i* flexible swan necks of the skin-spreading device.

lamp. Adjust tube 2 with its central light about 2 cm parallel to the surface of the thermal pad and perpendicular to the longitudinal axis of the mouse (**Fig. 19.2**).

5. Lift up a fold of the depilated dorsal skin. This fold, running from the sacrum to the neck of the mouse, should be located directly over, and parallel to, the spine. Adjust the fold under transillumination (tube 2) in such a way that both sides of the skinfold become congruent (*see* **Note 7**). Then, spread the skin in the upright position by fastening two holding threads (polypropylene 4-0; no knots but one baby mosquito each fixed at their ends) at the edge of the skinfold (*see* **Note 8**) and hanging them over two flexible swan necks of the custom-made skin-spreading device (Fig. 19.2; *see* **Note 9**).

6. Fix the first titanium frame of the chamber to the side of the skinfold facing away from the surgeon with two temporary holding sutures (polypropylene 4-0) through both of the borehole pairs, left and right of the apical screw (*see* **Notes 10–12**).

7. Remove the two holding threads completely and make two small incisions (using a pair of Cottle-Masing scissors or a no. 15 scalpel blade), perforating the entire skinfold to let the two lower screws of the titanium frame come through (*see* **Notes 13** and **14**). The fixation of the last two screws using the baby mosquitos helps to adapt the skinfold temporarily to the titanium frame, facilitating further surgical preparations. The bent tip of the baby mosquito should then point toward the mouse and not the surgeon.

8. Use the cranial baby mosquito to turn the back titanium frame in a perpendicular position to the transilluminating

light. The edge of the circular area projected through the central window of the back titanium frame to the skin facing the surgeon now can be easily marked with a dotted line using a fine black permanent pen.

9. Place the mouse in right lateral position under the dissecting microscope (~sixfold magnification; *see* **Note 15**) with the skinfold pointing to the surgeon.

10. Using delicate dissecting forceps (Micro-Adson) and microscissors (FD 103), remove all layers of the skin completely (epidermis, dermis, subcutis, cutaneous muscle, parts of the subcutaneous fatty tissue) along the marked dotted line (*see* **Notes 16** and **17**).

12. Stop possible bleeding along the edge of the wound gently using sterile Q-tips slightly moistened with isotonic saline. Now, allow 3–5 min to elapse to ensure that no further bleeding takes place. Meanwhile, to avoid drying out the operation field, perfuse this area with isotonic saline (*see* **Note 18**).

12. Absorb excess saline with dry Q-tips (*see* **Note 19**) and put the optical magnification on about 10-fold. *The next step is probably the most critical step of the chamber preparation*: Use microforceps (BD 331) and a pair of microscissors (Vannas) to carefully remove the last layer of subcutaneous fatty tissue that is connected to the underlying cutaneous muscle of the opposite skin (*see* **Notes 20–22**).

13. Close the chamber preparation like a sandwich with the second titanium frame: First, connect the lower two screws to the corresponding holes of the second frame and then the apical one (*see* **Note 23**). If no air bubbles are visible between the coverslip and the underlying cutaneous muscle, screw on the nuts to finally fix the two frames of the chamber together (*see* **Notes 24** and **25**).

14. Perform four holding sutures (polypropylene 4-0) to spread and fix the edges of the sandwiched skinfold to all four pairs of boreholes left and right from the apical screw in both chamber frames. After stitching but before closing the knots of the two central holding sutures, cut the old temporary holding sutures from **step 6** and remove them completely (*see* **Note 26**).

15. Place the operated mouse in its cage and leave it on a thermal pad outside the hood at least until the mouse has regained consciousness (*see* **Note 27**).

3.2. Tissue Implantation into the Chamber Preparation

In the following, the explantation of a solid tumor from a donor mouse and the consecutive implantation of a chunk of this tumor into the dorsal skinfold chamber preparation of a recipient mouse are described (*see* **Notes 28** and **29**).

1. Allow at least 48 h to elapse for the animals to recover completely from surgery before implantation of any tissue into the dorsal skinfold chamber preparation. Exclude all animals from further treatment and sacrifice them when there are signs of bleeding, inflammation, or any other irritation at the implantation site (*see* **Note 30**). Only chambers meeting criteria of intact microcirculation *(11)* should be used as sites for implantation.

2. Sacrifice the donor mouse bearing a solid subcutaneous tumor according to official *Guidelines for Care and Use of Experimental Animals.* For 2–3 min, completely insert the dead animal into a 70% alcohol solution for disinfection. Excise the desired tumor surgically under aseptic conditions in the hood and put it into a sterile Petri dish previously filled with cold (~6°C) Hank's balanced salt solution.

3. The dissecting microscope is only needed for this step of the protocol (magnification ~ 10-fold). Remove the capsule and all hemorrhagic or necrotic parts of the tumor with the help of microforceps and a pair of microscissors. Cut the remaining tumor into small chunks of a diameter no greater than about 0.5–1 mm (*see* **Note 31**).

4. Put the nonanesthetized recipient mouse in the slitted polycarbon tube (**Fig. 19.3**). An adhesive tape fixed across the slit right behind the chamber jutting out will prevent the animal escaping from the tube (*see* **Note 32**). Then, fix the chamber in a horizontal position in the special mounting stage, which will also serve as a stage to perform intravital microscopy of the implant at a later time (**Fig. 19.3**).

5. Remove the tension ring with the wrench. Use a 26-gauge needle as a lever to lift the coverslip a few millimeters, finally grasping and removing it with microforceps.

6. Transfer one of the tumor chunks with another set of sterile microforceps onto the cutaneous muscle in the center of the open chamber (*see* **Note 33**).

7. With a new, sterile coverslip, reclose the chamber preparation. Before inserting the tension ring to fix the coverslip in position, make sure that there are no persisting air bubbles. These air bubbles can be removed with a Q-tip (*see* **Notes 24** and **34**).

8. Cover the central window of the back titanium frame, which has no coverslip inserted, with a piece of adhesive hypoallergenic tape (*see* **Notes 35** and **36**) and release the mouse back into its cage after removing the other adhesive tape from the tube.

9. Intravital microscopy of the implanted tumor chunk can be performed now repeatedly in the conscious or anesthetized animal by means of normal light and transillumination or

epi-illumination from a mercury lamp and a fluorescent filter set in combination with appropriate fluorescent dyes injected intravenously into the animal.

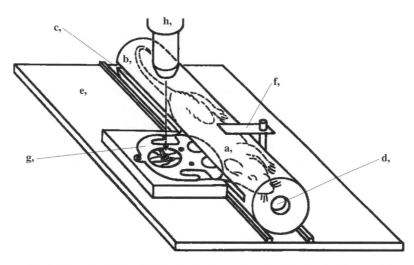

Fig. 19.3. A mouse fitted with a dorsal skinfold chamber inside a polycarbon tube fixed on the special mounting stage: *a* mouse, *b* slitted polycarbon tube, *c* slit, *d* breathing hole, *e* special mounting stage, *f* device to fix the polycarbon tube on the mounting stage, *g* dorsal skinfold chamber preparation, *h* objective of the intravital microscope.

4. Notes

1. For optimal quality of the chamber preparation, one of the basic requirements is that the area of the dorsal skin associated with the chamber preparation lacks any injuries, scars, or other irritations. Therefore, prior to surgery, only mice from one brood should be held together in one cage since mice from different broods tend to cause injury to one another. After chamber implantation, the animals must be housed separately in single cages; otherwise, they may destroy each other's chamber preparations by scratching and biting.

2. Work under a laminar flow hood and wear a surgical mask as well as rubber gloves to minimize the possibility of bacterial contamination of the chamber preparation. Between preparations of two different animals, all surgical instruments should be first cleaned mechanically with sterile nonwoven swabs soaked in alcohol and then sterilized with a dry sterilizer. The gloves should be washed with alcohol and changed from time to time.

3. To avoid cooling of the body temperature of the mouse, the anesthetized animal should be placed on a thermal pad (~37°C) whenever possible.

4. While the depilatory cream is taking effect, clean and prepare the surgical instruments for the following chamber preparation.

5. To avoid irritation of the skin, wipe gently but unsparingly with fresh water-soaked swabs.

6. It is important to use depilatory cream as needed to remove all hairs. Otherwise, during later transillumination microscopy the remaining hair roots will show up as dark shadows, decreasing the optical quality of the region of interest.

7. Points of reference are the larger vessels of the skin, which run symmetrically to each other on the left and right sides of the sagittal plane of the mouse.

8. The chamber should fit between the two holding sutures.

9. The weight of a baby mosquito fixed at the end of each holding thread is heavy enough to keep the skinfold in an upright position.

10. If possible, place the central window of this frame in such a manner that it lies centrally between the two main vascular trunks coming from caudal and cranial.

11. The skin between the two sutures should be unstressed, and the apical screw of the chamber frame should just jut over the upper edge of the skinfold.

12. Because the two holding sutures have to be removed at a later time (*see* **step 14**) and to avoid local skin necrosis, do not make these sutures too tight.

13. The location of larger skin vessels may be controlled with transilluminating light. These vessels should not be cut or damaged by the incisions.

14. There should not be any tension on the skin area between the two screws and holding sutures.

15. The two baby mosquitos may be used to adjust the skinfold and keep it level, parallel to the surface of the thermal pad.

16. Be sure to remove all macroscopic particles left inside or around the marked skin area before cutting. Loose hairs or small fibers of the nonwoven swab may be detected easily under the dissecting microscope and can be removed using microforceps.

17. Avoid hurting the underlying inside of the opposite skinfold. It may be advantageous to perform the initial incision in the center of the marked area. Then, continue cutting toward and along the dotted line.

18. Allow enough time for bleeding to stop. The last layer of subcutaneous fatty tissue still protects the underlying cutaneous muscle of the opposite skin, which will later serve as

the site of tissue implantation. After-bleeding at a later time onto the unprotected cutaneous muscle can easily destroy the chamber preparation.

19. Place the Q-tip close to the edge of the operation field. In doing so, touching of the vulnerable inside of the underlying skin is avoided.

20. Be sure to dissect the subcutaneous fatty tissue from the underlying cutaneous muscle by cutting and not by pulling it away. Too much pulling may lead to disruption of small vessels of the muscle and thus to uncontrolled microbleeding.

21. **Steps 12** and **13** have to be performed as free of interruption as possible to avoid drying and damage of the tissue layer (cutaneous muscle), which will later be used as a bed for implanting other tissues.

22. To save time, it may be advantageous to remove all the remaining subcutaneous fatty tissue *in toto*, starting from caudal and toward cranial if you are cutting right-handed.

23. At the time of closure of the chamber preparation, the coverslip should already be inserted and fixed in the second titanium frame with the tension ring.

24. Usually, the cutaneous muscle should stick to the coverslip solely by adhesion forces, automatically expelling remaining air. After closure of the two chamber frames, small, persisting air bubbles may be carefully "pushed out" of the chamber from behind through the central window of the first titanium frame with a dry Q-tip. If you fail to remove all air bubbles, open the slit between both titanium frames for some millimeters, insert a few drops of saline between the coverslip and the cutaneous muscle using a 26-gauge needle to drive out remaining air bubbles and close the chamber again.

25. The first nut should be screwed on the apical screw. Make sure not to tighten the nuts too much. This could result in local skin necrosis or deficient blood flow to and from the skin being part of the chamber preparation.

26. Make at least 6–8 knots in each of the four holding sutures since the mice sometimes try to chew through sutures.

27. It will take a maximum of 1–2 d for the mouse to get completely accustomed to its new "knapsack." To allow the mouse to eat easily during these first days after surgery, some food may be put directly on the floor of the cage. After this time period, the mouse should show normal behavior again (e.g., cleaning itself, eating, drinking, sleeping, playing, and climbing around in the cage).

28. To avoid immune reactions between the recipient animal and the tumor, use either isografted mouse carcinomas or

immunedeficient mice as recipients (e.g., severe combined immunodeficient [SCID] mice).

29. Only fast-growing tumors are suitable for implantation since the mice must be sacrificed (on average) less than 30 d after the initial chamber implantation. Stimulated by the weight of the chamber, new skin will grow and lead to a lateral tipping over of the chamber preparation, causing reduced blood flow to and from the skinfold sandwiched between the two titanium frames. As a rule of thumb, solid tumors reaching visible size within 1–3 wk after subcutaneous implantation may be suitable for implantation into the chamber preparation.

30. Daily weight monitoring may help to appraise the general state of health of the animal. After an initial loss of weight (less than 10%), mice should stabilize again within the first 48 h after surgery. When bearing a tumor, further loss of weight might be observed in these animals with increasing tumor volume over time.

31. To avoid warming of the tumor chunks before implantation in different animals, put the Petri dish on ice from time to time.

32. Since it may be stressful for mice to be inserted into a tube during experiments, it is recommended to leave a tube in their cage 1 or 2 wk prior to the chamber implantation so that they become accustomed to it.

33. To avoid drying of the cutaneous muscle and to facilitate air-bubble-free reclosure of the chamber, moisten it with few drops of saline.

34. Be sure that the tumor chunk implanted does not move away from its position in the center of the chamber.

35. The growing tumor may sometimes provoke an itching stimulus at the site of implantation. Using a tape may prevent injuries to the skin and implant of the mice caused by scratching.

36. Avoid direct contact between the tape and the skin of the mouse.

References

1.. Sandison, J. C. (1924) A new method for the microscopic study of living growing tissues by the introduction of a transparent chamber in the rabbit's ear. *Anat Rec* 28, 281–287.

2. Menger, M. D., Lehr, H. A. (1993) Scope and perspectives of intravital microscopy—bridge over from *in vitro* to *in vivo*. *Immunol Today* 14, 519–522.

3. Leunig, M., Messmer, K. (1995) Intravital microscopy in tumor biology: current status

and future perspectives [review]. *Int J Oncol* 6, 413–417.

4. Jain, R. K., Schlenger, K., Höckel, M., Yuan, F. (1997) Quantitative angiogenesis assays: progress and problems. *Nat Med* 3, 1203–1208.

5. Leunig, M., Yuan, F., Menger, M. D .et al ., (1992) Angiogenesis, microvascular architecture, microhemodynamics, and interstitial fluid pressure during early growth of

human adenocarcinoma LS174T in SCID mice. *Cancer Res* 52, 6553–6560.

6. Leunig, M., Yuan, F., Berk, D. A., Gerweck, L. E., Jain, R. K. (1994) Angiogenesis and growth of isografted bone: quantitative *in vivo* assay in nude mice. *Lab Invest* 71, 300–307.

7. Endrich, B., Asaishi, K., Goetz, A. E., Messmer, K. (1980) Technical report: a new chamber technique for microvascular studies in unanesthetized hamsters. *Res Exp Med* 177, 125–134.

8. Sckell, A., Leunig, M., Fraitzl, C. R., Ganz, R., Ballmer, F. T. (1999) The connective-tissue envelope in revascularization of patellar tendon grafts. *J Bone Joint Surg (Br)* 81-B, 915–920.

9. Leunig, M., Demhartner, T. J., Sckell, A . et al., (1999) Quantitative assessment of angiogenesis and osteogenesis after transplantation of bone: comparison of isograft and allograft bone in mice. *Acta Orthop Scand* 70, 374–380.

10. Vajkoczy, P., Menger, M. D., Simpson, E., Messmer, K. (1995) Angiogenesis and vascularization of murine pancreatic islet isografts. *Transplantation* 60, 123–127.

11. Sewell, I. A. (1966) Studies of the microcirculation using transparent tissue observation chambers inserted in the hamster cheek pouch. *J Anat* 100, 839–856.

Chapter 20

The Corneal Pocket Assay

Marina Ziche and Lucia Morbidelli

Abstract

Continuous monitoring of neovascular growth *in vivo* is required for the development and evaluation of drugs acting as suppressors or stimulators of angiogenesis. The cornea assay consists of the placement of an angiogenesis stimulus (tumor tissue, cell suspension, growth factor) into a micropocket produced in the cornea thickness to evoke vascular outgrowth from the peripherally located limbal vasculature. Neovascular development and progression can be modified by the presence of locally released or applied inhibitory factors or by systemically given antiangiogenic drugs. This assay has the advantage over other *in vivo* assays of measuring new blood vessels only since the cornea is initially avascular. The experimental details of the avascular cornea assay and its advantages and disadvantages in different species are discussed.

Key words: Angiogenesis, capillary, endothelial cell, fibroblast growth factor, immunohistochemistry, vascular endothelial growth factor.

1. Introduction

Continuous monitoring of angiogenesis *in vivo* is required for the development and evaluation of drugs acting as suppressors or stimulators of angiogenesis. In this respect, there are concerted efforts to provide an animal model for more quantitative analysis of *in vivo* angiogenesis *(1)*. The cornea assay consists of the placement of an angiogenesis inducer (tumor tissue, cell suspension, growth factor) into a micropocket produced in the cornea thickness to evoke vascular outgrowth from the peripherally located limbal vasculature. This assay has the advantage over other *in vivo* assays of measuring only new blood vessels since the cornea is initially avascular.

The corneal assay performed in New Zealand white rabbits was first described by Gimbrone et al. *(2)*. Modifications of the

S. Martin and C. Murray (eds.), *Methods in Molecular Biology, Angiogenesis Protocols, Second edition, Vol. 467*
© Humana Press, a part of Springer Science+ Business Media, LLC 2009
DOI: 10.1007/978-1-59745-241-0_20

original method have been reported by our group that allow the implanting of multiple samples, including cell suspensions and tissue fragments. It was chosen for the absence of a vascular pattern and for the easy manipulation and monitoring of the neovascular growth. This technique, extensively used during the years, has been substantially modified to fulfil different experimental requirements: characterization of angiogenesis inducers; evaluation of angiogenesis inhibitors; interaction between different factors; study of the cellular, biochemical, and molecular mechanisms of angiogenesis.

The experimental details of the avascular cornea assay and its advantages and disadvantages in different species (rabbit, mouse, rat) are discussed.

2. Materials

2.1. Rabbit Cornea Assay

The angiogenesis cornea assay is performed in albino rabbits (*see* **Note 1**) and requires the simultaneous presence of two operators (*see* **Note 2**).

2.1. Reagents, Tissues

1. Animals: New Zealand albino rabbits (Charles River, www.criver.com (1.5–2 kg (see **Note 3**).
2. Cells: Cell suspension (e.g., tumor cells) in medium with 10% foetal calf serum at a dilution of $2–5 \times 10^5$ cells/5 µL.
3. Tissue samples: Fresh tissue fragments (2–3 mg each) isolated from surgery specimens removed within 2 h from patients or animals and kept at 4°C in medium.
4. Recombinant growth factors or drugs in water or phosphate-buffered saline (PBS) or ethanol or methanol in highly concentrated solutions (0.1–1 mg/mL) (see **Note 4**).
5. Ethylene-vinyl-acetate copolymer (Elvax-40; DuPont de Nemours, Wilmington, DE, www.dupont.com) (*see* **Note 5**). Elvax-40 preparation and testing: Weigh 1 g Elvax-40, wash it in absolute alcohol for 100-fold at 37°C, and dissolve in 10 mL of methylene-chloride to prepare 10% casting stock solution. Leave Elvax-40 in methylene-chloride at 37°C for 30–60 min to speed up solubilization. Test the Elvax-40 preparation for its biocompatibility *(3)*. The casting solution is eligible for use if no implant performed with this preparation induces the slightest histological reaction in the rabbit cornea.
6. Sodium pentothal in saline solution (0.1 g/mL).
7. Benoxinate 0.4%.

8. Fixative: 4% paraformaldehyde in PBS, pH 7.4.

9. Isopentane.

10. Liquid nitrogen.

11. OCT tissue-teck medium or similar.

12. Acetone.

13. Haematoxylin and eosin.

14. Phosphate-buffered saline.

15. Hydrogen peroxide in PBS.

16. Bovine serum albumin (BSA).

17. Primary antibodies: For markers of neovascularization (anti-CD31 Ab, Dako); inflammation (anti-RAM11 Ab, Dako); and adhesion molecule (anti-α5β1 integrin Ab, Chemicon).

18. Goat antimouse immunoglobulin G (IgG; Sigma).

19. Mouse peroxidase antiperoxidase (PAP, Sigma).

20. 3,3′-Diaminobenzidine tetrahydrochloride (DAB, Sigma).

21. Aquatex medium (Merck).

2.2. Facilities and Equipment

1. Cell culture facility equipped with vertical laminal flow hood and autoclave.

2. Animal facility equipped with a sterile surgical room.

3. Disposable scalpel for ocular microsurgery (no. 10/11, Aesculap).

4. Sterile forceps, silver spatula, microsurgery scissors, Teflon plate (10 × 10 cm), microspatula.

5. 6-cm glass Petri dishes.

6. Latex dental dam for endodontic procedures (DentalTrey, www.dentaltrey.com).

7. Insulin syringes.

8. Slit-lamp stereomicroscope equipped with a digital camera.

9. Cryostat.

10. Slides and glassware for histology.

11. Microscope equipped with a digital camera.

3. Methods

3.1. Rabbit Cornea Assay

The material under test can be in the form of slow-release pellets incorporating recombinant growth factors, cell suspensions, or tissue samples.

3.1.1. Sample Preparation

1. Preparation of slow-release pellets: Recombinant growth factors are prepared as slow-release pellets by incorporating the substance under test in Elvax-40. For testing, a predetermined volume of Elvax-40 casting solution is mixed with a given amount of the compound to be tested and previously dried on a flat Teflon surface. The polymer and the compound are homogeneously mixed under a laminar flow hood. After drying, the film sequestering the compound is cut into 1 × 1 × 0.5 mm pieces. Empty pellets of Elvax-40 are used as negative controls, while vascular endothelial growth factor (VEGF) or fibroblast growth factor 2 (FGF-2)-containing pellets are considered positive controls (see **Note 6**).

2. Prepare a cell suspension by trypsinization of confluent cell monolayers or concentrated cell suspensions to a final dilution of $2–5 \times 10^5$ cells in 5 µL, depending on cell volume and cell type.

3. When tissue samples are tested, samples of 2–3 mg are obtained by cutting the fresh tissue fragments under sterile conditions.

3.1.2 Surgical Procedure

1. Anaesthetise animals with sodium pentothal (30 mg/kg iv) (see **Note 7**).

2. Isolate the eye with a dental dam. Wash with a few drops of local anaesthetic (0.4% benoxinate)

3. Under aseptic conditions, produce a micropocket (1.5 × 3 mm) using a pliable silver spatula 1.5-mm wide in the lower half of the cornea (*see* **Note 8**). A small amount (20–50 µL) of the aqueous humour can be drained from the anterior chamber with an insulin syringe when reduced corneal tension is required (i.e., for cell or tissue samples).

4. Locate the implant at 2.5–3 mm from the limbus to avoid false positives due to the mechanical procedure and to allow the diffusion of test substances in the tissue, with the formation of a gradient for the endothelial cells of the limbal vessels (**Figs. 20.1** and **20.2**). Implants sequestering the test material and the controls are coded and implanted in a double-masked manner.

5. By using a micropipette, introduce in each corneal pocket 5 µL containing $2–5 \times 10^5$ cells in medium supplemented with 10% serum in the corneal micropocket. When the overexpression of growth factors/inhibitors by stable transfection of specific complementary DNA (cDNA) is studied, one eye is implanted with transfected cells and the other with the wild-type or vector-transduced cell line. Suitable cell lines for these experiments are mammary carcinoma (MCF-7) cells, Chinese hamster ovary (CHO) cells, or lymphoma Burkitt's cells (DG75) *(4–6)*. It might be necessary to evaluate

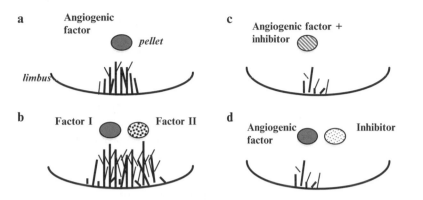

Fig. 20.1. Schematic representation of different experimental settings performed in the corneal micropocket assay. (**a**) Characterization of the angiogenic activity of a purified factor released in the corneal stroma from a slow-release device. (**b**)–(**d**) Study of the interaction among different factors. Two adjacent micropockets can be surgically produced in the same eye and the synergistic effect of two co-released factors (**b**) or inhibition of angiogenesis (**d**) monitored. Otherwise, the angiogenesis inducer and inhibitor can be incorporated in the same pellet (**c**).

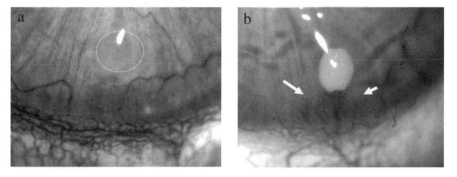

Fig 20.2. Examples of a negative (**a**) and positive (**b**) angiogenic responses. Pictures of an Elvax-40 empty pellet implant (**a**) and VEGF (200 ng/pellet)-induced angiogenesis were taken at d 7. The newly formed vessels (indicated by arrows) start from the limbal vasculature and progress toward the implanted stimulus.

the angiogenic potential of drug-treated cells. In these experiments, cell monolayers are treated before the implant (18–24 h). One eye is implanted with treated cells and the contralateral with control cells.

6. Tissue fragments are inserted in the corneal micropocket with small forceps. The angiogenic activity of tumor samples is compared with macroscopically healthy tissue *(7)*.

7. When drug solutions incompatible with Elvax polymerisation and genes transduced by viral vectors have to be locally tested, microinjection of concentrated solutions is performed using insulin syringes equipped with 30-gauge needles. After the removal of aqueous humour, a 10-µL volume is injected within the corneal stroma in the space between the limbus and the pellet implant.

3.1.3. Quantification of Angiogenesis

1. Subsequent daily observation of the implants is made with a slit-lamp stereomicroscope without anaesthesia. Angiogenesis, edema, and cellular infiltrate are recorded, and images are taken.

2. An angiogenic response is scored positive when budding of vessels from the limbal plexus occurs after 3–4 d and capillaries progress to reach the implanted pellet in 7–10 d (**Fig. 20.2**). Implants that fail to produce a neovascular growth within 10 d are considered negative, while implants showing an inflammatory reaction are discarded.

3. During each observation, the number of positive implants over the total implants performed is scored.

4. The potency of angiogenic activity is evaluated on the basis of the number and growth rate of newly formed capillaries, and an angiogenic score is calculated by the following formula: Vessel density × Distance from limbus *(4, 8)*. A density value of 1 corresponds to 0–25 vessels per cornea, 2 from 25 to 50, 3 from 50 to 75, 4 from 75 to 100 and 5 for more than 100 vessels. The distance from the limbus is graded with the aid of an ocular grid (**Fig. 20.3**).

3.1.4. Histological Examination and Immunohistochemical Analysis

Depending on the experimental design, histological or immunohistochemical analysis of corneal sections can be performed at fixed times during angiogenesis progression or at the end of the observations *(4)*.

1. Animals are sacrificed with a bolus of anaesthetic drug at d 15.

2. The corneas are removed, oriented, and marked (*see* **Note 9** and **Fig. 20.4**), immediately frozen in isopentane cooled in liquid nitrogen for 10 s, and stored at –80°C in OCT tissue-teck medium. If required, the cornea can be fixed in paraformaldehyde.

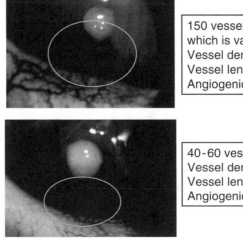

150 vessels to the pellet which is vascularized:
Vessel density: 5
Vessel length: 5
Angiogenic score=25

40-60 vessels, 2,5 mm:
Vessel density: 2
Vessel length: 3
Angiogenic score=6

Fig. 20.3. Examples of angiogenic score evaluation.

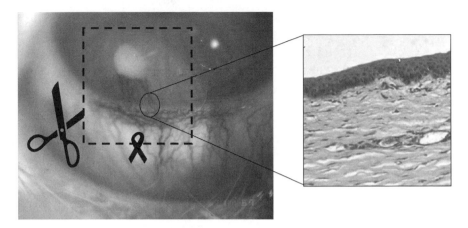

Fig 20.4. Schematic representation of cornea sampling and orientation for histological analysis. An example of the histological analysis of a positive angiogenic implant is reported in the right inset. Note the presence of newly formed vessels containing red blood cells within the corneal stroma.

3. Cryostat sections 7-μm thick are stained with haematoxylin and eosin, and adjacent sections are used for immunohistochemical staining. After fixation in absolute acetone at –20°C for 5 min, sections are washed in PBS and then treated with 1.5% hydrogen peroxide in PBS for 8 min to perform quenching of endogenous peroxidases. Aspecific binding sites are then blocked in 3% BSA in PBS for 45 min. Sections are incubated overnight with the primary antibodies diluted in 0.5% BSA in PBS.

4. Primary antibodies can be anti-CD31 Ab (Dako, 200 μg/mL) (marker of neovascularization), anti-RAM11 Ab (Dako, 1.2 μg/mL) (marker of inflammation), or anti-α5β1 integrin Ab (Chemicon, 1:50) (adhesion molecule expressed in epithelial and endothelial cells). For colocalization, serial and adjacent sections can be labelled with different antibodies.

5. Sections are extensively washed in 0.5% BSA in PBS and then incubated in goat antimouse IgG (Sigma, 1:40) for 1 h. After washing in 0.5% BSA in PBS, sections are incubated in mouse PAP (Sigma 1:35) for 45 min. Immunoreaction is developed in DAB for 8 min. Sections are then extensively rinsed in distilled water, counterstained in haematoxylin, and mounted in Aquatex medium (Merck).

6. Sections are observed at the microscope (at ×*10*–40), and digital images are taken.

3.2. Mouse Corneal Micropocket

The mouse cornea micropocket assay was first described by Muthukkaruppan and Auerbach *(9)*.

1. Anaesthetise animals with methoxyflurane.

2. Make corneal micropockets in both eyes, reaching within 1 mm of the limbus, and pellets containing substances to be tested coated with Hydron (Interferon Science, New Brunswick, NJ) are implanted.

3. Use Hydron as a casting solution (12% w/v) solution, prepared by dissolving the polymer in absolute alcohol at 37°C *(3)*. When peptides are tested, sucralfate (sucrose aluminium sulphate, Bukh Meditec, Copenhagen, Denmark) is added to stabilize the molecule and to slow its release from Hydron *(10, 11)*.

4. The vascular response, measured as the maximal vessel length and number of clock hours of neovascularization, is scored at fixed time intervals (usually on postoperative d 5 and 7) using slit-lamp biomicroscopy and photographed. To quantify the section of the cornea in which new vessels are sprouting from the preexisting limbal vessels, the circumference of the cornea is divided into the equivalent of 12 clock hours. The number of clock hours of neovascularization for each eye is measured during each observation.

3.3. Rat Corneal Assay

1. Purified growth factors are combined 1:1 with Hydron as described by Polverini and Leibovich *(12)*.

2. Pellets are implanted 1–1.5 mm from the limbus of the cornea of anaesthetised rats (sodium pentobarbital, 30 mg/kg ip).

3. Neovascularization is assessed at fixed days (usually 3, 5, and 7 d): Animals are perfused with colloidal carbon solution to label vessels; eyes are enucleated and fixed in 10% neutral buffered formalin overnight. The following day, corneas are excised, flattened, and photographed. A positive neovascularization response is recorded only if sustained directional growth of capillary sprouts and hairpin loops toward the implant are observed. Negative responses are recorded either when no growth is observed or when only an occasional sprout or hairpin loop showing no evidence of sustained growth is detected.

3.4. Advantages and Disadvantages in Different Species

3.4.1. Species

The rabbit size (2–3 kg) allows easy manipulation of the animal; the eye may be easily extruded from its location and surgically manipulated.

Rabbit cornea has been avascular in all strains examined so far. In some strains of rats, the presence of preexisting vessels within the cornea and the development of keratitis are serious disadvantages. Furthermore, rabbits are more docile and amenable to handling and experimentation than mice and rats. In case of inflammatory reactions, these are easily detectable in rabbits by stereomicroscopic examination as corneal opacity.

3.4.2. Measurements

In mice and rats, it is possible to obtain time-point results. The evolution of the angiogenic response in the same animal is not recommended because each time the cornea is observed the animal has to be anaesthetised. Experiments are made with a large number of animals, and vessel growth during the time can be visualised by perfusion with colloidal carbon solution in individual animals. Multiple observations are easily performed in rabbits, thus reducing the number of animals required for statistical evaluation. The use of slit-lamp stereomicroscopy and of awake animals allows the observation of newly formed vessels during times with prolonged monitoring, up to 1–2 mo.

3.4.3. Different Experimental Procedures

In the rabbit eye, due to its wide area, stimuli in different forms can be placed. In particular, the activity of specific growth factors can be studied in the form of slow-release pellets *(6, 13–15)* and of tumor or nontumor cell lines stably transfected for the overexpression of angiogenic factors *(4, 6, 16)*. Cells with double transfection can also be studied *(6, 17)*. The modulation of the angiogenic responses by different stimuli can be assessed in the rabbit cornea assay *(1)* by implanting single pellets releasing both the angiogenic stimulus and the inhibitor *(18–20)*, *(2)* by implanting in the same cornea two pellets placed in parallel micropockets and releasing different molecules *(21, 22)*, and *(3)* through the removal or addition of multiple pellets *(21)* (**Fig. 20.1**). The implant of tumor samples from different locations can be performed both in corneal micropockets and in the anterior chamber of the eye to monitor angiogenesis produced by hormone-dependent tissues or tumors (i.e., human breast or ovary carcinoma in female rabbits), and it allows the detection of both the iris and the corneal neovascular growth *(7, 23)*.

3.4.4. Treatment with Drugs

The effect of local drug treatment on corneal neovascularization may be studied in the form of ocular drops or ointment *(24)* or microinjection in the corneal thickness. The effect of systemic drug treatment on corneal angiogenesis may also be evaluated *(4, 7, 8, 25)*. However, when considering the size of the animals, systemic drug treatment in rabbits requires a higher amount of drugs than for smaller animals.

Interestingly, the use of nude mice allows the study of angiogenesis modulation in response to effectors produced and released by tumors or tumor cell lines of human origin growing subcutaneously. Treatment of mice with antiangiogenic or antitumor drugs allows the simultaneous measurement of tumor growth and metastasis and corneal angiogenesis.

4. Notes

1. Cornea has been avascular in all strains examined so far. In albino rabbits, the newly formed vessels are more visible on the faint pink background of the iris.

2. Operator skill for pellet manipulation, surgery, and monitoring of angiogenesis is required.

3. Body weight in the range of 1.8 to 2.5 kg allows for easy handling and prompt recovery from anaesthesia.

4. Sterility of materials and procedures is crucial to avoid non-specific responses. DMSO (dimethylsulphoxide) should be avoided since it is incompatible with Elvax-40 polymerization and handling.

5. Polyvinyl alcohol and Hydron can be used instead of Elvax-40. In our experience, a polymer of hydroxyethyl-methacrylate gave less-satisfactory results than Elvax-40.

6. Variability among growth factors in inducing angiogenesis has been found considering different angiogenic factors, different providers, and batch of preparation. Usually, the dose of VEGF or FGF-2 able to give a positive angiogenic response varies in the range 200–400 ng/pellet.

7. Immobilisation during the anaesthetic procedure and observation is important to avoid self-induced injury.

8. Make the cut in the cornea in correspondence with the pupil and orient the micropocket toward the lower eyelid for easy daily observation. When two factors are tested separately, make two independent and parallel micropockets (**Fig. 20.1**).

9. Before embedding in OCT tissue-teck medium, pellets should be removed and corneas cut and marked (i.e., with a cotton thread) for orientation at the cryostat (**Fig. 20.4**).

Acknowledgements

We would like to thank Dr. Raffaella Solito for technical assistance. The work was supported by the Italian Ministry of University (MIUR), the Italian Association for Cancer Research (AIRC), and the University of Siena.

References

1. Jain, R. K., Schlenger, K., Hockel, M., Yuan, F. (1997) Quantitative angiogenesis assays: progress and problems. *Nat Med* 3, 1203–1208.

2. Gimbrone, M., Jr., Cotran, R., Leapman, S. B., Folkman, J. (1974) Tumor growth and neovascularization: an experimental model using the rabbit cornea. *J Natl Cancer Inst* 52, 413–427.

3. Langer, R., Folkman. J. (1976) Polymers for the sustained release of proteins and other macromolecules. *Nature* 363, 797–800.

4. Ziche, M., Morbidelli, L., Choudhuri, R., et al. (1997) Nitric oxide-synthase lies downstream of vascular endothelial growth factor but not basic fibroblast growth factor induced angiogenesis. *J Clin Invest* 99, 2625–2634.

5. Marconcini, L., Marchio, S., Morbidelli, L., et al. (1999) c-fos-induced growth factor/vascular endothelial growth factor D induces angiogenesis *in vivo* and *in vitro*. *Proc Natl Acad Sci U S A* 96(17), 9671–9676.

6. Cervenak, L., Morbidelli, L., Donati, D., et al. (2000) Abolished angiogenicity and tumorigenicity of Burkitt lymphoma by interleukin-10. *Blood* 96, 2568–2573.

7. Gallo, O., Masini, E., Morbidelli, L., et al. (1998) Role of nitric oxide in angiogenesis and tumor progression in head and neck cancer. *J Natl Cancer Inst* 90, 587–596.

8. Ziche, M., Morbidelli, L., Masini, E., et al. (1994) Nitric oxide mediates angiogenesis *in vivo* and endothelial cell growth and migration *in vitro* promoted by substance P. *J Clin Invest* 94, 2036–2044.

9.. Muthukkaruppan, V., Auerbach, R. (1979) Angiogenesis in the mouse cornea. *Science* 206, 1416–1418.

10. Chen, C., Parangi, S., Tolentino, M.T., Folkman, J. (1995) A strategy to discover circulating angiogenesis inhibitors generated by human tumors. *Cancer Res* 55, 4230—4233.

11. Voest, E. E., Kenyon, B. M., O'Really, M. S., Truitt, G., D'Amato, R. J., Folkman, J. (1995) Inhibition of angiogenesis *in vivo* by interleukin 12. *J Natl Cancer Inst* 87, 581—586.

12. Polverini, P. J., Leibovich, S. J. (1984) Induction of neovascularization *in vivo* and endothelial cell proliferation *in vitro* by tumor associated macrophages. *Lab Invest* 51, 635–642.

13. Ziche, M., Jones, J., Gullino, P. M. (1982) Role of prostaglandin E1 and copper in angiogenesis. *J Natl Cancer Inst* 69, 475–482.

14. Taraboletti, G., Morbidelli, L., Donnini, S., et al. (2000) The heparin binding 25 kDa fragment of thrombospondin-1 promotes angiogenesis and modulates gelatinases and TIMP-2 in endothelial cells. *FASEB J* 14, 1674–1676.

15. Parenti, A., Morbidelli, L., Ledda, F., Granger, H. J., Ziche, M. (2001) The bradykinin/B1 receptor promotes angiogenesis by upregulation of endogenous FGF-2 in endothelium via the nitric oxide synthase pathway. *FASEB J* 15(8), 1487–1489.

16. Lasagna, N., Fantappiè, O., Solazzo, M., et al. (2006) Hepatocyte growth factor and inducible nitric oxide synthase are involved in multidrug resistance-induced angiogenesis in hepatocellular carcinoma cell lines. *Cancer Res* 66(5), 2673–2682.

17. Woolard, J., Wang, W. Y., Bevan, H. S., et al. (2004) VEGF165b, an inhibitory vascular endothelial growth factor splice variant: mechanism of action, *in vivo* effect on angiogenesis and endogenous protein expression. *Cancer Res* 64(21), 7822–7835.

18. Morbidelli, L., Donnini, S., Chillemi, F., Giochetti, A., Ziche, M. (2003) Angiosuppressive and angiostimulatory effects exerted by synthetic partial sequences of endostatin. *Clin Cancer Res* 9(14), 5358–5369.

19. Bagli, E., Stefaniotou, M., Morbidelli, L., et al. (2004) Luteolin inhibits vascular endothelial growth factor-induced angiogenesis; inhibition of endothelial cell survival and proliferation by targeting phosphatidylinositol 3 -kinase activity. *Cancer Res* 64(21), 7936–7946.

20. Donnini, S., Finetti, F., Lusini, L., et al. (2006) Divergent effects of quercetin conjugates on angiogenesis. *Br J Nutr* 95(5), 1016–1023.

21. Ziche, M., Alessandri, G., Gullino, P. M. (1989) Gangliosides promote the angiogenic response. *Lab Invest* 61, 629–634.

22. Cantara, S., Donnini, S., Morbidelli, L., et al. (2004) Physiological levels of amyloid peptides stimulate the angiogenic response through FGF-2. *FASEB J* 18(15), 1943–1945.

23. Federman, J. L., Brown, G. C., Felberg, N. T., Felton, S. M. (1980) Experimental ocular angiogenesis. *Am J Ophthalmol* 89(2), 231–237.

24. Presta, M., Rusnati, M., Belleri, M., Morbidelli, L., Ziche, M., Ribatti, D. (1999) Purine analog 6-methylmercaptopurine ribose inhibits early and late phases of the angiogenesis process. *Cancer Res* 59(10), 2417–2424.

25. Ziche, M., Donnini, S., Morbidelli, L., Parenti, A., Gasparini, G., Ledda, F. (1998) Linomide blocks angiogenesis by breast carcinoma vascular endothelial growth factor transfectants *Br J Cancer* 77(7), 1123–1129.

Chapter 21

Use of the Hollow Fibre Assay for Studies of Tumor Neovasculature

Steven D. Shnyder

Abstract

In vivo preclinical assays are required to screen potential agents that target the tumor vasculature. Here, a hollow fibre-based assay for the quantification of neovasculature in the presence or absence of an agent that potentially targets tumor neovasculature is described. The neovasculature is developed as a consequence of the presence of tumor cells encapsulated in hollow fibres, which are transplanted subcutaneously in the dorsal flanks of mice.

Keywords: Angiogenesis, cancer therapy, CD-31, hollow fibre assay, preclinical screening, vascular-targeting agents.

1. Introduction

The progression of novel agents that target tumor vasculature into the clinic is reliant on them demonstrating activity in suitable preclinical assays. With these assays, the analytical end point is either the detection of new vessel growth in an area where there is minimal vascularisation (e.g., Matrigel or sponge assays; *1, 2*), or the differentiation of neovasculature from existing vasculature (e.g., dorsal skin chamber; *3*). *In vivo* assays currently used in this field have distinct features and limitations (as reviewed in **ref.** *4*), and thus it is not ideal when screening to rely on just one assay. Hence, there is a requirement for further development of novel screening assays.

S. Martin and C. Murray (eds.), *Methods in Molecular Biology, Angiogenesis Protocols, Second edition, Vol. 467*
© Humana Press, a part of Springer Science+Business Media, LLC 2009
DOI: 10.1007/978-1-59745-241-0_21

As the basis for developing such a novel *in vivo* screening assay, we have adapted the hollow fibre assay that is currently used as part of the drug-screening programme at the National Cancer Institute (NCI) in the United States *(5)* for studies on vasculature. In the NCI assay, human tumor cell lines are loaded into biocompatible polyvinylidene fluoride (PVDF) hollow fibres with an internal diameter of 1 mm and a 500-kDa molecular weight exclusion, which allows for free passage of macromolecules and drugs but restricts passage of cells. The hollow fibres are cut into 2-cm lengths, which are heat sealed at both ends; using aseptic surgical procedures, these are then transplanted subcutaneously and intraperitoneally into anaesthetised mice. Animals are treated with the test agent for 4 consecutive days from d 3 and the studies terminated on d 7 or 10, with cells then analysed for viability using the 3-(4,5-Dimethylthiazol-2-yl)-2,5-diphenylterazolium bromide (MTT) cell viability assay, with the amount of viable cells in hollow fibres from the treated animals compared with untreated controls.

The rationale for adapting this assay is from initial studies carried out in this laboratory in which assay times were extended beyond the standard 6-d period, and extensive vascular networks were observed surrounding subcutaneously implanted fibres *(6)*.

We have subsequently modified the NCI assay to make it more suitable for investigating antiangiogenic and other vascular-targeting strategies by only transplanting subcutaneously, and reducing the size and number of fibres transplanted to minimise inflammation surrounding the fibres due to damage inflicted by the fibre seals on the surrounding tissue *(7)*. We have also incorporated immunohistochemical analysis of the neovascularisation surrounding the fibres as an end point for quantifying agent effect *(7, 8)*.

2. Materials

2.1. Hollow Fibre Preparation and Loading

1. PVDF hollow fibres with an internal diameter of 1.0 mm and a molecular weight cut-off point of 500 kDa (Spectrum Medical, Houston, TX, USA): These are supplied dehydrated in 180-mm lengths and in different colors to enable different cell lines to be placed in the same animal.

2. 70% ethanol solution in distilled water.

3. Sterile 5- and 10-mL syringes with 21-gauge blunt needles.

4. 600-mm glass chromatography column or similar.

5. Stainless steel tray (approximate dimensions 220 × 150 mm).

6. Autoclavable stainless steel box and lid (approximate dimensions 250 × 50 × 40 mm).

7. Sterile distilled water.

8. Autoclave tape.

9. Autoclavable bags for sterilising box containing fibres and instruments.

10. Tumor cell lines obtained from the European Collection of Cell Cultures (Porton Down, UK) or LGC Promochem (Teddington, UK) (*see* **Note 1**).

11. Tissue culture medium, such as RPMI 1640 supplemented with 10% or 20% foetal calf serum (FCS), 2 mM sodium pyruvate, and 2 mM L-glutamine (all from Sigma, Poole, UK), stored at 4°C.

12. Cell culture consumables, including 75-cm^2 culture flasks, 10-mL pipettes, six-well plates (all Corning), 30-mL polypropylene tubes (Sarstedt).

13. Hank's balanced salt solution (HBSS) (Sigma), stored at room temperature.

14. 0.25% trypsin-EDTA (ethylenediaminetetraacetic acid) solution (Sigma), stored at –20°C.

15. Haemocytometer.

16. Polystyrene box containing ice.

17. "Fireboy plus" rechargable mobile Bunsen burner (Integra Biosciences, Chur, Switzerland) with CV360 butane cylinder.

18. Sterilised demarcated stainless steel tray (approximate dimensions 450 × 140 mm) with demarcations every 1.5 cm.

19. Sterilised instrument tray (approximate dimensions 220 × 150 mm) and instruments, including smooth-jawed needle holders, scissors, and forceps.

2.2. In Vivo Transplantation of Hollow Fibres and Treatment

1. Mice (*see* **Note 2**).

2. Polystyrene box to transport fibres to operating theatre.

3. Anaesthesia equipment.

4. Sterile instruments: scissors, two forceps, 3-mm diameter trocar.

5. Histoacryl tissue adhesive (Braun, Sheffield, UK). Store at 4°C.

6. Autoclips (IMS, Edenbridge, UK).

7. Sterile 1-mL syringe plus either a 25-gauge needle for intraperitoneal or a 26-gauge needle for intravenous administration of compound.

2.3. Removal and Processing of Hollow Fibres

1. CO_2 euthanasia system.

2. Dissecting board with pins.

3. Instruments: scissors and forceps.

4. Fuji Finepix digital camera and stand.

5. Stiff card cut to 22 × 22 mm.

6. Processing cassettes, 28 × 32 mm.

7. Zinc fixative (BD Pharmingen, San Diego, CA, USA). Store at room temperature.

8. 50-mL polystyrene sample pots (Greiner Bio-One, Stonehouse, UK).

9. Graded ethanols: 70%, 90%, and 100%.

10. Histoclear.

11. Paraffin wax.

2.4. Assessment of Effects on the Vasculature

1. Xylene.

2. Distilled water.

3. 50-mL Coplin jars.

4. 1% hydrogen peroxide solution: Make immediately before use by adding 3.33 mL of 30% hydrogen peroxide solution (stored at 4°C) to 96.67mL of distilled water.

5. Phosphate-buffered saline (PBS), 0.01M, pH 7.4: Make by dissolving a sachet of powder (Sigma) in 1 L distilled water. Store at room temperature.

6. Wax pen (Zymed Laboratories, San Francisco, CA, USA).

7. Normal rabbit serum (NRS) (Vector Laboratories, Burlingame, CA, USA). Make by adding 150 µL of stock solution to 9.95 mL of PBS. Make up fresh before every run-through.

8. Purified rat anti-mouse CD-31 (PECAM-1, platelet endothelial cell adhesion molecule-1) monoclonal antibody (BD Pharmingen). Store stock at 4°C. Dilute 1:100 in normal rabbit serum prior to use.

9. Secondary antibody: Biotinylated rabbit anti-rat immunoglobulin G (IgG) (H+L) (Vector Laboratories). Store stock at 4°C. Dilute 1:400 in PBS prior to use.

10. ABC-peroxidase reagent kit (Vector Laboratories): Components provided in dropper bottles; store at 4°C. Make up 30 min before use by adding 1 drop of "bottle A" plus 1 drop of "bottle B" to 5 mL of PBS, mixing thoroughly between each addition.

11. Peroxidase substrate kit (Vector Laboratories): Components provided in dropper bottles; store at 4°C. Make up immediately before use by adding 2 drops of "buffer pH 7.5" plus 4 drops of "DAB (3,3-diaminobenzidine)", plus 2 drops of "hydrogen peroxide" to 5 mL of distilled water, mixing thoroughly between each addition.

12. Sheet of white paper.

13. 10% bleach in tap water solution for disposal of DAB solutions.

14. Harris's haematoxylin solution (Sigma). Store at room temperature.

15. Scott's tap water substitute: Make by dissolving 3.5 g of sodium bicarbonate and 20 g of magnesium sulphate in 1 L distilled water. Store at room temperature.

16. DPX mountant for microscopy (BDH).

17. 22 × 50 mm glass coverslips.

3. Methods

The evaluation of the effects of agents on the neovasculature surrounding hollow fibres loaded with tumor cells comprises four procedures: rehydration and sterilisation of the hollow fibres followed by loading of the tumor cells; transplantation of the fibres in recipient mice and treatment with agents; removal of the fibres and processing for analysis; and the analysis itself.

3.1. Hollow Fibre Preparation and Loading

1. Ensure that any closed, cut ends of the fibre are opened by applying gentle pressure and flush each fibre through with 70% ethanol using the syringe attached to a blunt needle.

2. Place the fibres in a tray containing more 70% ethanol, and when all fibres are flushed through, transfer the fibres to the glass chromatography column and fill with more 70% ethanol so that the fibres are well covered; seal.

3. Store vertically for at least 72 h (*see* **Note 3**).

4. To sterilize the fibres, first transfer them to a tray.

5. Place 100 mL of distilled water into the autoclavable box.

6. Flush each fibre through with fresh 70% ethanol using the syringe and needle.

7. Flush each fibre through with distilled water, and with the syringe and needle still attached, place the fibre in the box filled with distilled water and flush to eliminate trapped air from the fibre. Remove the syringe and needle (*see* **Note 4**).

8. Once the required number of fibres have been rehydrated, place the lid on the box and secure with autoclave tape.

9. Carefully place the box in an autoclave bag, seal, label, and place into an autoclave.

10. Autoclave at 131°C for 5 min (*see* **Note 5**).

11. Once fibres have been sterilised, they can be stored at 4°C in their container until required.

12. Prepare conditioned medium by decanting the medium from one of the flasks of cells to be harvested into a 30-mL polypropylene tube and centrifuge at 700*g* for 10 min. Decant the supernatant into a second tube and label as conditioned medium.

13. Use monolayers of tumor cells approaching confluence in 75-cm² flasks. Wash twice in HBSS, followed by trypsinization in the trypsin/EDTA solution at 37°C for between 2 and 10 min (dependent on the cell line) and resuspend the cells in a mixture of 95% growth medium containing 20% FCS and 5% conditioned medium.

14. Count the cells using a haemocytometer and adjust to the required cell density using a mixture of 95% growth medium containing 20% FCS and 5% conditioned medium (*see* **Note 6**).

15. Store the cell suspension on ice while the class II cabinet is prepared for loading the fibres. Spray the cabinet with 70% ethanol and wipe. Lay out the equipment for loading the cells into the hollow fibres as suggested in **Fig. 21.1a**.

16. Position the Bunsen burner to the rear right-hand side of the cabinet (*see* **Note 7**). Place the demarcated tray at the front of the cabinet with the instrument tray to the right of it; the small steel tray is placed to the left of it.

17. Add 2 mL of growth medium containing 20% FCS to each well of a six-well plate, and also fill a 30-mL polypropylene tube with the medium; store both at 4°C.

18. Open the packaging of the syringe and needles without touching the inside of the package.

19. Bring the tube containing the medium into the cabinet and loosen the lid. Using the sterile 10-mL syringe, take up approximately 7–8 mL of medium, attach a sterile blunt needle, and place on the small steel tray.

20. Mix the tube containing the cell suspension by inversion and bring into the cabinet and loosen the lid. Using the sterile 5-mL syringe, take up approximately 3–4 mL of medium, attach a sterile blunt needle, and place on the instrument tray.

21. Place the box containing the sterilized fibres on the furthest left of the cabinet and open the lid.

22. Manipulate the fibre using one set of forceps and flush through with fresh medium onto the demarcated tray.

23. While holding the fibre with a second set of forceps, flush the fibre through with cell suspension onto the demarcated tray (*see* **Note 8**). Keep the syringe and needle attached to the fibre.

24. Heat the needle holders in the Bunsen flame for 3 s and clamp the loose end of the fibre to heat-seal (*see* **Fig. 21.1b**).

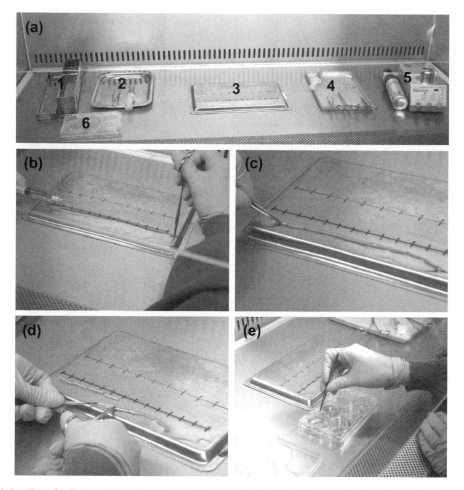

Fig. 21.1. Loading of cells into hollow fibres. (a) Suggested layout of equipment in the class II cabinet: *1* box containing sterilised fibres, *2* small tray for instruments in use, *3* demarcated tray, *4* instrument tray, *5* "Fireboy plus" rechargable Bunsen burner, *6* six-well plate containing medium. (b) Heat-sealing one end of the fibre after loading with cells. (c) Fibre following heat-sealing at 1.5-cm intervals along its length. (d) Dividing the fibre into segments by cutting across the heat-seals. (e) Placing the fibre segments into a six-well plate.

Dispense further cell suspension into the fibre until it is evidently full and heat-seal the needle end. Remove the needle and syringe and place on the instrument tray.

25. Lay the cell-filled fibre along the demarcated tray, lining up the first sealed end with one of the marks on the tray, and cover the fibre with medium. Heat-seal at 1.5-cm intervals (*see* **Fig. 21.1c**) and then, using scissors, cut across the seals and separate the segments (*see* **Fig. 21.1d**). For ease of handling, each segment should have a 2-mm tail (*see* **Note 9**).

26. Wash off any cells from the outside of the fibres by immersing the fibres in a series of wells of a six-well plate filled with medium.

27. Place the washed segments into fresh six-well plates and incubate at 37°C in a 5% CO_2 atmosphere overnight before proceeding to implantation.

3.2. In Vivo Transplantation of Hollow Fibres and Treatment

All animal procedures adhere to guidelines issued by the UK Co-ordinating Committee on Cancer Research *(9)*.

1. To minimise fluctuations in temperature during transport to the operating theatre, the six-well plates that hold the fibres are transported in an insulating polystyrene box.

2. Following induction general inhalation anaesthesia with 2% isofluorane in a 2% O_2 atmosphere, anaesthesia is maintained at the same rate using a nose cone, with the recipient mouse placed on its abdomen.

3. A small incision is made in the skin near to the left leg on the left dorsal flank, and a fibre is placed subcutaneously using a trocar. The incision is sealed with tissue glue, and an autoclip is applied (*see* **Note 10**). The procedure is repeated to load another fibre in the right flank, with the fibres lying as seen in **Fig. 21.2a**.

4. Mice are then transferred to a box containing a heating pad maintained at 37°C and are monitored until they have recovered from the anaesthesia. They are then returned to their cages and frequently monitored for any deleterious effects until the time of sacrifice.

3.3. Removal and Processing of Hollow Fibres

1. At various times following implantation (studies have been carried out up to 32 d), mice are sacrificed by CO_2 inhalation and are pinned out on a dissecting board with their dorsal flank exposed.

2. An incision is made in the skin near the base of the left fibre, which is then extended parallel with the fibre toward the centre of the body, and then up along the top of the fibre, such that there is at least a 5-mm margin of skin surrounding the fibre on three sides. The skin flap is then carefully pulled back and pinned to the dissecting board. The same procedure is repeated for the right fibre.

3. The dissection board is then placed on the photographic stand, and an image of the fibres *in situ* is captured using the digital camera (*see* **Fig. 21.2b**, **Fig. 21.2c**) to show the extent of the neovascularisation.

4. The skin flap surrounding the fibre is then totally dissected, laid flat on a piece of card, and placed into a histology processing cassette. This in turn is placed into a histology pot containing zinc fixative.

5. The samples are fixed overnight and then processed through a graded alcohol series and Histoclear before paraffin wax embedding using the automatic tissue processor.

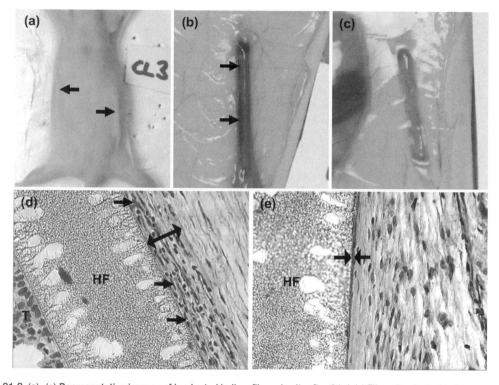

Fig. 21.2. (**a**)–(**c**) Representative images of implanted hollow fibres *in situ* after 21 d. (**a**) Fibres implanted subcutaneously in the right and left dorsal areas with no visible signs of inflammation (*arrows*). (**b**) Image of a fibre loaded with cells exposed *in situ* following dissection. The area directly adjacent to the fibre is heavily vascularised (*arrows*). (**c**) Similar image of an unloaded fibre showing lack of vascularisation surrounding the fibre in comparison. (**d**) and (**e**) Photomicrographs at 21 d of sections of tissue surrounding implanted fibres immunostained with anti-CD31 antibody. (**d**) Fibre (*HF*) loaded with MCF-7 mammary tumor cells (*T*) with large numbers of vascular profiles positive for CD31 (*arrows*) in a thick layer of granulation tissue (*double-headed arrow*). (**e**) Area next to an unloaded fibre (*HF*) with minimal granulation tissue (*arrows*) and negligible CD31 immunopositivity is seen.

3.4. Assessment of Effects on the Vasculature

All incubations in this section are carried out at room temperature unless otherwise stated, with 100 µL of solution applied to each section.

1. 5 µm thick sections of the fibre in longitudinal profile are cut using a microtome and collected onto 3-aminopropyltriethoxysilane (APES)-coated slides (*see* **Note 11**). Sections are then de-waxed in xylene and hydrated using a graded alcohol series and finally distilled water.

2. Slides are then placed in Coplin jars containing freshly made 1% H_2O_2 solution in distilled water for 30 min to block any endogenous peroxidase activity.

3. Following a wash in PBS for 5 min, slides are carefully blotted dry with a tissue, and a circle is drawn around the section using a wax pen, which ensures that solutions remain over the area of the section when applied and thus that a smaller volume of reagent can be used.

4. NRS is then applied as a blocking serum for 20 min, after which the excess liquid is blotted off by carefully tilting the slide and allowing the edge of a folded tissue to come into contact with the solution to remove the majority of it.

5. The α-CD31 primary antibody is then applied to the sections, and these are incubated for 90 min (*see* **Note 12**), followed by three washes of 5 min each in PBS.

6. The biotinylated rabbit α-rat IgG (H+L) secondary antibody is then applied for 30 min. At this stage, the ABC peroxidase reagent is made up in a 30-mL polypropylene tube.

7. After three washes of 5 min each in PBS, the ABC peroxidase reagent is applied for 30 min.

8. Slides are then washed a further three times in PBS. During the third wash, the DAB substrate is prepared.

9. The DAB substrate is then applied to the slides, which are placed on a piece of white paper and are closely observed until brown staining is evident (approximately 3 min). Once colour is observed, the reaction is stopped by gently running distilled water over the whole of the section using a wash bottle, with the waste collected into a 500-mL beaker containing 10% bleach solution (*see* **Note 13**).

10. The sections are then placed in a Coplin jar full of distilled water for 5 min.

11. Sections are then counterstained with Harris' haematoxylin for 5 min followed by washing in running tap water for 2 min, immersion in Scott's tap water substitute for 1 min, a further rinse in tap water for 30 s, and then dehydration through the graded series of alcohols and xylene before mounting the sections with DPX mounting medium and addition of a 22 × 50 mm coverslip.

12. Sections are then analysed by capturing digital images from a light microscope with a ×40 magnification objective lens, which is connected to a 3-CCD video camera, using image capture software (*see* **Fig. 21.2d, Fig. 21.2e**).

13. Using a 21 × 17 cm graticule overlayed on the image on the personal computer monitor, measurements are taken in the area toward the centre of the fibre, with the area of granulation tissue surrounding the fibre and the amount of CD-31 immunolabelling of the neovasculature recorded. Measure 20 fields for each section, with three sections are evaluated for each fibre.

14. A figure for the mean percentage density of neovasculature in the granulation tissue surrounding the fibre can then be calculated by dividing the mean number of CD31-positive intersects on a section by the mean area of the granulation tissue and multiplying by 100.

15. Statistical analysis of differences in neovascularisation surrounding treated and control fibres can then be carried out using a one-way analysis of variance (ANOVA).

4. Notes

1. Cell lines are chosen for their ability to stimulate angiogenesis (e.g., high vascular endothelial growth factor [VEGF] expression). This can be confirmed by enzyme-linked immunosorbent assay (ELISA) analysis of VEGF levels in *in vitro* hollow fibre experiments of the same duration as the *in vivo* experiments.

2. We have demonstrated that for the longer implantation times required for studying vascular effects, human tumor cell lines must be placed in immunocompromised nude mice, whereas if murine tumor cell lines are used, then they can be used with either immunocompetent or immunocompromised mice *(10)*.

3. Fibres can be stored like this indefinitely as long as they do not dry out. If this occurs, then the procedure must be repeated.

4. When the fibres are being rehydrated with water, the trapping of air bubbles may occur. To prevent this happening, ensure that the fibre is fully immersed in the autoclavable box before removing the needle.

5. Fibres melt at 143°C and should not be autoclaved above 131°C.

6. Cell density depends on the growth characteristics of the cell line to be used, but the typical range is for densities between 2×10^5 and 1×10^7 cells per millilitre.

7. Positions are for a right-handed person. Reverse if left-handed.

8. Keep the fibre as near to horizontal as possible while loading and sealing.

9. Seals must be properly formed or they may break open, resulting in leakage of cells. It is important to practice not only the sealing procedure, but also the segmentation, as both procedures require steady hands and good eyesight. It may be helpful to employ a magnifying lens during this stage.

10. Once the incision wound has healed, then the clips can be removed. This normally occurs around 7 d post-implantation.

11. Discard initial sections until the full lumen of the hollow fibre is exposed. Then, take serial sections at 200-µm intervals.

12. For the immunostaining procedure, control sections are included in the protocol where NRS is applied rather than

the primary antibody, and also in addition to omitting the primary antibody, PBS is then applied instead of the secondary antibody.

13. DAB is a known carcinogen; therefore, all unused DAB solution should be poured into 10% bleach, which can then be disposed of by pouring down the drain in the presence of a large excess of running tap water.

Acknowledgements

I would like to thank Tricia Cooper for her advice and practical input and Shofiq Al-Islam for his additional technical assistance. This work was supported by Cancer Research UK Programme grant C7589/A5953.

References

1. Passaniti, A., Taylor, R. M., Pili, R., et al. (1992) A simple, quantitative method for assessing angiogenesis and antiangiogenic agents using reconstituted basement membrane, heparin, and fibroblast growth factor. *Lab Invest* 67, 519–528.

2. Mahadevan, V., Hart, I. R., Lewis, G. P. (1989) Factors influencing blood supply in wound granuloma quantitated by a new *in vivo* technique. *Cancer Res* 49, 415–419.

3. Algire, G. H. (1945) An adaptation of the transparent chamber technique to the mouse. *J Natl Cancer Inst. U S A* 4, 1–11.

4. Hasan, J., Shnyder, S. D., Bibby, M., Double, J. A., Bicknel, R., Jayson, G. C. (2004) Quantitative angiogenesis assays *in vivo*—a review. *Angiogenesis* 7, 1–16.

5. Hollingshead, M. G., Alley, M. C., Camalier, R. F., et al. (1995) *In vivo* cultivation of tumor cells in hollow fibres. *Life Sci* 57, 131–141.

6. Phillips, R. M., Pearce, J., Loadman, P. M. et al. (1998) Angiogenesis in the hollow fibre tumor model influences drug delivery to tumor cells: implications for anticancer drug screening programs. *Cancer Res* 58, 5263–5266.

7. Shnyder, S. D., Hasan, J., Cooper, P. A., et al. (2005) Development of a modified hollow fibre assay for studying agents targeting the tumor neovasculature. *Anticancer Res* 25, 1889–1894.

8. Hasan, J., Shnyder, S. D., Clamp, A. R., et al. (2005) Heparin octasaccharides inhibit angiogenesis *in vivo*. *Clin Cancer Res* 11, 8172–8179.

9. Workman, P., Twentyman, P., Balkwill, F., et al. (1998) United Kingdom Co-ordinating Committee on Cancer Research (UKCCCR) Guidelines for the Welfare of Animals in Experimental Neoplasia (second edition). *Br J Cancer* 77, 1–10.

10. Shnyder, S. D., Cooper, P. A., Scally, A. J., Bibby, M. C. (2006) Reducing the cost of screening novel agents using the hollow fibre assay. *Anticancer Res* 26, 2049–2052.

Chapter 22

The Cranial Bone Window Model: Studying Angiogenesis of Primary and Secondary Bone Tumors by Intravital Microscopy

Axel Sckell and Frank M. Klenke

Abstract

The successful treatment of primary and secondary bone tumors in a huge number of cases remains one of the major unsolved challenges in modern medicine. Malignant primary bone tumor growth predominantly occurs in younger people, whereas older people predominantly suffer from secondary bone tumors since up to 85% of the most frequently occurring malignant solid tumors, such as lung, mammary, and prostate carcinomas, metastasize into the bone. It is well known that a tumor's course may be altered by its surrounding tissue. For this reason, reported here is the protocol for the surgical preparation of a cranial bone window in mice as well as the method to implant tumors in this bone window for further investigations of angiogenesis and other microcirculatory parameters in orthotopically growing primary or secondary bone tumors using intravital microscopy. Intravital microscopy represents an internationally accepted and sophisticated experimental method to study angiogenesis, microcirculation, and many other parameters in a wide variety of neoplastic and nonneoplastic tissues. Since most physiologic and pathophysiologic processes are active and dynamic events, one of the major strengths of chronic animal models using intravital microscopy is the possibility of monitoring the regions of interest *in vivo* continuously up to several weeks with high spatial and temporal resolution. In addition, after the termination of experiments, tissue samples can be excised easily and further examined by various *in vitro* methods such as histology, immunohistochemistry, and molecular biology.

Key words: Angiogenesis, bone tumor, cranial window, intravital microscopy, microcirculation.

1. Introduction

Intravital microscopy represents a sophisticated method to study angiogenesis, microcirculation, and many other parameters in a wide variety of neoplastic and nonneoplastic tissues (for reviews, *see* refs. 1–3). Since most physiologic and pathophysiologic processes

S. Martin and C. Murray (eds.), *Methods in Molecular Biology, Angiogenesis Protocols, Second edition, Vol. 467*
© Humana Press, a part of Springer Science+ Business Media, LLC 2009
DOI: 10.1007/978-1-59745-241-0_22

are active and dynamic events, one of the major strengths of chronic animal models using intravital microscopy is the possibility of monitoring the regions of interest *in vivo* continuously up to several weeks with high spatial and temporal resolution. In addition, after the termination of experiments, tissue samples can be excised easily and further examined by various in vitro methods such as histology, immunohistochemistry, and molecular biology.

Up to 85% of the most frequently occurring malignant solid tumors, such as lung and prostate carcinomas, metastasize into the bone, and thus, like primary solid malignant bone tumors, they still represent one of the great challenges of modern medicine due to their high mortality. A tumor's course may be altered by its surrounding tissue. For this reason, our research team developed a novel model for the investigation of orthotopically growing primary and secondary bone tumors using intravital microscopy *(4, 5)*. The model presented is a development of a preexisting model in which a similar cranial window preparation enabled observation of the brain tissue and implants onto the brain tissue *(6, 7)*.

The advantages of using mice as experimental animals are, for instance, the availability of a large number of different well-defined mouse strains, including transgenic or knockout mice, and the wide variety of commercially generated agents suitable for mice, such as monoclonal antibodies, nanoparticles, and single-gene products.

This chapter describes the protocol for the surgical preparation of the cranial bone window *(4)* in mice as well as the method to implant tumors in this bone window for further investigations of angiogenesis and other microcirculatory parameters. In brief, resect a circular area of skin and all underlying soft tissues above the calvaria. As the site for tumor implantation, drill a groove (~2.0 mm × 1.0 mm × 0.5 mm) in one of the parietal bones close to the sagittal sinus. Implant a small tumor chunk (volume ~ 1 mm^3) into this groove and finally seal the cranial bone window preparation airtight with bone cement. From this point, intravital microscopy can be performed for monitoring angiogenesis and other parameters such as tumor growth, microvascular perfusion index, microcirculation, and leukocyte endothelium interaction. However, the application of the cranial bone window model is not limited to the investigation of neoplastic tissues. Also applications like the implantation and investigation of bone substitute materials such as hydroxylapatite or β-tricaliumphosphate are conceivable.

2. Materials

Except for commonly used devices, all materials necessary for the preparation of the chronic cranial bone window in mice are listed with a detailed manufacturer's record. All materials given are only suggestions and may be modified for personal preferences.

2.1. Cranial Bone Window Preparation

2.1.1. Facilities and Apparatus

1. Laminar flow hood (HPH 12, Merck Eurolab GmbH, Bruchsal, Germany).

2. Dry sterilizer (model IS-350, Inotech, Intergra Biosciences, Wallisellen, Switzerland).

3. Dissecting microscope (Leica MZ75, ×6.3–50, Leica Microsystems AG, Heerbrugg, Switzerland).

4. Two or more flat, custom-made, self-controlling thermal pads (Silicon, Therm TSW 3, Isopad GmbH, Heidelberg, Germany).

5. Halogen lamp with two flexible swan-neck light transmission tubes (Intralux, 150H, Volpi AG, Urdorf-Zürich, Switzerland).

6. Custom-made stereotactic device (*see* **Fig. 22.1**) to fix the head of the animal in an optimal position for the surgical preparation as well as the investigations using intravital microscopy at a later time (Workshop, Department of Experimental Surgery, University of Heidelberg, Germany).

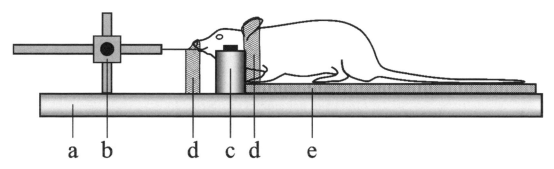

Fig. 22.1. Setup for the surgical preparation inside the hood showing a mouse fixed on the custom-made stereotactic device: **a** ground plate, **b** head mount with three-dimensional movable metal loop for the upper incisor teeth, **c** neck mount consisting of 2 two-dimensional movable anatomically shaped units left and right of the animal (depending on anatomy of the mouse, the distance between the two units can be changed), **d** adhesive tape to fix the head of the animal to the ground plate of the stereotactic device, **e** flat, custom-made, self-controlling thermal pad.

2.1.2. Drugs

1. Isotonic sodium chloride: 0.9% NaCl solution injectable.

2. Anesthesia: Mixture consisting of isotonic sodium chloride, ketamine hydrochloride (Ketalar®, Parke-Davis, Morris Plains, NJ), and xylazine (XylaJect®, Phoenix Pharmaceutical, St. Joseph, MO).

2.1.3. Cranial Bone Window

1. Sterile transparent circular glass coverslips (0.16- to 0.19-mm thick; depending on the size of the animal's head, 7- or 8-mm diameter, circular, Assistent, Sontheim, Germany).

2. Bone cement to fix the glass coverslip to the calvaria of the mouse: Mixture consisting of instant ethyl cyanoacrylate glue (Krazy Glue, original formula, Elmer's Products, Columbus, OH) and methyl methacrylate polymer thickener (GC Ostron®-Powder, GC Europe, Leuven, Belgium).

2.1.4. Surgical Instruments

1. Electric hair clipper (Electra®II, GH 204 or 201; Aesculap®, Aesculap, Tuttlingen, Germany) equipped with a 1/20-mm cutting head (GH 700, Aesculap).

2. One delicate dissecting forceps (Micro-Adson, BD 220, Aesculap).

3. Two microforceps (BD 331, Aesculap).

4. One pair of dissecting scissors, fine patterns (Cottle-Masing, sharp, OK 365, Aesculap).

5. One pair of microscissors (spring type) with round handles (FD 103, Aesculap).

6. One scalpel handle (BB 73, Aesculap).

7. Sterile scalpel blades (no. 15, BB 515, Aesculap).

8. One air turbine dental drill (motor system consisting of economic control unit GD 641, EC-micro motor GD 622, foot control unit GD 614, angel piece connector 2:1 GB 119, and angel handpiece 1:2 GB 114, Aesculap).

9. Spherical drill tips (0.5-mm diameter).

2.1.5. Other Materials

1. Mouse (25–30 g body weight, 6–12 wk); depending on the research goal, inbreed, outbreed, immunecompetent, immunedeficient, and so on.

2. One cage per animal (*see* **Note 1**).

3. Adhesive tape (Transpore™ 3M, hypoallergenic, ≈1.2 cm × 9.1 m, 3M Medical-Surgical Division, St. Paul, MN, USA).

4. Syringes (1 mL).

5. 26- and 30-gauge needles.

6. Sterile nonwoven swabs (5 × 5 cm).

7. Sterile Q-tips (cotton pads on wooden sticks; *see* **Note 2**).

8. Surgical masks.

9. Rubber gloves.

10. 70% alcohol to disinfect skin of the mouse, surgical instruments, and rubber gloves.

2.1.6. Additional Equipment Necessary for Tissue Implantation

1. Hank's balanced salt solution (H 9269, 100 mL, Sigma-Aldrich, Irvine, CA, USA); store at 6°C.

2. Sterile Petri dishes (≈100 × 20 mm diameter).

3. Methods

3.1. Surgical Preparation of the Cranial Bone Window

All surgical procedures should be performed under aseptic conditions (*see* **Note 3**).

1. Prior to the start of surgical preparations of cranial bone windows, the preparation of small chunks of tumor tissue for implantation into the cranial bone window has to be finished for organizing reasons (*see* **Subheading 3.2**).

2. Anesthetize the mouse by an injection of a mixture of ketamine (100 mg/kg body weight) and xylazine (10 mg/kg body weight) intramuscularly into the limb (*see* **Notes 4** and **5**).

3. Carefully shave the scalp hair of the anesthetized mouse from the neck to the supraorbital arch with the electric hair clipper outside the hood to avoid contamination of the hood with loose hairs. Thoroughly disinfect the scalp with alcohol after mechanically cleaning of the shaved skin (*see* **Notes 6** and **7**).

4. Move the mouse inside the hood and fix it in a prone position with its head to the custom-made stereotactic device (*see* **Fig. 22.1**). The upper incisor teeth have to hang in the three-dimensional movable metal loop, and the caudal parts of the lower jawbone have to be placed onto the two-dimensional, movable, anatomically shaped neck mount. After positioning the calvaria parallel to the surface of the thermal pad by manipulating the three-dimensional movable metal loop, the head of the animal has to be fixed with adhesive tape to the ground plate of the stereotactic device. In detail, one adhesive tape piece has to fix the snout to the metal loop and another piece of adhesive tape has to fix caudal parts of the lower jawbone and the neck to the anatomically shaped neck mount (*see* **Fig. 22.1** and **Note 8**). The longitudinal axis of the mouse (with its head lined up to the left-hand side of the surgeon) should be parallel to the frontal plane of the surgeon (*see* **Note 9**). Adhesive tape pieces are suitable to cover the area directly around the operation field to avoid hairs contaminating the operation field. The body of the mouse should be covered with sterile nonwoven cotton swabs to avoid cooling.

5. Using delicate tissue forceps and dissecting scissors, remove the scalp skin in an approximately circular area. This circle should be about 10–12 mm in diameter. Its central point should be over the middle of the sagittal suture (*see* **Fig. 22.2A** and **Note 10**).

6. Turn the stereotaxic device with the mouse clockwise by 90° and place it under the dissecting microscope. Now, the head of the mouse is pointing away from the surgeon. Carefully

scrape off all the periosteum of the accessible calvaria with a scalpel (no. 15 blade; *see* **Note 11**). While doing this procedure, the caudal and the lateral margins should be about 2–3 mm caudal from the lambdoid suture and 5–6 mm lateral from the sagittal suture, respectively, whereas the frontal margin should be demarcated by the upper edges of the two orbits. Then, clean the surface of the exposed calvaria with dry sterile Q-tips to identify possible bleeding. If there is bleeding, stop it by mechanical compression. After this, always avoid complete drying of soft and osseous tissues by moistening with isotonic NaCl.

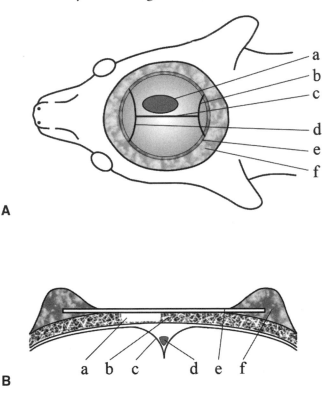

Fig. 22.2. Construction plan of the cranial bone window preparation. (**A**) View from above: *a* groove that serves as site of tumor implantation, *b* lambdoid suture, *c* sagittal suture, *d* coronal suture, *e*, transparent round glass coverslip, *f* bone cement. (**B**) Frontal cross section: *a* groove that serves as site of tumor implantation, *b* calvaria, *c* meninx, *d* sagittal sinus, *e* transparent round glass coverslip, *f* bone cement.

7. The goal of this step is to drill a small groove into the parietal bone that later will serve as the site for tumor implantation (*see* **Fig. 22.2**). The location of this groove is in the center of the preparation at the medial margin of one of the parietal bones parallel to the sagittal suture (*see* **Notes 12** and **13**). Hold the dental drill in one hand like a pencil with its drill tip always perpendicular to the calvaria. In the other hand, hold a 1-mL syringe filled with 0.9% NaCl and a 26-gauge needle on its tip. Under permanent superfusion of the drill tip with

NaCl for cooling (*see* **Note 14**), drill a groove in one of the parietal bones parallel to the sagittal suture in the center of the preparation, with the groove approximately 2.0 mm long, 1.0 mm wide, and 0.5 mm deep; move the drill tip back and forth continuously, never resting directly over one location (*see* **Note 15**). The groove should be deepened slowly and homogeneously (*see* **Note 16**).

8. From time to time, clean the groove and the surrounding bone with a dry sterile Q-tip and check the depth and length of the groove, first optically and then also mechanically with the drill tip (*see* **Note 17**). Continue with drilling under NaCl superfusion as long as necessary. The drilling of the groove is finished when on the one hand the groove is as deep as possible but on the other hand the inner layer of the compact tissue of the calvaria is just not perforated. Then, clean the groove again with dry sterile Q-tips to identify possible bleeding. If there is bleeding, stop it by mechanical compression with dry sterile Q-tips.

9. Now, the groove may serve as the site for implantation of small chunks of tumor tissue that have been prepared prior to preparation of cranial bone windows (compare with **step 1** and *see* **Subheading 3.2**); no active bleeding in and around the operation field (compare with **steps 6** and **8**) may occur. Turn the stereotaxic device with the mouse clockwise by 90° until the head of the mouse is pointing toward the surgeon (*see* **Note 18**). Place the Petri dish with the adequate tumor chunks (volume ~ 1 mm³) in it next to the stereotaxic device (*see* **Note 19**).

10. Dry the complete operation field up to the cutting edges of the skin with dry sterile Q-tips. Then, take one of the chunks using microforceps and place it into the central, almost dry groove of the calvaria (*see* **Note 20**). Due to adhesion forces, the tumor chunk will become perfectly attached to the osseous bottom inside the groove.

11. If the tumor chunk is located where desired in the central area of the groove, carefully and slowly superfuse the groove, including the tumor tissue in its center, with NaCl using a syringe equipped with a 30-gauge needle. Interrupt superfusion immediately when the tumor chunk seems to move (*see* **Note 21**). Finally, stop superfusion when NaCl reaches areas approximately 1–2 mm outside the groove. By doing so, moistening the peripheral areas of the calvaria close to the margins of surgical resection of the surrounding soft tissues is avoided (*see* **Note 22**). Take a sterile, round glass coverslip (diameter depending on the size of the animal's head, 7 or 8 mm) with a microforceps and carefully place it with its center located directly over the center of the cranial bone

window preparation and parallel to the horizontal plane of the animal's head onto the calvarium. During this maneuver, the tumor chunk has to be observed attentively to ensure that it is not displaced out of the groove. The round glass coverslip should complete the upper margins of the groove like a tangent to the calvaria with no space between the upper margins of the groove and the central part of the glass coverslip.

12. The entire space outside the groove between the round glass coverslip and the calvaria beneath should be filled with NaCl using a syringe equipped with a 30-gauge needle until the peripheral edge of the round glass coverslip is reached in all directions. Due to adhesion forces, the round glass cover-slip now will float on the "NaCl bolster" parallel to the surface, being attached to the calvaria itself only in the center where the groove with the tumor chunk inside is located (*see* **Note 23**).

13. Place an uncovered clean and sterile Petri dish next to the stereotaxic device, put a small pile of the GC Ostron-Powder in it (~6 × 6 mm diameter) and form a volcano-like hollow in the pile with the tip of a sharpened wooden stick (*see* **Notes 2** and **24**). Fill up the hollow with instant ethyl cyanoacrylate glue. Using the wooden stick, mix the glue with the powder to form a homogeneous gelatinous glue mixture.

14. If necessary, again dry the external surface of the calvarian bone with a dry sterile Q-tip.

15. Now, the cranial window has to be sealed and fixed airtight to the dry calvaria using a sharpened wooden stick (*see* **Note 2**) to distribute the glue mixture from **step 13** around the glass coverslip (*see* **Note 25**). During this procedure, it might be helpful to turn the stereotactic device with the mouse clockwise or counterclockwise whenever necessary to facilitate the handling and to be in optimal position to seal the preparation airtight all over.

16. After a few minutes when the glue will be completely dried and stable, the preparation of the chronic cranial window is finished (*see* **Note 26** and **Fig. 22.2**).

17. Place the operated mouse in its cage and leave it on a thermal pad outside the hood at least until the mouse has regained consciousness (*see* **Notes 1** and **27**).

18. Intravital microscopy of the implanted tumor chunk can be performed now repeatedly in the anesthetized animal by means of epi-illumination from a normal light source (e.g., for measuring the two-dimensional tumor surface or to detect the first appearance of newly formed blood vessels and hemorrhages) or by means of epi-illumination from a mercury lamp and a fluorescent filter set in combination with appropriate fluorescent dyes injected intravenously into the animal (e.g., for investigation of microcirculation and leukocyte endothelial interaction). For general advice, please also *see* **Note 28**.

3.2. Preparation of Small Chunks of Tumor Tissue for Implantation into the Cranial Bone Window

In the following, the explantation of a solid tumor from a donor mouse and its preparation into small chunks adequate for consecutive implantation into cranial bone windows in recipient mice are described. This procedure has to be performed prior to preparation of the cranial bone windows (*see* **Notes 29** and **30**).

1. Sacrifice the donor mouse bearing a solid subcutaneous tumor according to *Official Guidelines for Care and Use of Experimental Animals*. For 2–3 min, completely insert the dead animal into a 70% alcohol solution for disinfection. Excise the desired tumor surgically under aseptic conditions in the hood and put it into a sterile Petri dish previously filled with cold (~6°C) Hank's balanced salt solution.

2. The dissecting microscope is only needed for this step of the protocol (magnification ~10-fold). Remove the capsule and all hemorrhagic or necrotic parts of the tumor with the help of microforceps and a pair of microscissors. Cut the remaining tumor into small chunks with a diameter no greater than about 0.5–1 mm using microforceps and a pair of microscissors (*see* **Note 31**).

3. Start with the surgical preparation of the cranial bone window as described in **Subheading 3.1.**

4. Notes

1. For standardized reproducible experiments, one of the basic requirements is that animals lack any injuries, scars, or other irritations. Therefore, prior to surgery only mice from one brood should be held together in one cage since mice from different broods tend to cause injury to one another. After the bone window implantation, the animals must be housed separately in single cages in every case; otherwise, they may destroy each other's bone window preparations by scratching and biting.

2. Prior to the surgery, sharpen the end of few wooden sticks as in toothpicks. Later, these sticks will be used as a tool to mix the Krazy Glue with the Coe tray Plastic-powder and finally to seal the cranial window preparation with this glue mixture.

3. Work under a laminar flow hood and wear a surgical mask as well as rubber gloves to minimize the possibility of bacterial contamination of the bone window preparation. Between preparations of two different animals, all surgical instruments should be first cleaned mechanically with sterile nonwoven swabs soaked in alcohol and then sterilized with

a dry sterilizer. The gloves should be washed with alcohol and changed from time to time.

4. After initiation of the anesthesia with the suggested dose, if needed, the maintenance dose will be single doses of one-third of the initial dose.

5. To avoid cooling of the body temperature of the mouse, the anesthetized animal should be placed on a thermal pad (~37°C) whenever possible.

6. Clean the skin thoroughly and remove any clipped hairs by using small pieces of adhesive tape. Loose hairs will eventually contaminate the operation field and could induce irritation of tissues within the cranial bone window preparation.

7. Avoid alcohol contamination of the eyes of the animal. This could cause serious irritation and pain. Eyes should be protected from drying with a perfume-free eye salve (Bepanthen® eye and nose salve, Bayer Healthcare, Germany).

8. During and after fixing the animal to the stereotactic device, control the airway of the animal. The mouth has to be open, and the trachea may not be pressurized indirectly by the adhesive tape that fixes the neck to the anatomically shaped neck mount. Ideally, the head of the animal is tightly fixed to the stereotactic device in a manner that movements caused by breathing or the heartbeat are not transferred to the head of the animal. This is also a prerequisite for future intravital microscopy since otherwise movements of the head will cause bad quality of video recordings making it impossible to analyse the images.

9. All instructions for positioning of the animal in this chapter are suggestions only. Every researcher should place the animal in an individually adequate position considering personal preferences such as special habits, handedness, and other things.

10. A transparent, round glass coverslip held with a microforceps over the operation field can be used as reference. The diameter of the area remaining after resection of the skin should be approximately 2–3 mm longer than the diameter of the glass coverslip.

11. If any small parts of the periosteum are not removed from the calvaria during this step of the preparation, it will later weaken the glue seal between the calvaria and the coverslip. The adhesion between periosteum and the underlying bone is less strong than a direct glue-to-bone adhesion.

12. It does not matter whether the left or right parietal bone is used for site of tumor implantation. However, for reproducibility of experiments, in all experiments always the same side and location should be used.

13. Drilling too close at the sagittal sinus might be dangerous since harming/opening this sinus mechanically or due to

friction heat caused by the drill tip will lead to uncontrollable bleeding and exclusion of this preparation from experiments.

14. Do not spare superfusion of NaCl. This will avoid thermal injury of the underlying brain tissue and the sagittal sinus as well as wash out particles of bone dust caused by drilling.

15. This also helps to prevent thermal injury.

16. Take enough time for this step of the procedure. The consequences of rushing while drilling the bone are thermal injuries or mechanical damage to the underlying brain tissue caused by the drill tip.

17. It is important not to bring too much pressure on the bone with the drill tip while testing to avoid unexpected penetrating of the bone with consecutive mechanical damage to the underlying brain tissue. A lot of practice will be necessary to get the right "feeling" for the drill and how to use it appropriately during this critical procedure. The optimal bone thickness at the bottom of the groove is reached when the rose-colored brain tissue is dimly visible through the groove and thus when only a small bone plate remains between the groove and the meninx.

18. In this position, the correct horizontal position of the glass coverslip that will be used to close the cranial window can be assessed the best.

19. The following steps (**Subheading 3.1, steps 10–16**) describing how to close and seal the cranial window have to be done quickly and routinely without any unnecessary delays. Otherwise, the open areas of the surgical preparation will become too dry or the bone cement will harden before the sealing of the cranial window is completed.

20. At the moment of implantation, the groove should be almost dry. This guaranties firm initial "attachment" of the tumor chunk to the underlying osseous tissue of the groove. Otherwise, overfilling of the groove with NaCl may displace the chunk out of the groove toward the periphery of the surgical preparation.

21. If the chunk disappears out of the groove unexpectedly, it is better to remove the chunk completely and start the procedure with a new chunk as described beginning with **step 10**.

22. This circular peripheral area of the preparation has to remain dry to enable airtight sealing of the cranial bone window preparation with bone cement as described in **step 15**.

23. Do not overfill the space under the round glass coverslip. If the distance between the glass coverslip and the upper edges of the groove becomes too wide, there may be a chance for the tumor chunk to get sluiced out of its site of implantation within the groove.

24. For this, the tip of the wooden stick has to be used as a tool to distribute the glue mixture, to fix the glass coverslip to the calvaria, and to seal the cranial window preparation.

25. A prerequisite for sealing of the bone window preparation with bone cement is the lack of air bubbles inside the preparation under the transparent glass coverslip. If there are any bubbles visible, they have to be removed, usually by repeating the procedure of placing the coverslip on the calvaria (*see* **steps 11–14**). If the tumor chunk becomes sliced out of the groove by this maneuver, **step 10** also has to be repeated.

26. Residual dried glue on top of the transparent glass coverslip preventing clear visibility of the inside of the cranial bone window preparation easily may be scratched away using a sterile scalpel blade (no. 15).

27. After recovery from anesthesia, the mouse should show normal behavior within a few hours (e.g., cleaning itself, eating, drinking, sleeping, playing, and climbing around in the cage).

28. Daily weight monitoring may help to appraise the general state of health of the animal. After an initial loss of weight (less than 10%), mice should stabilize again within the first 48 h after surgery. When bearing a tumor, further loss of weight might be observed in these animals with increasing tumor volume over time.

29. To avoid immune reactions between the recipient animal and the tumor, use either isografted mouse carcinomas or immunedeficient mice as recipients (e.g., severe combined immunodeficient [SCID] mice).

30. The durability of the cranial bone window preparation is determined primarily by the growth rate and the local invasiveness of the tumor. Fast and infiltrative growing tumors may detach the transparent glass coverslip from the calvaria by displacement, thus destroying the bone window preparation, or induce neurological side effects by mechanical pressure to or infiltration of the brain. If this happens, mice have to be excluded from experiments and sacrificed. Experience with some human tumors (Saos-2 osteosarcoma, A 549 lung carcinoma, and PC 3 prostate carcinoma) were described by Klenke et al. *(4, 5)*.

31. To avoid warming of the tumor chunks before implantation in different animals, put the Petri dish on ice from time to time.

References

1. Menger, M. D., Lehr, H. A. (1993) Scope and perspectives of intravital microscopy - bridge over from *in vitro* to *in vivo*. *Immunol Today* 14, 519–522.

2. Leunig, M., Messmer, K. (1995) Intravital microscopy in tumor biology: current status and future perspectives (review). *Int J Oncol* 6, 413–417.

3. Jain, R. K., Schlenger, K., Höckel, M., Yuan, F. (1997) Quantitative angiogenesis assays: progress and problems. *Nat Med* 3, 1203–1208.

4. Klenke, F. M., Merkle, T., Fellenberg, J., et al. (2005) A novel model for the investigation of orthotopically growing primary and secondary bone tumors using intravital microscopy. *Lab Anim* 39, 377–383.

5. Klenke, F. M., Gebhard, M. M., Ewerbeck, V., Abdollahi, A., Huber, P. E., Sckell, A. (2006) The selective Cox-2 inhibitor Celecoxib suppresses angiogenesis and growth of secondary bone tumors: an intravital microscopy study in mice. *BMC Cancer* 6, 9.

6. Yuan, F., Salehi, H. A., Boucher, Y., Vasthare, U. S., Tuma, R. F., Jain, R. K. (1994) Vascular permeability and microcirculation of gliomas and mammary carcinomas transplanted in rat and mouse cranial windows. *Cancer Res* 54, 4564–4568.

7. Sckell, A., Safabakhsh, N., Dellian, M., Jain, R. K. (1998) Primary tumor size-dependent inhibition of angiogenesis at a secondary site: an intravital microscopic study in mice. *Cancer Res* 58, 5866–5869.

INDEX

Printed in the United States of America